AF356610

Vertical-Axis Wind Turbine

Vertical-Axis Wind Turbine

Editors

Yutaka Hara
Yoshifumi Jodai

Basel • Beijing • Wuhan • Barcelona • Belgrade • Novi Sad • Cluj • Manchester

Editors

Yutaka Hara
Tottori University
Tottori
Japan

Yoshifumi Jodai
Kagawa College
Takamatsu
Japan

Editorial Office
MDPI
St. Alban-Anlage 66
4052 Basel, Switzerland

This is a reprint of articles from the Special Issue published online in the open access journal *Education Sciences* (ISSN 2227-7102) (available at: https://www.mdpi.com/journal/energies/special_issues/Vertical_Axis_Wind_Turbine).

For citation purposes, cite each article independently as indicated on the article page online and as indicated below:

Lastname, A.A.; Lastname, B.B. Article Title. *Journal Name* **Year**, *Volume Number*, Page Range.

ISBN 978-3-7258-0259-3 (Hbk)
ISBN 978-3-7258-0260-9 (PDF)
doi.org/10.3390/books978-3-7258-0260-9

Cover image courtesy of Tadakazu Tanino

Contents

About the Editors

Yutaka Hara

Yutaka Hara graduated in Electronic-Mechanical Engineering Course from Nagoya University in 1987 and finished a master's degree in 1989. He proceeded to complete a doctoral course and obtained his D. Eng. from Nagoya University in 1993. He worked at Nagoya University from April 1992 to April 1995 as a Research Associate and from May 1995 to March 1997 as an Assistant Professor. He moved to Tottori University as an Associate Professor in April 1997. He was an Overseas Research Scholar of the Education Ministry of Japan at Syracuse University, USA, from March 1998 to January 1999. He became a Professor of the Faculty of Engineering at Tottori University in April 2019.

Yutaka Hara mainly studied Laser measurement techniques and their application at Nagoya University. After moving to Tottori, he started studying wind turbines. He invented the Butterfly Wind Turbine in 2012. Currently, he is developing a low-cost Butterfly Wind Turbine made out of an aluminum extrusion in collaboration with Nikkeikin Aluminium Core Technology Company, Ltd.

Yoshifumi Jodai

Yoshifumi Jodai graduated with a B.Sc. in Creative Designing Engineering in 1991, a M.Sc. in Mechanical Systems Engineering in 1993 from the Nagaoka University of Technology, and a Ph.D. in Macro Control Engineering from Tokushima University in 2008. He worked at Mitsubishi Heavy Industries, Ltd., from April 1995 to December 1998. Currently, he is working at the National Institute of Technology (KOSEN), Kagawa College, as Head of the Mechanical Engineering Department. He has more than 25 years of experience as a college teacher, including one year of experience at Delft University of Technology as a visiting researcher. Professor Jodai has written five textbooks on fluid mechanics. He has received an Industry-Academia-Government Collaboration Honorary Award in 2023 from the National Institute of Technology, a JSME Hatakeyama Award in 1993, and an Industrial Education Award in Shimane Prefecture in 1991.

Preface

As the expectations for renewable energy increase toward the realization of carbon neutrality, wind power is taking on a leading role as the primary power source for electricity generation. Currently, large propeller-type horizontal-axis wind turbines (HAWTs) have become mainstream, and their development is progressing toward further increasing their size, which is not easy. For floating offshore wind turbines, vertical-axis wind turbines (VAWTs), in which the tilt of the axis of rotation is not an issue, could be superior to HAWTs due a large float not being a requirement. There have also been proposals to increase the output power of small VAWTs via proximity arrangement, which could lead to the development of a small VAWT wind farm utilizing the land more effectively. Furthermore, owing to the inherent characteristics of VAWTs, i.e., no wind direction dependence resulting in a simple structure, low-cost wind power generation equipment can be developed, regardless of the size and application of the VAWT. As we move toward a carbon-free society, it is important to investigate the various capabilities of VAWTs. Therefore, this Special Issue collected original papers on various topics related to VAWTs. The collected papers are classified by the wind turbine type into four groups, including seven papers on the lift-type Darrieus rotor (including straight-bladed Darrieus and butterfly wind turbines), one on the drag-type Savonius rotor, one on the cross-flow rotor, and one on the hybrid type of Darrieus and Savonius rotors. As depicted by research content, four papers are related to the interaction (output enhancement) among rotors in wind turbine clusters (including paired wind turbines), four papers related to the optimization of the rotor's shape/construction (using machine learning, deflection or end plates, and the gap between the main blade and arm), one paper to the effects of the rotor axis's inclination on the offshore floating VAWT performance, and one paper to the life cycle assessment. The Guest Editors would like to express their gratitude to all the authors who have contributed to this Special Issue, and anticipate that this Special Issue will make a contribution to the practical application of vertical-axis wind turbines.

Yutaka Hara and Yoshifumi Jodai
Editors

 *energies*

Article

Integrated Surrogate Optimization of a Vertical Axis Wind Turbine

Marco A. Moreno-Armendáriz [1], Eddy Ibarra-Ontiveros [1], Hiram Calvo [1],* and Carlos A. Duchanoy [2]

[1] Instituto Politécnico Nacional, Centro de Investigación en Computación, Av. Juan de Dios Bátiz s/n, Ciudad de Mexico 07738, Mexico; mam_armendariz@cic.ipn.mx (M.A.M.-A.); gabriel040595@gmail.com (E.I.-O.)

[2] Gus Chat, Av. Paseo de la Reforma 26-Piso 19, Ciudad de Mexico 06600, Mexico; carlos.duchanoy@gus.chat

* Correspondence: hcalvo@cic.ipn.mx

Abstract: In this work, a 3D computational model based on computational fluid dynamics (CFD) is built to simulate the aerodynamic behavior of a Savonius-type vertical axis wind turbine with a semi-elliptical profile. This computational model is used to evaluate the performance of the wind turbine in terms of its power coefficient (Cp). Subsequently, a full factorial design of experiments (DOE) is defined to obtain a representative sample of the search space on the geometry of the wind turbine. A dataset is built on the performance of each geometry proposed in the DOE. This process is carried out in an automated way through a scheme of integrated computational platforms. Later, a surrogate model of the wind turbine is fitted to estimate its performance using machine learning algorithms. Finally, a process of optimization of the geometry of the wind turbine is carried out employing metaheuristic optimization algorithms to maximize its Cp; the final optimized designs are evaluated using the computational model for validating their performance.

Keywords: vertical axis wind turbine; CAE model; computational fluid dynamics; machine learning; surrogate model; optimization; evolutionary algorithms

Citation: Moreno-Armendáriz, M.A.; Ibarra-Ontiveros, E.; Calvo, H.; Duchanoy, C.A. Integrated Surrogate Optimization of a Vertical Axis Wind Turbine. *Energies* **2022**, *15*, 233. https://doi.org/10.3390/en15010233

Academic Editors: Yutaka Hara and Yoshifumi Jodai

Received: 3 November 2021
Accepted: 27 December 2021
Published: 30 December 2021

Publisher's Note: MDPI stays neutral with regard to jurisdictional claims in published maps and institutional affiliations.

1. Introduction

The issue of renewable energy is a topic that has gained relevance in recent years, mainly due to the depletion of traditional energy sources and the environmental impact. Consequently, more and more countries are interested in developing a sustainable energy industry based on renewable energy sources, such as solar, hydro, or wind energy.

The conventional way of harnessing the energy of the wind is through wind turbines. According to the orientation of its axis of rotation to the direction of the wind flow, two types are distinguished: the so-called Horizontal Axis Wind Turbine (HAWT) and those of Vertical Axis Wind Turbine (VAWT).

VAWT represents a good option for implementation in small-scale applications for self-consumption purposes, in areas with limited space and changing wind conditions, such as urban and rural environments [1]. In particular, the Savonius VAWT has a simple design that allows it to operate regardless of wind direction and at relatively low wind speeds. In addition, it has the advantage of having high capacity for self-starting [2] and low cost of construction and maintenance. However, the main disadvantage of this wind turbine resides in its low efficiency in the use of the energy available in the wind, being the design with the lowest theoretical efficiency of all [3–5].

Despite the inherent advantages of the Savonius VAWT for its implementation in small-scale power generation applications, its low efficiency makes it an unprofitable option for its practical implementation. Therefore, its use in practical applications is not widely spread.

A possible solution with which to solve this problem is the optimization of the design of the Savonius wind turbine, seeking to maximize its efficiency. In recent years, there has been growing interest in the design and optimization of vertical axis wind turbines [6,7]. For instance, in [8] the authors performed a study focused on finding the optimal values of the design parameters of Darrieus VAWT blade airfoil. The authors also studied the effect of varying the number of blades on the rotor, concluding that a four-blade configuration has better efficiency than a three-blade configuration. Such an optimization study was carried out by using the QBlade simulation software for wind turbine blade design. Similarly, in [9], the authors performed an optimization process on the blade profile of a Savonius wind turbine. The authors built a 2D computational model for the Savonius wind turbine by employing the software Ansys Fluent, and later they surrogated the computational model by means of a Kriging model. Finally, the authors employed the surrogate model along with a PSO algorithm to optimize the Savonius blade profile, claiming to reach a maximum C_p of 0.262 with the optimized design. On the other hand, in [10] a fluid-structure interaction analysis was carried out to study the stresses occurring on the blade of a small horizontal axis wind turbine designed to operate urban areas. This study employed a 2D computational model built on Ansys Fluent, which was validated with wind tunnel tests, showing that the computational model tends to underestimate the wind turbine performance but has an acceptable level of matching between simulation and experimental results. Another interesting example is found in [11], where the authors study the effect of varying the helix angle of blades for a Darrieus VAWT, analyzing its influence on the aerodynamic performance of the wind turbine. The authors employ a 3D computational model to simulate the wind turbine's performance and find the optimal helix angle from a small set of helix angles. In [12], a review of the rotor parameters on the performance of Savonius wind turbine is presented. The review focuses on analyzing a series of works concerning the design parameters modifications of Savonius wind turbines and their effect on the wind turbine's performance. The authors highlight some interesting remarks about the Savonius wind turbine configurations, for instance, better performance is expected from turbines with two blades rather than turbines with three blades, operating at low wind speeds. Besides, it is suggested to employ end plates to cover the turbine blades in order to increase the air volume acting on the advancing blade. Moreover, the existence of an overlap region between the turbine blades has shown to have a positive effect on the Savonius wind turbine performance. Finally, it is remarked that the use of curtains or wind deflectors to guide the airflow to the advancing blade is a promisingly alternative to significantly improve the Savonius wind turbine performance. A more detailed review about different proposals of wind deflectors is presented in [13], where the authors analyze different kinds of wind deflectors designed for a variety of VAWT's, including some Savonius and Darrieus wind turbines. It is highlighted that flat wind deflectors are more suitable for Savonius type wind turbines with two blades; meanwhile, aerodynamic deflector profiles are more suitable for Darrieus-type wind turbines, even having the possibility of using one single wind deflector on arrangements of at least two vertical axis wind turbines. Finally, in [14] the authors present a proposal to improve the performance of VAWT, based on the design and implementation of an improved wind duct to increase the wind speed. The optimum parameters of the wind duct are found by means of numerical simulations and a surrogate model based on DOE and the Kriging method. The authors simulate the behavior of an H-Darrieus VAWT placed at the inside of the improved wind duct, and the results show that the VAWT power coefficient can be enhanced up to 2.9 times.

In this work, we propose optimizing the design of a vertical wind turbine type Savonius to maximize its efficiency in terms of its energy conversion capacity through meta-heuristic algorithms and using a surrogate model of the Savonius wind turbine obtained through techniques of machine learning.

By performing the optimization process using a surrogate model of the wind turbine, it intended to reduce the computational cost of the optimization process and, therefore, reduce the time spent in the design and optimization process of the Savonius wind turbine.

2. State of the Art

2.1. Design and Optimization Based on Experimental Tests

The design methodology based on experimental tests in controlled environments is one of the classic approaches most used by researchers due to the fidelity of the data collected concerning the physical phenomenon being studied. The most used technique in this type of test consists of wind tunnels to carry out the different experiments associated with the design and optimization process [15–22].

For example, in [23] a series of experimental tests was carried out to design a new blade profile for the Savonius wind turbine based on progressive modifications on the Bach and Benesh blade profiles (2013–2015). During this process, the authors sought to achieve an optimal design for the geometry of the wind turbine blades to maximize the power coefficient of the wind turbine as a function of the tip speed ratio or TSR. Roy and Saha's profile was tested using a prototype with a frontal area of 0.0529 m^2 in an open section wind tunnel. Several other profiles reported in the literature show an increase in the power coefficient of up to 34.8% compared to the classic semicircular profile, reaching a maximum C_p of up to 0.31 at a TSR of 0.82 with a Reynolds number of 1.2×10^5 and a wind speed of 6 m/s. The work reported by the authors also considers a correction factor for their experimental data due to the blocking effect produced by the construction of the wind tunnel. In these works, the authors carried out the optimization process by trial and error. The number of prototypes analyzed was small, and therefore the authors explored a limited search space during the optimization process. The sample was small due to the time and cost required for construction and experimentation with the prototypes.

On the other hand, work in [23] tests small prototypes, with a frontal area of only 0.0377 to 0.0529 m^2, obtaining a meager amount of total energy available in the wind. Furthermore, the efficiency values achieved with designs in this order of dimension are not necessarily maintained when scaling up to larger prototypes since these tend to decrease as the dimension of the wind turbines increases.

2.2. Design and Optimization Based on Computational Models

An alternative methodology to experimental tests in controlled environments is that which makes use of computational models. This methodology has become quite popular in recent years because it allows for the modeling of many physical phenomena through computational tools with relatively high precision, thus reducing the costs and times associated with the design and optimization process. It reduces the number of experimental tests required for this process. The most widely used technique in this field is computational fluid dynamics or CFD, which solves the complex equations that describe the behavior of fluids using finite element or finite volume methods [24–28].

In 2016, Alom et al. [29] used a CFD model of a Savonius wind turbine with a semi-elliptical profile and a frontal area of 0.286 m^2 to optimize this wind turbine's geometry. For this, dynamic simulations are carried out in a 2D space following the sliding mesh method to simulate the rotation of the wind turbine. The study analyzed the torque and power coefficients of the wind turbine as a function of the TSR, while the angle Theta varied. They evaluated four different cut angles. As a result, they obtained an optimal cutting angle of 47.5°, reaching a maximum power coefficient of up to 0.33 at $TSR = 0.80$, considering a wind speed of 6.2 m/s. Subsequently, steady-state simulations were performed in 3D space to evaluate the wind speed and pressure contours on the wind turbine blades. Finally, in [30], the authors validated their optimized design through dynamic simulations in a 3D space and experimental tests in a wind tunnel, using a prototype with a frontal area of 0.046 m^2 and keeping the wind speed constant at 6.2 m/s, which resulted in significantly lower performance values than previously estimated in 2D studies and reported in [29,30]. Through 3D simulations, a maximum C_p of 0.1288 was estimated at a $TSR = 0.80$, while through experimental wind tunnel tests, a C_p maximum of 0.134 at a $TSR = 0.49$; this indicates that the 2D computational model can overestimate the maximum value of C_p by a factor of up to 2.4 times, while the 3D model estimated values were closer to those obtained through experimental tests.

They also highlight the differences between the power coefficient values estimated by the computational models and the values obtained through experimental tests, where the 2D computational models can overestimate up to a factor of up to 2.4 times the value of the maximum C_p [29,30]. In contrast, the 3D models show a better correspondence with the experimental data [30]. This improvement is because neglecting the rotor height phenomenon leaves out a crucial factor in the design of the Savonius wind turbine, which directly affects the precision of the estimates made by such 2D models.

2.3. Design and Optimization Based on Surrogate Models

The use of surrogate models in the Savonius wind turbine design and optimization processes has begun to gain popularity among researchers in recent years [31–33]. This interest is mainly because these models offer the possibility of modeling complex systems at a low computational cost and without the need to thoroughly understand the intrinsic relationships between the inputs and the outputs of the system. In addition, they have the particularity of being built from data collected from experimental tests and computer simulations.

In 2018, in Mohammadi et al. [33], the experimental data on the performance of a Savonius wind turbine reported in 2009 in [34] were taken as a basis. They built a computational model using CFD software to emulate the experimental results reported in [34], where wind tunnel tests are carried out with Savonius wind turbine prototypes with a twisted semicircular profile and frontal areas of around 0.062 m^2. The computational model contemplates a 3D space and uses the sliding mesh method to simulate the rotation of the wind turbine. Such a model shows an acceptable correspondence with the data reported in [34]. Subsequently, they built a surrogate model based on artificial neural networks (ANN) to replace the computational model of the wind turbine, having as input the geometric parameters of the wind turbine and the characteristic speed and as output the power coefficient of the wind turbine. This surrogate model was a multilayer perceptron or MLP and showed an excellent correspondence with the data collected through simulations using the computational model. Next, an optimization process was carried out through genetic algorithms using the surrogate model of the computational model. This process evaluated 100 possible designs for the wind turbine in each generation of the genetic algorithm, which would demand a high computational cost for the 3D computational model. However, using the surrogate model during the optimization process makes it possible to evaluate a more significant number of possible designs for the wind turbine with a significantly lower computational cost. Finally, the authors assessed the optimized design of the Savonius wind turbine using the original computational model. This model showed good correspondence with the performance estimate produced by the surrogate model and obtained a significant improvement with the base design reported in [34], reaching a C_p maximum up to 0.222 with a wind speed of 8.75 m/s.

From the works presented in this section, the advantage of substituting computational models or experimental schemes in wind tunnels with surrogate models stands out; these allow one to carry out the optimization process in a considerably shorter period, without losing precision or generality with the results obtained by the original models.

3. Methodology

This work uses the following proposed methodology to optimize the design of a Savonius wind turbine, using a surrogate model based on data collected through simulations of a wind turbine's computational model and metaheuristic optimization algorithms to maximize the power coefficient of the wind turbine.

Figure 1 shows the flow diagram of the methodology to be followed in this work to optimize the Savonius wind turbine that includes five stages. Each one is detailed later. However, in general, they consist of the following:

- (A) Geometric model of the wind turbine: a blade profile is chosen for the wind turbine. Using computer-aided design software (CAD), a 3D model of the wind turbine is built. The design parameters and their respective limits and restrictions are defined.
- (B) A computational model of the wind turbine: through the use of computer-aided engineering (CAE) software, a computational model of the wind turbine capable of simulating the rotor's interaction with the wind flow is built, allowing one to calculate its performance.
- (C) Design of experiments: a set of experiments to simulate a representative sample of the solution space corresponding to the wind turbine design is designed. The necessary data are collected to form a representative data set on the wind turbine phenomenon.
- (D) Surrogate model of the wind turbine: using the data set collected in the previous stage, a surrogate model of the wind turbine is built using machine learning algorithms.
- (E) Wind turbine optimization: the optimization process for the wind turbine design using the surrogate model and metaheuristic optimization algorithms is executed.

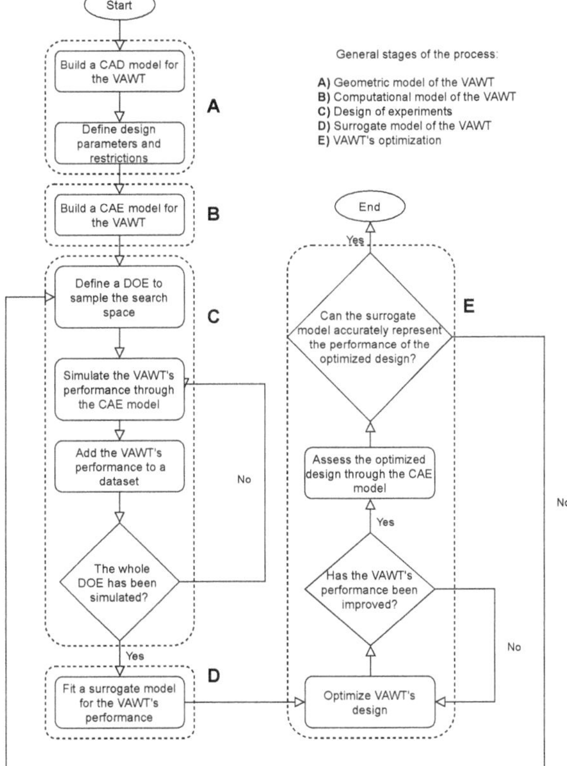

Figure 1. Proposed methodology.

3.1. Geometry

As shown in [8], there are several VAWT configurations and geometries, which include Darrieus VAWT and different Savonius blade profiles. In order to have a better guidance on the choice of VAWT type and blade profile, in Table 1 advantages and disadvantages for some VAWT types and blade profiles are summarized. From such Table 1, it is clear that Savonius VAWT type is more suitable for self-consumption applications in urban and rural areas, where normally the wind flow speed is lower and the power demand is not higher than 1 kW. Moreover, Savonius VAWT with semi circular blade profiles represent one of the simplest geometry designs, which, however, still have the potential to improve their capabilities with modifications on their geometric parameters that lead to more promising designs, such as the semi-elliptical blade profile.

Table 1. Comparison of different VAWT Types.

VAWT Type	Advantages	Disadvantages
Darrieus type	• Independence of wind flow direction • Higher rated power coefficient • Tolerance to turbulence wind flow • Scalable to higher dimensions • Suitable to medium scale power generation	• Higher cut-in speed • Higher rated wind speed • Poor self-starting capability • Complex geometry and construction
Savonius Semi-circular and semi-elliptical blade profiles	• High self starting capability • Independence of wind flow direction • Low cut-in speed • Lower rated wind speed • Tolerance to turbulent wind flow • Simple geometry and construction • Suitable for small scale power generation	• Lower rated power coefficient • Not suitable for larger-scale power generation
Savonius twisted blade profiles	• High self-starting capability • Independence of wind flow direction • Low cut-in speed • Lower rated wind speed • Tolerance to turbulent wind flow • Suitable for small scale power generation	• Lower rated power coefficient • More complex geometry and construction • Not suitable for larger scale power generation

The profile of the wind turbine blade proposed in this work is semi-elliptical and is obtained by taking the upper section of the sectional cut made on an ellipse, as shown in Figure 2a. The geometric parameters that define the semi-elliptical profile are the largest radius of the ellipse (OA), a minor radius of an ellipse (OB), distance to cut point (OM), and cut angle (*theta*).

The rotor configuration of the wind turbine proposed in this work includes two blades with an overlap between them, without a central axis, and an upper cover and a lower cover for the rotor. Figure 2b shows the parameters that define the geometry of the wind turbine rotor and correspond to the overlap distance between blades (e), the length of the blade chord (d), the rotor height (H), the diameter (D), and the diameter of the rotor caps (D_0). Additionally, Equation (1) defines the dimensionless parameter known as the overlap ratio (O_r).

$$O_r = \frac{e}{d} \tag{1}$$

The design parameters selected in this work and their respective restrictions for the optimization process are in Table 2. On the other hand, the geometric parameters that remain constant are in Table 3.

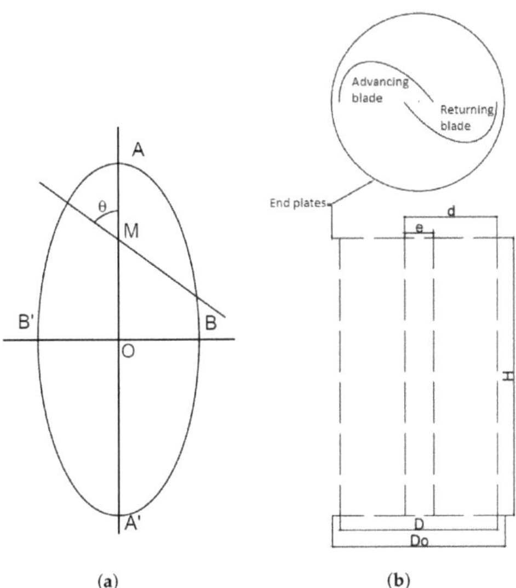

(a)   (b)

Figure 2. Geometric aspects of the Savonius wind turbine, (**a**) Sectional cut of an ellipse, (**b**) Savonius rotor geometry.

Table 2. Design parameters of the Savonius wind turbine.

Design Parameter	Restriction	Units
Cutting angle	$40 \leq \theta \leq 90$	$^{\circ}$
Overlap ratio	$0.1 \leq O_r \leq 0.3$	–
Major radius of the ellipse	$140 \leq OA \leq 200$	mm

Table 3. Constant parameters of the geometry of the Savonius wind turbine.

Parameter	Value	Units
Rotor height (H)	500	mm
Rotor diameter (D)	500	mm
Caps diameter (D_o)	$1.1 \times D$	mm
Distance to cut-off point (OM)	$0.54 \times OA$	mm

3.2. Computational Model of the Wind Turbine

The study of the aerodynamic characteristics of the Savonius wind turbine is carried out following the CFD approach and making use of the finite element analysis method to solve the transport equations that describe the interaction of the wind turbine with a wind flow at a constant speed. A computational model of the Savonius wind turbine using the CAE multiphysics simulation software Comsol Multiphysics is built.

3.2.1. Definition of Parameters

A set of global parameters is defined to drive the computational model, shown in Table 4.

Table 4. Global parameters of the computational model in Comsol Multiphysics.

Name	Value	Unit	Description
V	6	m/s	Wind speed
TSR	[0.1:0.1:0.8]	-	Characteristic speed
R	0.25	m	Rotor radius
N	$(TSR * V)/R$	rad/s	Rotor angular velocity
Cutting angle	θ	°	Cutting angle
Overlap	O_r	-	Overlap relationship
Major radius	OA	mm	Major radius of the ellipse

3.2.2. Geometry Construction

The simulation space has two computational domains: a rectangular prismatic domain to limit the total domain of the air within, called the stationary domain, and a cylindrical domain that surrounds the geometry of the Savonius wind turbine, called the rotating domain. Figure 3 illustrates the dimensions corresponding to these domains as a function of the rotor diameter and the respective contours intended for the inlet and outlet of the airflow. The simulation is set for the wind to flow from the inlet contour at constant speed and interacting with the Savonius VAWT on its way to the outlet contour, while the symmetry contours are set simply as symmetric side walls dedicated to contain the airflow within the stationary domain. It is worth mentioning that the height of both domains corresponds to four times the height of the rotor.

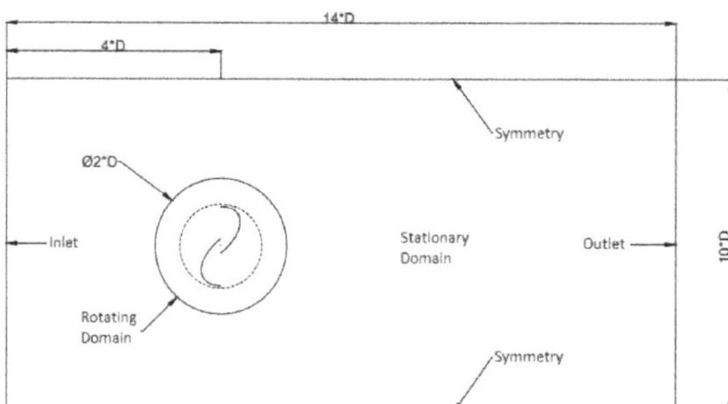

Figure 3. 2D view of the computational domain of the CAE model of the Savonius VAWT.

3.2.3. Physics Configuration

The selected physical interface corresponds to a single-phase and laminar flow without considering any turbulence model. This interface is applied to both the stationary and rotating domains, adding an inlet boundary condition to set the airflow inlet constant V velocity.

3.2.4. Sliding Mesh

We use the moving mesh method to simulate the rotation of the wind turbine at a constant angular speed N. At the same time, it is exposed to airflow at a constant speed V; this configuration is applied on the rotary domain. The angular speed is $-N/(2.pi)$ revolutions per second, where N is the angular velocity in rad/s.

3.2.5. Study Configuration

Two study steps are set up: a stationary step called a frozen rotor, and a temporary step. The function of the frozen rotor study is to approximate the solution to the dynamic problem of wind turbine rotation through a stationary study, to later use this approximate solution as initial conditions for the temporary study.

3.2.6. Definition of Study Variables

The total torque τ exerted on the Savonius wind turbine is defined as the difference in forces exerted between the forward blade and the return blade, as shown in Equation (2).

$$\tau = \tau_{av} - \tau_{ret} \tag{2}$$

where τ_{av} is the torque exerted by the forward blade and τ_{ret} is the torque exerted by the return blade.

To calculate this magnitude in Comsol Multiphysics, firstly, a series of integration operators are defined: the surface of the concave face of the advance blade (*intop*1), the surface of the convex face of the advance blade (*intop*2), the surface of the concave face of the return blade (*intop*3), and the surface of the convex face of the return blade (*intop*4). Subsequently, the variables in Equations (3) and (4) are defined in Comsol Multiphysics:

$$F_1 = x * (Stress_y) - y * (Stress_x) \tag{3}$$

$$F_2 = y * (Stress_x) - x * (Stress_y) \tag{4}$$

where x and y denote the position of a point within the Comsol coordinate frame. $Stress_x$ and $Stress_y$ refer to the total stresses, which include the contributions of the pressure force and the viscous force exerted on a point on axes x and y, respectively. In both cases, these variables are created automatically in the Comsol simulation environment.

Finally, in Equation (5), a variable for the total torque is defined.

$$Torque = intop_1(F_1) - intop_2(F_2) - intop_3(F_2) + intop_4(F_1). \tag{5}$$

3.3. Design of Experiments: Full Factorial

A full factorial design of experiments is designed with three design parameters and three levels for each one (see Table 5), and the total number of experiments required is defined as $N_{exp} = 3^3$, which gives a total of 27 experiments that correspond to the 27 combinations of the levels of the design parameters.

Table 5. Design parameter levels.

Design Parameter	Level 1	Level 2	Level 3	Units
Cutting angle (θ)	40	65	90	$^{\circ}$
Overlap ratio (O_r)	0.1	0.2	0.3	–
Major radius of the ellipse (OA)	140	170	200	mm

Performance Metrics for the Design of Experiments

The performance of each Savonouis geometry is measured in terms of the power coefficient and TSR as defined in Equations (6) and (7).

$$C_p = \frac{P_r}{P_v} \tag{6}$$

$$TSR = \frac{N \cdot R}{V} \tag{7}$$

Furthermore, P_r and P_v are the rotor and total wind power, respectively, defined in Equations (8) and (9).

$$P_r = T \cdot N \tag{8}$$

$$P_v = \frac{1}{2}\rho \cdot A \cdot V^3 \tag{9}$$

where T is the average torque of the wind turbine in a time interval, ρ is the density of the wind, and A is the frontal area of the wind turbine (defined as $A = D \cdot H$).

Using the computational model of the wind turbine, we obtain Equation (10), which corresponds to the total torque exerted on the wind turbine rotor during the time interval from t_0 to t_f, where t_f is the total simulation time. With this, the dynamic behavior of the total torque over time is computed.

Now, the average torque (T) is obtained by integrating the total torque obtained from (5) in the period t_1 to t_2, which is the time interval of the last complete rotation of the rotor, as shown in Equation (10).

$$T = \frac{\int_{t_1}^{t_2} Torque\, dt}{t_2 - t_1} \tag{10}$$

Thus, the rotor power is obtained using (10) in (8), and the total wind power is obtained using $\rho = 1.225$ in (9). Finally, the value of C_p is obtained by substituting (8) and (9) in (6). This calculation is repeated for all the TSR values indicated in the parametric sweep defined in Comsol Multiphysics. We obtained the behavior of the C_p as a function of the TSR.

3.4. Integration of Computing Platforms

The process of evaluating the performance of the different geometries of the Savonius wind turbine is carried out automatically through the integration of the different computational platforms used for the construction and execution of the computational model of the wind turbine.

For this, a client-server type link is established between Matlab and Comsol Multiphysics and a direct link between Comsol Multiphysics and SolidWorks; in this way, the three platforms are perfectly integrated, with Matlab being the base platform from which the others are driven. Figure 4 shows a diagram of the interaction between the different integrated computing platforms.

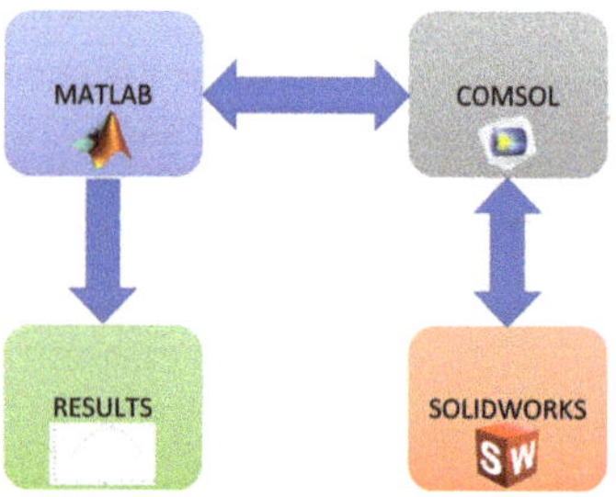

Figure 4. Integration of computing platforms.

3.5. Surrogate Model of the Wind Turbine

The dataset of the Savonius wind turbine obtained through the DOE is made up of 216 values, with the input vector $X = [\theta, Or, OA, TSR]$ and as a target $Y = C_p$, see Figure 5. This data set is separated into 80% data for the training stage and 20% for the testing stage, during the surrogate model fitting process.

The aim is to fit a surrogate model that receives the input data X (design parameters of the wind turbine) and estimates the output Y (power coefficient of the wind turbine). Figure 1 shows the diagram that illustrates this process. For this purpose, machine learning algorithms are used for regression, following the supervised learning approach.

The algorithms that are selected to build the surrogate model are the following:

- Support vector machine (SVM);
- Random forest (RFR);

- Bayesian ridge regression (BR);
- Multilayer perceptron (MLP);

The hyperparameters of each of these algorithms are adjusted by means of a grid search, using the cross-validation method. The metric used during this process is the root mean square error (MSE), see Equation (11).

$$MSE = \frac{1}{n} \sum_{i=1}^{n} (y_i - \bar{y}_i)^2 \tag{11}$$

where n is the total number of data, y_i is the correct label, and $\bar{y}_i$ corresponds to the estimated label.

In Table 6, the selected values of the hyperparameters of each algorithm and its performance are shown.

Table 6. Selected hyperparameters for each machine learning algorithm.

Algorithm	Selected Hyperparameters	Performance (MSE)
SVM	kernel = *linear*; C = 0.01; gamma = 1	9.40×10^{-4}
RFR	max_depth = *None*; max_features = *log2*; min_samples_leaf = 1; min_samples_split = 2, n_estimators = 1000	9.95×10^{-5}
BR	alpha_1 = 0.0001; alpha_2 = 1×10^{-7}; lambda_1 = 1×10^{-7}; lambda_2 = 0.0001	7.55×10^{-4}
MLP	activation = *tanh*; hidden_layer_sizes = (30, 30)	6.82×10^{-4}

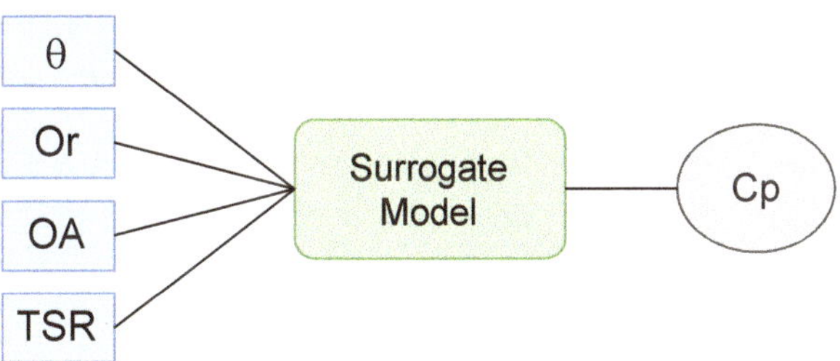

Figure 5. Surrogate model diagram.

3.6. Wind Turbine Optimization

The wind turbine optimization problem is approached as a single objective optimization problem in which the objective function to be maximized is the power coefficient of the wind turbine, while the design parameters to be optimized are the cut angle (*theta*), the overlap ratio (O_r), and the ellipse's largest radius (OA), and the TSR. So, the problem is stated in Equation (12)

$$f_{ob}(X) = C_p, \tag{12}$$

where $Cp \in \mathbb{R}$, $X \in \mathbb{R}^{1 \times 4}$, $X = [\theta, O_r, OA, TSR]$, and it seeks to maximize the value of f_{ob}.

Finally, the restrictions of each design parameter are shown in Table 2. Additionally, we establish that $0.595 \leq TSR \leq 0.605$.

The optimization process uses metaheuristic algorithms, specifically population algorithms based on hive intelligence and evolutionary algorithms. The algorithms selected to carry out this task are the following:

- Artificial Bee Colony (ABC);
- Genetic Algorithm (GA);
- Particle Swarm Optimization (PSO).

Hyperparameters of Optimization Algorithms

The hyperparameters associated with each optimization algorithm are adjusted manually, seeking to balance each algorithm's exploration and exploitation capabilities. Tables 7–9 list the hyperparameters selected for each algorithm during the wind turbine optimization process.

Table 7. Selected hyperparameters for the ABC algorithm

Hyperparameter	Value
Number of employed bees	10
Number of Onlookers	10
Selection method	*Roulette*
Number of steps	150
Number of cycles	100

Table 8. Selected hyperparameters for the GA algorithm

Hyperparameter	Value
Number of individuals	10
Encoding method	*Binary*
Selection method	*Tournament [2 Ind.]*
Cross method	*Crosses at 3 points*
Mutation method	*Binary [threshold = 0.01]*
Number of iterations	50

Table 9. Selected hyperparameters for the PSO algorithm

Hyperparameter	Value
Number of individuals	81
Cognitive constant	1.3
Social constant	0.7
Inertia range	$[0, 1.2]$
Decay of inertia	0.95 every 10 Iter.
Number of iterations	40

4. Experiments and Results

4.1. Computational Model Validation

The CAE computational model is validated based on two criteria: with a mesh independence study, where the model's behavior is analyzed based on the number of mesh elements, and through a direct comparison of the data on the performance of the Savonius VAWT calculated with the model against experimental performance data reported in state of the art to analyze the level of correspondence between the simulated and experimental data.

4.1.1. Mesh Convergence Study

The mesh convergence study is carried out through an iterative mesh refinement process. This process consists of the following steps:

1. Build different models with increasingly finer mesh elements.
2. Analyze the variations in the output result of the model as a function of the number of mesh elements used.
3. When no significant variations are observed, the model has reached convergence in terms of the mesh density.
4. Finish the process since the precision of the model is already totally independent of the number of mesh elements used, given that better results will not be achieved with finer meshes.

Figure 6 shows a variation in the model's output of less than 0.3% in terms of Cp as the number of mesh elements increases considerably. Therefore, the model has reached convergence.

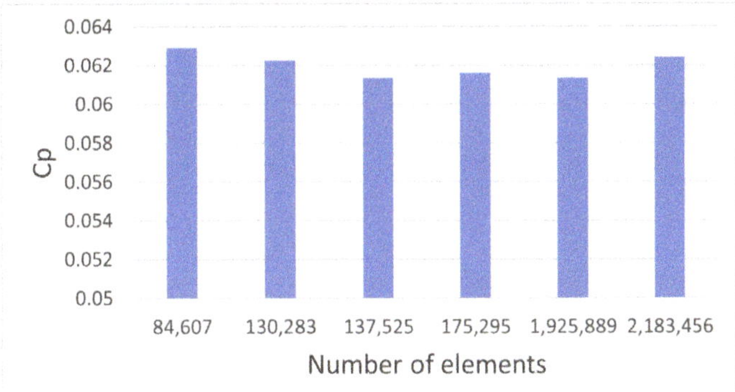

Figure 6. Output values of the CAE model as a function of the number of mesh elements.

4.1.2. Comparison with Experimental Data

Figure 7 shows a direct comparison between the experimental data reported by [30], referring to the performance of a Savonius wind turbine with a semi-elliptical profile and the data calculated by the CAE model proposed in this work for a Savonius wind turbine with the same reported design at [30]. From this Figure, the good shape of the $C_p - TSR$ curve and good operating range of the wind turbine are observed. However, there is no precise correspondence in the C_p since there is an average error of 0.0548; this indicates a tendency for the CAE model to underestimate the value of C_p. However, since the CAE model presents good convergence in terms of the mesh, the CAE model, although built correctly, could be leaving out some physical phenomenon, such as the level of turbulence present in the flow of wind, which could cause this tendency to underestimate the C_p.

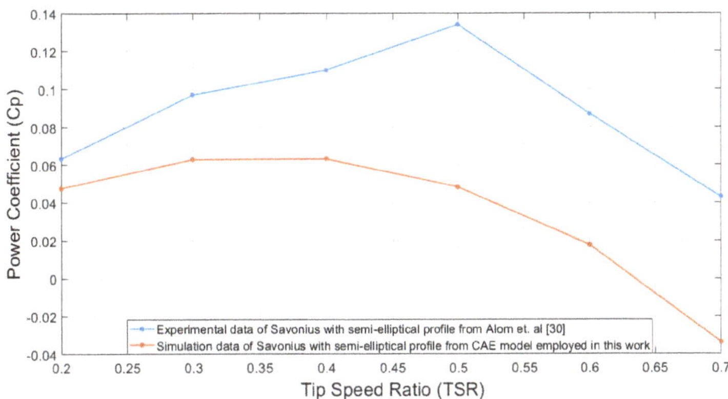

Figure 7. Comparison between data calculated using the CAE model and data collected experimentally.

4.2. Validation of Surrogate Models

The coefficient of determination (R^2) metric is used to test the performance of surrogate models previously adjusted using machine learning algorithms. The coefficient of determination is a widely used metric to measure the performance of regression algorithms and is defined in Equation (13), where SSR and SST are determined in (14) and (15).

$$R^2 = 1 - \frac{SSR}{SST} \tag{13}$$

$$SSR = \sum_{i=1}^{n}(y_i - \bar{y}_i)^2 \tag{14}$$

$$SST = \sum_{i=1}^{n}(y_i - \mu)^2 \tag{15}$$

where n is the total number of data, y_i is the target, $\bar{y}_i$ is the estimated label, and μ is the average of the data.

By applying this metric on the surrogate models adjusted with the hyperparameters of Table 6, Table 10 is obtained. This Table shows that the surrogate model with the best performance is the Random Forests algorithm, with a coefficient of determination of 0.97 and 0.98 in the training and testing stages, respectively.

Table 10. Performance of the surrogate models of the wind turbine based on the coefficient of determination.

Algorithm	R^2 in Training	R^2 in Testing
SVM	-4.19×10^{30}	0.0
RFR	0.973	0.981
BR	-6.03	-6.73
MLP	-9.61	-13.63

4.3. Savonius Wind Turbine Optimization

Table 11 groups the design parameters delivered by each optimization algorithm at the end of this process. It is important to note that, although the TSR does not correspond to a design parameter of the wind turbine geometry, it is relevant to indicate the maximum performance point of a given design. However, the performance of each optimized design must be evaluated over the entire operating range of the TSR (0.1 to 0.8). Figure 8 shows the C_p vs. TSR curves corresponding to the optimized designs in Table 11, estimated by the RFR surrogate model. It is observed that the performance of the three designs is practically identical, as estimated by the surrogate model.

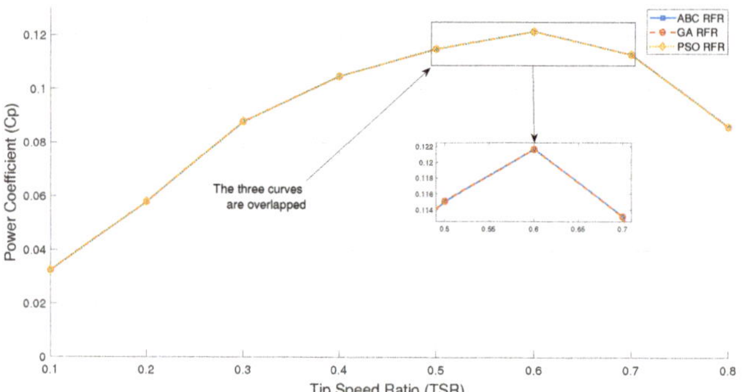

Figure 8. C_p curves estimated by the RFR model.

Table 11. Design parameters delivered by each optimization algorithm at the end of the process.

Algorithm	Design Parameters $[\theta, O_r, OA, TSR]$	Estimated Maximum Value of C_p
ABC	$[47.077, 0.135, 192.96, 0.6]$	0.1216873
GA	$[43.72, 0.14, 185.87, 0.59]$	0.121364
PSO	$[40.00005, 0.1000001, 199.999, 0.599]$	0.1216873

It is important to note that during the optimization process, around 8980 evaluations were carried out. The three algorithms evaluated 8980 possible designs in total, which consumed around one hour of computational time using the surrogate model. Considering that each evaluation requires between 1 to 1.5 h of computational time in the CAE model, the optimization process would have required between 12.47 to 18.7 months to complete. This meaningful amount of time highlights one of the significant advantages of using the surrogate model regarding the computation time required to complete the optimization process.

In Figure 9, the evolution of the population through generations of the GA is shown, where each dot represents an individual or candidate solution and each color represents a different generation of the algorithm execution. It is clear that during the first generations the population is more spread; this happens because at the beginning the algorithm is doing more exploration on the search space, and this is illustrated by the blue, red, and pink dots on Figure 9, which are more separate from each other. However, from generation 20 onwards the individuals start concentrating at some local optima near the circled region on Figure 9 (green, cyan, and dark red dots), where finally a global optima is found at the last generation (yellow dots).

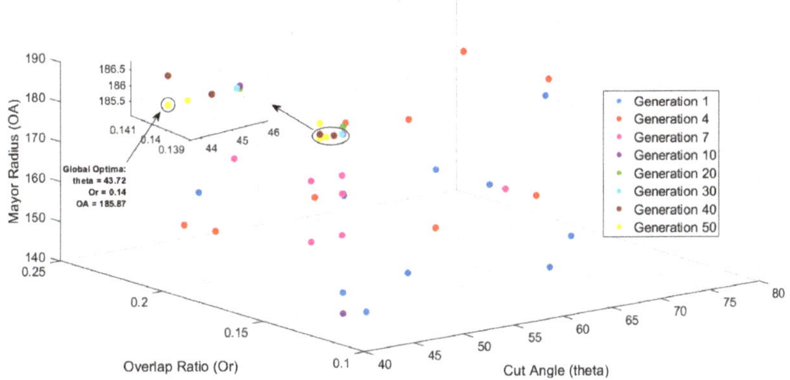

Figure 9. Evolution of population in genetic algorithm.

Optimized Design Validation

The CAE model evaluates the designs in Table 11 to re-estimate their performance in terms of C_p, and then they are compared with the estimation of the surrogate model. Figure 10 shows the comparison between the C_p curves estimated by the RFR surrogate model and the curves calculated by the CAE model, corresponding to the optimized designs in Table 11, where a good correspondence between the general shape of the curves is observed. However, in all cases, the surrogate model underestimates the value of C_p for all TSR points more significant than 0.1. Specifically, Table 12 shows a comparison between the maximum C_p values estimated by the RFR model and those calculated by the CAE model, and the difference or error between them. Finally, notice that the absolute error does not exceed 0.008; considering that C_p is measured in a range from 0 to 100%, it is equivalent to a margin of error of less than 1%.

Table 13 shows the MSE and R^2 metrics to measure the performance of the RFR model in the region of the optimized models. It is observed that the value of R^2 remains in a range between 0.80 and 0.95, which indicates that the regression model adjusted by the RFR algorithm has a reasonably acceptable performance in the region of the search space that corresponds to the optimal design. Since regression models tend to fail precisely at critical points in the search space, such as where the optimal design is found, in our case, the RFR model's acceptable performance in this region can be considered as a further test of the generality of the fitted regression model.

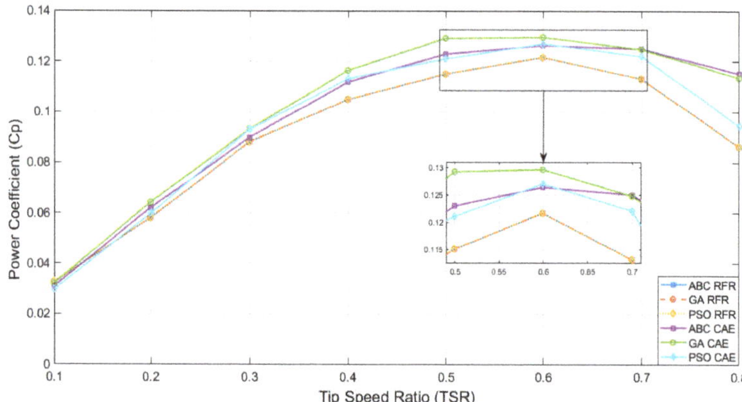

Figure 10. C_p curves estimated by RFR and CAE.

Table 12. Maximum C_p values estimated by the RFR model and calculated by the CAE model for the optimized designs with $TSR = 0.6$.

Optimized by	Maximum C_p (RFR)	Maximum C_p (CAE)	Absolute Error in C_p
ABC	0.1216	0.1264	0.0048
GA	0.1216	0.1296	0.008
PSO	0.1216	0.1270	0.0054

Table 13. Error metrics for C_p values of optimized designs.

Optimized by	MSE	R^2
ABC	0.000141	0.831
GA	0.000167	0.801
PSO	3.97×10^{-5}	0.952

From the same Figure 10 and Table 12, it is observed that the design with the best overall performance in terms of C_p is the one corresponding to the design delivered by the GA algorithm, which has a maximum C_p of 0.1296, while the design delivered by the ABC algorithm reaches a maximum C_p of 0.1264 and the design delivered by the PSO algorithm reaches a maximum C_p of 0.127, all at a TSR of 0.6, as calculated with the CAE model.

Finally, Figure 11 shows a comparison between the performance of the optimized design using the GA algorithm and the design with the best performance found during the DOE, which reached a maximum C_p of 0.1266 at a TSR of 0.6 with a design of $\theta = 40°$, $O_r = 0.10$, and $OA = 200$ mm. In this Figure, observe that the optimized design using the GA algorithm has superior performance to the best design found during the DOE throughout all the TSR points. In particular: (1) there is an increase in C_p of up to 9.49% at a TSR of 0.5, (2) an increase of up to 21.78% at a TSR of 0.8, and (3) an increase of 2.36% in the maximum C_p at a TSR of 0.6.

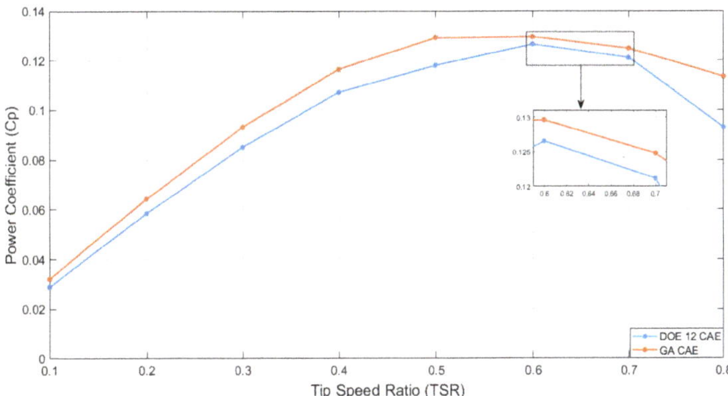

Figure 11. Performance comparison between GA-optimized design and best performance achieved during DOE.

5. Discussion

5.1. Optimized Geometry Analysis

The optimized geometry of the wind turbine blade, as can be seen in Figure 12a–d, obtains the following results: 1. it favors the concentration of the airflow directly towards the tips of the blades; 2. it increases the pressure that is exerted on the tips, which in turn helps to increase the torque exerted on them; and 3. the pressure lines exerted on the blades are shown in the instants of time when the torque is more outstanding.

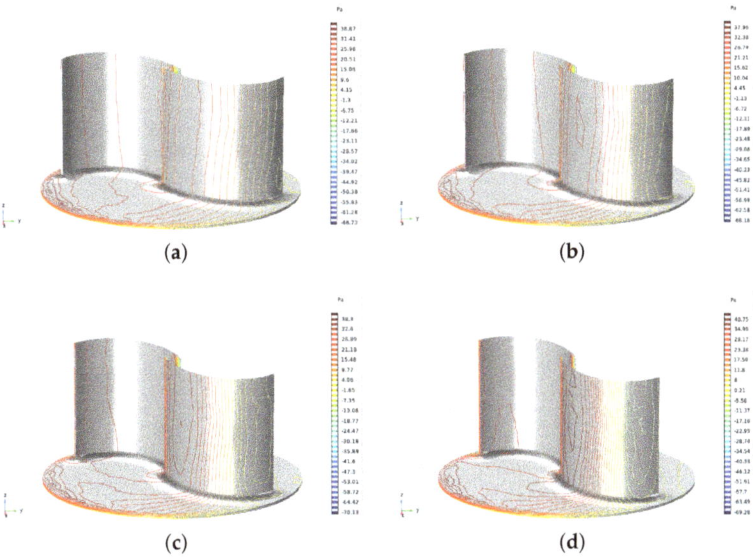

Figure 12. Pressure lines (Pa) on the lower half of the geometry of the GA-optimized Savonius wind turbine, (**a**) Instant pressure at $t = 1.2217$, (**b**) Instant pressure at $t = 1.2305$, (**c**) Instant pressure at $t = 1.239$, (**d**) Instant pressure at $t = 1.248$.

In the same way: (1) The overlap between blades is kept at an intermediate value. This value indicates that the redirection of flow from the forward blade to the return blade through the overlap between blades is a vital characteristic to reduce the negative effect of the return blade; (2) overlap should not be too wide so as not to cause air to escape freely through that space. This phenomenon is illustrated in Figure 13a–d, where it is observed that the wind flows through the overlap between blades with a speed of up to 4 m/s and that same wind flow runs through the entire concave surface of the return blade, providing it with an extra boost. We provide all necessary material to reproduce these results (https://tinyurl.com/68sn3ftc (accessed on 26 December 2021)).

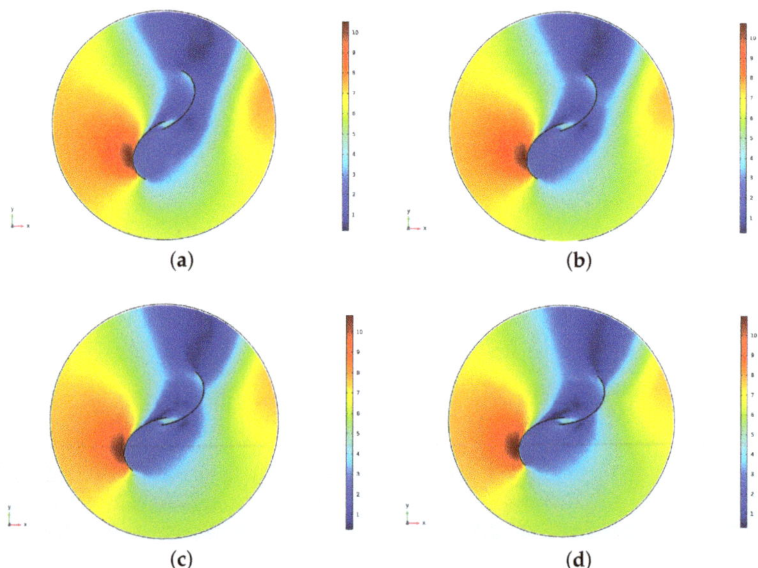

Figure 13. Wind speed (m/s) over the profile of the GA optimized Savonius wind turbine, (**a**) Instant speed at $t = 1.2217$, (**b**) Instant speed at $t = 1.2305$, (**c**) Instant speed at $t = 1.239$, (**d**) Instant speed at $t = 1.248$.

5.2. Comparison with the State of the Art

Due to the different approaches and methodologies followed by different studies over the years on the study of Savonius VAWT, it is complicated to establish a direct comparison between previous works and the present work. Specifically, such comparison is difficult because of the variety of blade profiles present in the state of the art and the different rotor dimensions considered for their prototypes, as well as the wind speeds considered on their tests. All these factors have been shown to have a strong impact on the final results, complicating the comparison task. Therefore, a fair comparison must consider works with similar features on the Savonius blade profile, the rotor dimensions, and the flow wind speed. The most suitable work complying with these characteristics is the one developed by Alom et al. on [30], and a comparison between this work and the present work is presented below.

The optimized design of this work reaches a maximum C_p of 0.1296 at $TSR = 0.6$; this is a marginally better performance than observed in the state of the art [30]. Alom et al., through a trial-and-error optimization process, report a maximum C_p of 0.1288 at a $TSR = 0.8$ for a Savonius wind turbine with a semi-elliptical profile, as estimated by their 3D computational model for the wind turbine. However, the performance of the design obtained in this work using metaheuristic optimization algorithms is below to that obtained by [30] through experimental tests in a wind tunnel, where they report a maximum C_p of 0.134 for their Savonius wind turbine with a semi-elliptical profile. However, it is essential to note the tendency of underestimating the C_p values concerning the experimental measurements reported in [30] that involves the CAE 3D model built in this work. Therefore, it is necessary to validate the results of this work through experimental tests to determine with certainty to what degree an improvement in the performance of the wind turbine was achieved or not, concerning that reported in state of the art. This experimental validation is proposed below, along with some other considerations only as future work.

6. Conclusions and Future Work

The design of a Savonius wind turbine was optimized through a GA. This design obtained improved performance in C_p, showing an increase of 21.78% with $TSR = 0.8$ and up to 2.36% in the maximum C_p with $TSR = 0.6$. However, the optimal design resulting from this work was not superior to that found in the state-of-the-art Savonius wind turbines with the same profile.

From the optimized geometry analysis, it is concluded that Savonius VAWT has better performance with blade profiles, which tend to concentrate the wind flow at the tips of the blade, exerting more pressure and, therefore, generating more torque on the blades. For this to be accomplished, the most relevant parameter is the cut angle, which must keep values between $40°$ and $50°$ (see Figure 9). Another relevant feature of the optimized blades is the overlap region between blades, which allow the wind to flow all along the concave side of the returning blade and, therefore, reduce the negative torque exerted by such blade. From Figure 9, it is observed that the overlap ratio should be kept at values between 0.14 and 0.15.

These results are valid for the operating conditions considered in this work, i.e, a wind speed of 6 m/s; a tip speed ratio from 0.1 to 0.8; and a Savonius VAWT with the following dimensions: diameter = 0.5 m and height = 0.5 m. Any change to these main conditions may yield slightly different final results on Savonius VAWT performance.

As future work, we plan to: (1) incorporate a turbulence model into the computational model of the wind turbine in order to capture the interactions of turbulent fluids inside the wind turbine and thus improve the accuracy of the model, (2) design a more extensive DOE to collect a more significant amount of data representing a more representative sample of the problem search space, (3) fit a new surrogate model with the extended data set to obtain a more accurate model and possibly free of local optimum, and (4) build and implement a test bench to test the optimized Savonius wind turbine prototype to measure and practically validate its performance.

Author Contributions: Conceptualization, methodology, M.A.M.-A. and C.A.D.; investigation and resources, M.A.M.-A., C.A.D., H.C., E.I.-O. and M.A.C.; software, visualization, and data curation, E.I.-O.; validation H.C.; formal analysis, M.A.M.-A., C.A.D. and H.C.; writing–original draft preparation, M.A.M.-A., E.I.O. and H.C.; writing–review and editing, H.C., C.A.D. and M.A.M.-A.; supervision, project administration and funding acquisition, M.A.M.-A. All authors have read and agreed to the published version of the manuscript.

Funding: This work has been possible thanks to the support of the Mexican government through the FORDECYT-PRONACES program of Consejo Nacional de Ciencia y Tecnología (CONACYT) under grant APN2017 − 5241; the SIP-IPN research grants SIP 2083, SIP 20210169, and SIP 20210189; and IPN-COFAA and IPN-EDI.

Institutional Review Board Statement: Not applicable.

Informed Consent Statement: Not applicable.

Data Availability Statement: The authors are committed to providing access to all the necessary information so that readers can fully reproduce the results presented in this work. For this, all the necessary information is available in the following repository at https://tinyurl.com/68sn3ftc (accessed on 26 December 2021).

Conflicts of Interest: The authors declare no conflict of interest.

References

1. Riegler, H. HAWT versus VAWT: Small VAWTs find a clear niche. *Refocus* **2003**, *4*, 44–46.
2. Shankar, P.N. Development of vertical axis wind turbines. *Proc. Indian Acad. Sci. Sect. C Eng. Sci.* **1979**, *2*, 49–66.
3. Akwa, J.V.; Vielmo, H.A.; Petry, A.P. A review on the performance of Savonius wind turbines. *Renew. Sustain. Energy Rev.* **2012**, *16*, 3054–3064. [CrossRef]
4. Mohamed, M.H.; Janiga, G.; Pap, E.; Thévenin, D. Optimization of Savonius turbines using an obstacle shielding the returning blade. *Renew. Energy* **2010**, *35*, 2618–2626. [CrossRef]
5. Schaffarczyk, A.P. *Introduction to Wind Turbine Aerodynamics*; Springer: Berlin/Heidelberg, Germany, 2014.
6. Alom, N.; Saha, U.K. Four decades of research into the augmentation techniques of Savonius wind turbine rotor. *J. Energy Resour. Technol.* **2018**, *140*, 050801. [CrossRef]
7. Alom, N.; Saha, U.K. Evolution and progress in the development of Savonius wind turbine rotor blade profiles and shapes. *J. Sol. Energy Eng.* **2019**, *141*, 030801. [CrossRef]
8. Altmimi, A.I.; Alaskari, M.; Abdullah, O.I.; Alhamadani, A.; Sherza, J.S. Design and Optimization of Vertical Axis Wind Turbines Using QBlade. *Appl. Syst. Innov.* **2021**, *4*, 74. [CrossRef]
9. Zhang, B.; Song, B.; Mao, Z.; Tian, W.; Li, B.; Li, B. A novel parametric modeling method and optimal design for Savonius wind turbines. *Energies* **2017**, *10*, 301. [CrossRef]
10. Lipian, M.; Czapski, P.; Obidowski, D. Fluid–Structure Interaction Numerical Analysis of a Small, Urban Wind Turbine Blade. *Energies* **2020**, *13*, 1832. [CrossRef]
11. Divakaran, U.; Ramesh, A.; Mohammad, A.; Velamati, R.K. Effect of helix angle on the performance of Helical Vertical axis wind turbine. *Energies* **2021**, *14*, 393. [CrossRef]
12. Fanel Dorel, S.; Adrian Mihai, G.; Nicusor, D. Review of Specific Performance Parameters of Vertical Wind Turbine Rotors Based on the Savonius Type. *Energies* **2021**, *14*, 1962. [CrossRef]
13. Rajpar, A.H.; Ali, I.; Eladwi, A.E.; Bashir, M.B.A. Recent Development in the Design of Wind Deflectors for Vertical Axis Wind Turbine: A Review. *Energies* **2021**, *14*, 5140. [CrossRef]
14. Ranjbar, M.H.; Rafiei, B.; Nasrazadani, S.A.; Gharali, K.; Soltani, M.; Al-Haq, A.; Nathwani, J. Power Enhancement of a Vertical Axis Wind Turbine Equipped with an Improved Duct. *Energies* **2021**, *14*, 5780. [CrossRef]
15. Alexander, A.J.; Holownia, B.P. Wind tunnel tests on a Savonius rotor. *J. Wind Eng. Ind. Aerodyn.* **1978**, *3*, 343–351. [CrossRef]
16. Kamoji, M.A.; Kedare, Shireesh, B.; Prabhu, S.V. Experimental investigations on single stage modified Savonius rotor. *Appl. Energy* **2009**, *86*, 1064–1073. [CrossRef]
17. Wenehenubun, F.; Saputra, A.; Sutanto, H. An experimental study on the performance of Savonius wind turbines related with the number of blades. *Energy Procedia* **2015**, *68*, 297–304. [CrossRef]
18. Blackwell, B.F.; Feltz, L.V.; Sheldahl, R.E. *Wind Tunnel Performance Data for Two-and Three-Bucket Savonius Rotors*; Sandia Laboratories: Albuquerque, NM, USA, 1977.
19. Saha, U.K.; Thotla, S.; Maity, D. Optimum design configuration of Savonius rotor through wind tunnel experiments. *J. Wind Eng. Ind. Aerodyn.* **2008**, *96*, 1359–1375. [CrossRef]
20. Roy, S.; Mukherjee, P.; Saha, U.K. Aerodynamic performance evaluation of a novel Savonius-style wind turbine under an oriented jet. In Proceedings of the ASME 2014 Gas Turbine India Conference, New Delhi, India, 15–17 December 2014.
21. Roy, S.; Saha, U.K. Investigations on the effect of aspect ratio into the performance of Savonius rotors. In Proceedings of the ASME 2013 Gas Turbine India Conference, Bangalore, Karnataka, India, 5–6 December 2013.
22. Roy, S. Aerodynamic Performance Evaluation of a Novel Savonius-Style Wind Turbine through Unsteady Simulations and Wind Tunnel Experiments. Ph.D. Thesis, Indian Institute of Technology Guwahati, Guwahati, Assam, India, 2014.
23. Roy, S.; Saha, U.K. Wind tunnel experiments of a newly developed two-bladed Savonius-style wind turbine. *Appl. Energy* **2015**, *137*, 117–125. [CrossRef]
24. Alom, N.; Saha, U.K. Influence of blade profiles on Savonius rotor performance: Numerical simulation and experimental validation. *Energy Convers. Manag.* **2019**, *186*, 267–277. [CrossRef]
25. Hosseini, A.; Goudarzi, N. Design and CFD study of a hybrid vertical-axis wind turbine by employing a combined Bach-type and H-Darrieus rotor systems. *Energy Convers. Manag.* **2019**, *189*, 49–59. [CrossRef]
26. Kacprzak, K.; Liskiewicz, G.; Sobczak, K. Numerical investigation of conventional and modified Savonius wind turbines. *Renew. Energy* **2013**, *60*, 578–585. [CrossRef]
27. Kacprzak, K.; Sobczak, K. Numerical analysis of the flow around the Bach-type Savonius wind turbine. *J. Physics Conf. Ser.* **2014**, *530*, 012063. [CrossRef]
28. Ramadan, A.; Yousef, K.; Said, M.; Mohamed, M.H. Optimization and experimental validation of a drag vertical axis wind turbine. *Energy* **2018**, *151*, 839–853. [CrossRef]
29. Alom, N.; Kolaparthi, S.C.; Gadde, S.C.; Saha, U.K. Aerodynamic design optimization of elliptical-bladed Savonius-style wind turbine by numerical simulations. In Proceedings of the ASME 2016 35th International Conference on Ocean, Offshore and Arctic Engineering, Busan, Korea, 19–24 June 2016.
30. Alom, N.; Saha, U.K. Performance evaluation of vent-augmented elliptical-bladed Savonius rotors by numerical simulation and wind tunnel experiments. *Energy* **2018**, *152*, 277–290. [CrossRef]

31. García, D.; Pineda J.; Paz, S. Modelo de Generador eólico Vertical Usando Redes Neuronales. Bachelor's Thesis, Escuela Superior de Cómputo IPN, Ciudad de Mexico, Mexico, 2019.
32. Moreno, T.; Juan, C. Multi-Objective Optimization of a Vertical Axis Wind Turbine Rotor for Self-Supply Applying Metaheuristics. Master's Thesis, Centro de Investigación en Computación IPN, Ciudad de Mexico, Mexico, 2019.
33. Mohammadi, M; Lakestani, M; Mohamed, MH. Intelligent parameter optimization of Savonius rotor using artificial neural network and genetic algorithm. *Energy* **2018**, *143*, 56–68. [CrossRef]
34. Kamoji, M.A.; Kedare, S.B.; Prabhu, S.V. Performance tests on helical Savonius rotors. *Renew. Energy* **2009**, *34*, 521–529. [CrossRef]

Article

Analytical Model for Phase Synchronization of a Pair of Vertical-Axis Wind Turbines

Masaru Furukawa [1,*], **Yutaka Hara** [1] **and Yoshifumi Jodai** [2]

[1] Faculty of Engineering, Tottori University, 4-101 Koyama-Minami, Tottori 680-8552, Japan; hara@tottori-u.ac.jp
[2] Department of Mechanical Engineering, Kagawa National Institute of Technology (KOSEN), Kagawa College, 355 Chokushi, Takamatsu 761-8058, Japan; jodai@t.kagawa-nct.ac.jp
[*] Correspondence: furukawa@tottori-u.ac.jp; Tel.: +81-857-31-5731

Abstract: The phase-synchronized rotation of a pair of closely spaced vertical-axis wind turbines has been found in wind tunnel experiments and computational fluid dynamics (CFD) simulations. During phase synchronization, the two wind turbine rotors rotate inversely at the same mean angular velocity. The blades of the two rotors pass through the gap between the turbines almost simultaneously, while the angular velocities oscillate with a small amplitude. A pressure drop in the gap region, explained by Bernoulli's law, has been proposed to generate the interaction torque required for phase synchronization. In this study, an analytical model of the interaction torques was developed. In our simulations using the model, (i) phase synchronization occurred, (ii) the angular velocities of the rotors oscillated during the phase synchronization, and (iii) the oscillation period became shorter and the amplitude became larger as the interaction became stronger. These observations agree qualitatively with the experiments and CFD simulations. Phase synchronization was found to occur even for a pair of rotors with slightly different torque characteristics. Our simulation also shows that the induced flow velocities influence the dependence of the angular velocities during phase synchronization on the rotation directions of the rotors and the distance between the rotors.

Keywords: vertical-axis wind turbine; phase synchronization; analytical model; Bernoulli's law

Citation: Furukawa, M.; Hara, Y.; Jodai, Y. Analytical Model for Phase Synchronization of a Pair of Vertical-Axis Wind Turbines. *Energies* **2022**, *15*, 4130. https://doi.org/10.3390/en15114130

Academic Editor: Davide Astolfi

Received: 2 May 2022
Accepted: 31 May 2022
Published: 4 June 2022

Publisher's Note: MDPI stays neutral with regard to jurisdictional claims in published maps and institutional affiliations.

1. Introduction

A pair of closely spaced vertical-axis wind turbines (VAWTs) can yield more power than two isolated VAWTs [1]. This idea was extended to a wind farm where many pairs of small-sized VAWTs were placed in a limited area in order to yield a high power density [2]. VAWTs accept wind from all directions; thus, it is possible to place them in proximity to each other. In order to investigate the performance and characteristics of such closely spaced VAWTs, wind tunnel experiments [3–8] and computational fluid dynamics (CFD) simulations of a pair of VAWTs were performed in two dimensions [9–14], as well as in three dimensions [6,15].

In Ref. [4,5], wind tunnel experiments of a pair of two-bladed H-type Darrieus turbines were reported, and synchronization of the rotations was found in both counter-down (CD) and counter-up (CU) layouts. Note that, in the CD layout, the blades of the two rotors move in the downwind direction in the gap region between the rotors. In the CU layout, the blades move in the upwind direction in the gap region. It was found that the rotational speeds equalize and the power output increases. Let us call this phenomenon phase synchronization. It was also found that the phase difference between the rotors oscillates around a mean value, and converges to it during the phase synchronization period. However, the mechanism of phase synchronization was not discussed in their papers.

Jodai et al. performed wind tunnel experiments on a pair of closely spaced, small-sized rotors made by a 3D printer and found that phase synchronization occurs when the distance between the rotors is sufficiently small in the CD layout [7,8]. It was found that the rotational speed with phase synchronization was 13% greater than that for a single rotor under the extremely small gap condition; the gap distance was 10% of the rotor diameter. This was the first report on the steep rise of the rotational speed under phase synchronization of a pair of closely spaced VAWTs with the extremely small gap. The oscillation of the phase difference was not reported in their papers.

Hara et al. performed a two-dimensional CFD analysis [13,14] to simulate the experiments by Jodai et al. They adopted a dynamic fluid–body interaction (DFBI) model that enabled the angular velocities of the rotors to change dynamically. The simulation results showed that phase synchronization also occurred in the CFD analysis for both the CD and CU layouts. They also found that the angular velocities of the rotors oscillated around the mean value during phase synchronization. Note that the oscillation of the angular velocities around the mean value means the same as the oscillation of the phase difference. From the observations of the CFD simulation results, it was proposed that phase synchronization and the oscillation of the angular velocities occur because of the interaction torques generated by pressure fluctuations in the gap between the rotors. In fact, they found that the velocity increases and the pressure decreases according to Bernoulli's law when blades of the two rotors come closer together in the gap region.

In this study, we developed an analytical model for the interaction torques that can be included in evolution equations of the angular velocities of rotors considered as solid bodies. Such a model is useful to understand the physics of the observed phenomena. Here, we present the details of the derivation, as well as numerical results showing the phase synchronization and the oscillation of the angular velocities. We also perform a simulation of a pair of rotors with slightly different torque characteristics and show that phase synchronization also occurs if interaction torques exist.

This paper is organized as follows. The details of the model, including the derivation of the interaction torques, are explained in Section 2. Then, Section 3 shows the numerical results based on the model. In particular, in Section 3.2, we report that the phase synchronization and oscillation of the angular velocities occur when using our model. We also report the dependence of the angular velocities in the phase synchronization regime on the gap in Section 3.3. The oscillations of the difference in angular velocities are shown in Section 3.4. The conclusions are given in Section 4. Appendix A shows the verification of our model based on comparison with CFD as well as experimental results, while Appendix B summarizes normalized expressions of our model.

2. Model

Let us consider a pair of VAWTs. The geometry of the layout is shown in Figure 1. The upstream wind flows from the left of the figure. The rotation directions of the rotors are shown by the arrows. Co-rotating, Counter Down, and Counter Up layouts are written as CO, CD, and CU, respectively. In the CD layout, the blades of the rotors move in the downwind direction in the gap region between the rotors. In contrast, in the CU layout, the blades move in the upwind direction.

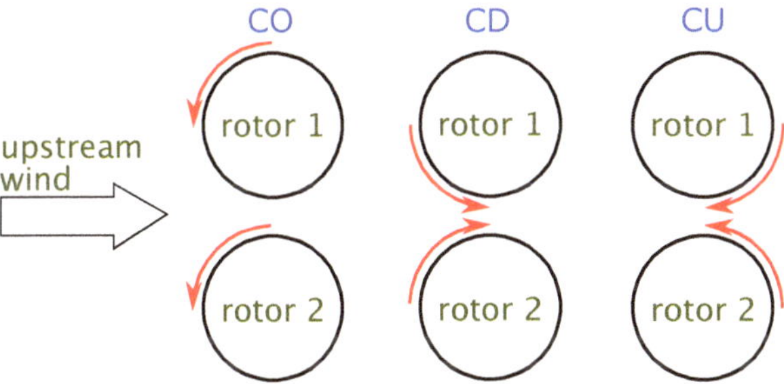

Figure 1. Layouts of rotors.

Figure 2 explains the definitions of variables. The diameter and the radius of a rotor i are denoted by D_i and R_i, respectively. The gap between rotors i and j is given by g_{ij}. In this study, we only considered a pair of rotors. A rotor i has n_i blades, and the position of a blade k is expressed as (x_{ik}, y_{ik}) in the two-dimensional plane, of which the origin is at the center of rotor 1. An azimuthal angle of a blade k of a rotor i is expressed by ψ_{ik}, which is zero in the direction of the y axis and increases in the counterclockwise direction. The blades of a rotor are equally spaced in the azimuthal angle, and thus the angle between the neighboring blades is $2\pi/n_i$. The chord length and the area projected along the rotation direction of a blade k of a rotor i are denoted c_{ik} and S_{ik}, respectively.

The distance between the blade k of the rotor i and the blade ℓ of the rotor j is written as $W_{ik,j\ell}$, which varies in time due to the rotation. As we explain later in this section, we adopted a model to describe a change in the flow velocity according to the temporal change of $W_{ik,j\ell}$, leading to a pressure fluctuation between the blades according to Bernoulli's law. The center of the rotor j is in the direction of an angle ϕ_{ij} seen from the center of rotor i. As with the azimuthal angle ψ_{ik}, the angle ϕ_{ij} is also zero in the direction of the y axis and increases in the counterclockwise direction. The angle of the center of the rotor i seen from the center of the rotor j is ϕ_{ji}, although only ϕ_{12} with $i = 1$ and $j = 2$ is shown in Figure 2.

The wind flows in the direction of the x axis. The upstream speed is denoted by V. The flow velocity experienced by the rotor differs for each rotor. The effective flow velocity immediately in front of rotor i is written as V_i.

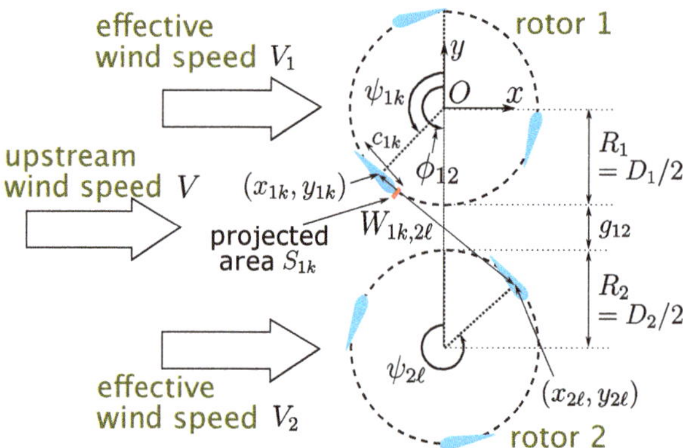

Figure 2. Definitions of variables.

As mentioned above, we only considered a pair of rotors. The evolution equations governing the rotation are

$$I_i \frac{d\omega_i}{dt} = Q_i - L_i + Q_{\mathrm{p}i},\tag{1}$$

$$\frac{d\psi_i}{dt} = \omega_i,\tag{2}$$

where $i = 1, 2$ represent the two rotors, ω_i is the angular velocity, $\psi_i := \psi_{i1}$ is the azimuthal angle of a representative blade, I_i is the moment of inertia, Q_i is the rotor torque, L_i is the load torque, and $Q_{\mathrm{p}i}$ is the torque due to the interaction with the other rotor $j\,(\neq i)$ through pressure fluctuation in the gap between them. Note that ψ_i is not necessarily ψ_{i1} but can be another ψ_{ik} with $k \neq 1$; the meaning is the same for any choice of $k = 1, \cdots, n_i$.

In the following section, we explain our torque models. First, we take the rotor torque Q_i as

$$Q_i = -\frac{3\sqrt{3}Q_{\max}(V_i)}{2\omega_0^3(V_i)}\,\omega_i(\omega_i + \omega_0(V_i))(\omega_i - \omega_0(V_i))F_{n_i}(\psi_i).\tag{3}$$

This torque, without the last term $F_{n_i}(\psi_i)$, simulates the torque characteristics in the two-dimensional CFD by Hara et al., as shown in Figure 2 of Ref. [14]. Note that the rotor is operated at an angular velocity, in a sense of average over phase, where the torque in Equation (3) is balanced by a load torque introduced below. Since the balanced state occurs at an angular velocity larger than that at the maximum torque in the present choice of the load torque for a given flow velocity, the rotor torque characteristics at small angular velocity regime is not essential. Here, $F_{n_i}(\psi_i)$ expresses a modulation in the rotor torque depending on the angle of the blades, which we explain shortly. Without this modulation, or setting $F_{n_i}(\psi_i) \equiv 1$, Q_i is a cubic function of ω_i for a given V_i and takes a maximum value $Q_{\max}(V_i)$ when $\omega_i = \omega_0(V_i)/\sqrt{3}$ on the positive ω_i side. When $\omega_i = \omega_0$ at the positive ω_i side, the rotor torque becomes zero. Thus, ω_0 is sometimes called no-load angular velocity.

Here, we take

$$Q_{\max}(V_i) = c_1 V_i^2, \tag{4}$$

$$\omega_0(V_i) = c_2 V_i, \tag{5}$$

where c_1 and c_2 are constants. As shown in Appendix A, we chose these expressions by observing the CFD [13,14], as well as the experimental [7,8] data. The actual values used in our simulation are given in the next section. Note that normalized expressions of Equations (3)–(5), as well as equations to appear below are given in Appendix B, where it is shown that the normalized rotor torque Q_i is expressed by a tip–speed ratio $\lambda_i := R_i \omega_i / V_i$ for the rotor i and normalized $Q_{\max}$ and ω_0 only.

The average of $F_{n_i}(\psi)$ over ψ is taken to be unity. As shown in Appendix A, it is known that the rotor torque by a single blade is finite at the upstream side and becomes maximal when a blade comes to the position $\psi_i = \pi/2$. It decreases to almost zero at the downstream side when the rotor solidity $\sigma_i := n_i c_{ik}/(\pi D_i)$ is large. We assume that

$$F_1(\psi) = \begin{cases} 4\sin^2 \psi & (0 \leq \psi \leq \pi) \\ 0 & (\pi < \psi < 2\pi) \end{cases} \tag{6}$$

for a single blade. Figure 3 shows the azimuthal angle ψ dependence of $F_n(\psi)$. In the figure, $n = 1$ plots Equation (6). The curve of $n = 3$ plots a summation of the modulation function for $n = 1$ over three blades equally spaced in the azimuthal angle and normalized such that $\int_0^{2\pi} F_3(\psi)\mathrm{d}\psi/(2\pi) = 1$. Note that the function for the modulation can be rather flexibly chosen since it is enough to simulate qualitative aspects of the torque modulation. Any function with its maximum at $\psi = \pi/2$ and almost zero in the downstream half may be acceptable.

For the effective wind speed, we use

$$V_i = (1 - a_i)V + \sum_{j \neq i} \frac{\Gamma_j}{2\pi(R_i + R_j + g_{ij})} \cos \phi_{ij}, \tag{7}$$

where a_i is a coefficient of self-induced velocity, and Γ_j is the circulation of rotor j. The second term expresses the velocity induced by another rotor. This assumption, assigning a circulation for a rotor, is similar to the one in Ref. [16]. We take

$$\Gamma_j = c_3 V_j, \tag{8}$$

where c_3 is a constant. This dependence was also obtained by the CFD results as shown in Appendix A. By using the effective wind speed V_i, the rotor torque on rotor i can be calculated. Note that V_i in Equation (7), especially its mutually induced velocity in the second term of the right-hand side, is evaluated by using the coordinate of the center of rotor i. In reality, the effective flow velocity is different for each blade, and the summation of the torques on every blade of a rotor determines the rotor torque. However, in this study, the effective velocity V_i is used for calculating the rotor torque as an average of the effective velocities for all blades in the rotor, and we assume that V_i is the flow velocity immediately in front of rotor i. Instead of considering the torques on each blade, we take into account the torque modulation of the rotor torque via $F_{n_i}(\psi)$. The modulation expresses the fact that the blade experiences significantly smaller flow than the upstream when it is in the downstream half of the rotation.

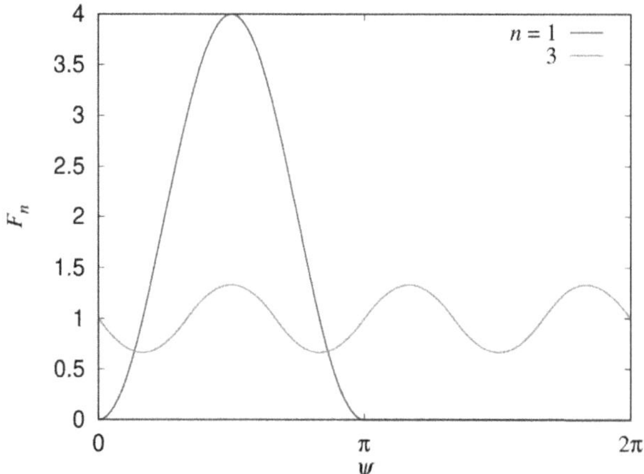

Figure 3. Modulation function $F_n(\psi)$ of the rotor torque on the azimuthal angle of the blade. The curves $n = 1$ and $n = 3$ plot the modulation for a single blade and a summation over three blades, respectively. The average of $F_n(\psi)$ over ψ is taken to be unity.

Secondly, we take the load torque as

$$L_i = c_4 \omega_i^2. \tag{9}$$

The dependence, square of angular velocity, is known as the ideal load torque to obtain highest power at each instance [17]. The CFD simulations by Hara et al. [13,14] also adopted the ideal load torque of the same form. Within our torque model, it is shown that the load torque in Equation (9) works to keep the highest power as follows. According to the rotor torque in Equation (3), the power $Q_i \omega_i$ takes a maximum value when $\omega_i = \omega_0(V_i)/\sqrt{2}$, and the corresponding rotor torque is $Q_i = 3\sqrt{3}Q_{\max}(V_i)/4\sqrt{2}$. By using Equations (4) and (5), these are written such that the maximum power is obtained when $\omega_i = c_2 V_i/\sqrt{2}$ and the corresponding torque is $Q_i = 3\sqrt{3}c_1 V_i^2/4\sqrt{2}$. The ideal operation of the turbine is achieved by balancing the rotor torque Q_i by the load torque L_i for whatever V_i. This is realized by setting $L_i \propto \omega_i^2$, which is obtained by eliminating V_i in Q by using $\omega_i = c_2 V_i/\sqrt{2}$. The coefficient c_4 is determined so that the load torque takes 95% of Q_i when the power becomes maximal for a given V_i in Section 3.1.

Finally, we come to the torque caused by pressure fluctuations between the blades. The positions of the blades of rotors 1 and 2 are given by

$$x_{1k} = -R_1 \sin\psi_{1k}, \tag{10}$$
$$y_{1k} = R_1 \cos\psi_{1k}, \tag{11}$$
$$x_{2\ell} = -R_2 \sin\psi_{2\ell}, \tag{12}$$
$$y_{2\ell} = R_2 \cos\psi_{2\ell} - (R_1 + R_2 + g_{12}). \tag{13}$$

The distance $W_{1k,2\ell}$ between two blades at (x_{1k}, y_{1k}) and $(x_{2\ell}, y_{2\ell})$ is

$$W_{1k,2\ell} = \sqrt{(x_{1k} - x_{2\ell})^2 + (y_{1k} - y_{2\ell})^2}. \tag{14}$$

Here, we consider the flow to be incompressible and assume that

$$V_{\mathrm{av}}(R_1 + R_2 + g_{12}) = (V_{\mathrm{av}} + \delta V)W_{1k,2\ell}, \tag{15}$$

where $V_{\mathrm{av}} := (V_1 + V_2)/2$ is the average flow velocity, and δV expresses the change in the flow velocity between the two blades due to a change in distance between them. Note that the calculation of δV may be improved by taking not only the x component of the flow velocity but also the y component, or taking the distance in the y direction between the blades k and ℓ instead of the distance $W_{1k,2\ell}$ itself. Let us leave this refinement as a future issue. We roughly estimate the change of velocity δV in this study. Then, we obtain the pressure fluctuation δp using Bernoulli's law

$$\frac{1}{2}\rho V_{\mathrm{av}}^2 + p = \frac{1}{2}\rho(V_{\mathrm{av}} + \delta V)^2 + p + \delta p, \tag{16}$$

as

$$\delta p = -\frac{1}{2}\rho\left(2V_{\mathrm{av}}\delta V + (\delta V)^2\right). \tag{17}$$

Here, the air pressure is written as p. Note that we neglected the effects of viscous force and unsteadiness of the flow. We adopted the Bernoulli's law to explain interactions between the blades due to pressure fluctuation through the increase in the flow velocity in the x direction observed in the CFD results, as shown in Figure 20 of Ref. [14]. The force on a blade due to this pressure fluctuation is calculated by integrating $-\nabla\delta p$ with the blade volume. The necessary force component is that along the rotation direction $-(\partial\delta p/\partial\psi_{ik})/R_i$ for the blade k of the rotor i. As an example, for $i = 1$, this is calculated as

$$\begin{aligned}
\frac{\partial\delta p}{\partial\psi_{1k}} &= -\rho(V_{\mathrm{av}} + \delta V)\frac{\partial\delta V}{\partial\psi_{1k}} \\
&= -\rho\frac{R_1 + R_2 + g_{12}}{W_{1k,2\ell}}V_{\mathrm{av}}\frac{\partial\delta V}{\partial\psi_{1k}}.
\end{aligned} \tag{18}$$

By using Equation (15),

$$\begin{aligned}
\frac{\partial\delta V}{\partial\psi_{1k}} &= -\frac{R_1 + R_2 + g_{12}}{W_{1k,2\ell}^2}V_{\mathrm{av}}\frac{\partial W_{1k,2\ell}}{\partial\psi_{1k}} \\
&= -\frac{R_1 + R_2 + g_{12}}{W_{1k,2\ell}^3}V_{\mathrm{av}}\left[(x_{1k} - x_{2\ell})\left(\frac{\partial x_{1k}}{\partial\psi_{1k}} - \frac{\partial x_{2\ell}}{\partial\psi_{1k}}\right)\right. \\
&\qquad\qquad\left. +(y_{1k} - y_{2\ell})\left(\frac{\partial y_{1k}}{\partial\psi_{1k}} - \frac{\partial y_{2\ell}}{\partial\psi_{1k}}\right)\right].
\end{aligned} \tag{19}$$

Then, we can obtain

$$\begin{aligned}
\frac{\partial\delta p}{\partial\psi_{1k}} &= \rho V_{\mathrm{av}}^2\frac{(R_1 + R_2 + g_{12})^2}{W_{1k,2\ell}^4}R_1[R_2\sin(\psi_{1k} - \psi_{2\ell}) \\
&\qquad\qquad -(R_1 + R_2 + g_{12})\sin\psi_{1k}],
\end{aligned} \tag{20}$$

$$\begin{aligned}
\frac{\partial\delta p}{\partial\psi_{2\ell}} &= -\rho V_{\mathrm{av}}^2\frac{(R_1 + R_2 + g_{12})^2}{W_{1k,2\ell}^4}R_2[R_1\sin(\psi_{1k} - \psi_{2\ell}) \\
&\qquad\qquad -(R_1 + R_2 + g_{12})\sin\psi_{2\ell}].
\end{aligned} \tag{21}$$

The torque on a blade due to pressure fluctuations may be approximated by multiplying the blade volume $c_{ik}S_{ik}$ by the rotor radius R_i. Now, we also assume that the total torque on a rotor can be obtained by adding contributions from all blades. Then, we obtain

$$Q_{p1} = -\alpha \sum_{k=1}^{n_i} \sum_{\ell=1}^{n_j} c_{1k}S_{1k}\rho V_{av}^2 \frac{(R_1 + R_2 + g_{12})^2}{W_{1k,2\ell}^4} R_1 [R_2 \sin(\psi_{1k} - \psi_{2\ell})$$

$$- (R_1 + R_2 + g_{12}) \sin \psi_{1k}], \tag{22}$$

$$Q_{p2} = \alpha \sum_{\ell=1}^{n_j} \sum_{k=1}^{n_i} c_{2\ell}S_{2\ell}\rho V_{av}^2 \frac{(R_1 + R_2 + g_{12})^2}{W_{1k,2\ell}^4} R_2 [R_1 \sin(\psi_{1k} - \psi_{2\ell})$$

$$- (R_1 + R_2 + g_{12}) \sin \psi_{2\ell}]. \tag{23}$$

Here, a coefficient α is included to control the strength of the interaction between the blades due to the pressure fluctuation, since the summation over all blades may be rather rough. Readers may think that it is strange to add a contribution from a blade at the opposite side of the gap between the rotors; it is natural to count only contributions from blades in the gap region. We also considered a model where rotors are taken as solid cylinders and only considered a narrowed channel between the gap region. This may not be a bad choice for high solidity rotors; however, the flow goes into the region surrounded by the blades of a rotor in reality. Thus, we decided to consider all combinations of blades and to add contributions from all pairs of blades in our model as an average in a sense, although big contributions must arise when two blades come to the gap region, and thus, other contributions may not play major roles. From a numerical computation viewpoint, this assumption, summation over all blades, is far simpler than taking the azimuthal angle of each blade into account to judge whether we should add its contribution or not. The on/off nature of the contributions can yield abrupt changes in torque, leading to a strange time evolution. Thus, we avoid this confusion.

3. Numerical Results

3.1. Characteristics of Rotors

We only consider a pair of rotors and assume that they have completely the same characteristics, except when otherwise stated. The following parameters are chosen. These are the same as those used in the CFD simulations in Hara et al. [13,14]. The number of blades per rotor is $n_1 = n_2 = 3$. The radius of the rotor is $R_1 = R_2 = 25\,\text{mm}$. The chord length is $c_{1k} = c_{2\ell} = 20\,\text{mm}$, and the area projected along the rotation direction is $S_{1k} = S_{2\ell} = 43.4\,\text{mm} \times 3.8\,\text{mm}$ for $k, \ell = 1, 2, 3$. The moment of inertia is $I_1 = I_2 = 5.574 \times 10^{-6}\,\text{kg}\,\text{m}^2$. Note that the previous CFD simulation [13,14] was two-dimensional, where the rotor height was 1 m instead of 43.4 mm, as used for the three-dimensional model, and the moment of inertia was set at $I_1 = I_2 = 1.284 \times 10^{-4}\,\text{kg}\,\text{m}^2 (= 5.574 \times 10^{-6}\,\text{kg}\,\text{m}^2 \times (1\,\text{m}/43.4\,\text{mm}))$.

From the CFD simulation for a single rotor with the above parameters, under a fixed angular velocity condition, it was found that the maximum power is obtained when $\omega = 366.52\,\text{rad/s}$ for $V = 10\,\text{m/s}$. The time-averaged rotor torque in this case is $Q = 0.525\,\text{mN}\cdot\text{m}$. By using these values, we determined $Q_{\max}(V_i)$ and $\omega_0(V_i)$ for the rotor torque, Equation (3), and c_4 for the load torque, Equation (9). We further determined that c_1 and c_2 from $Q_{\max}(V_i)$ and $\omega_0(V_i)$ via Equations (4) and (5).

In the case of our rotor torque (3), the power $Q_i\omega_i$ becomes maximal at $\omega_i = \omega_0(V_i)/\sqrt{2}$ for a given V_i. The rotor torque at this angular velocity is $Q_i = 3\sqrt{3}Q_{\max}(V_i)/4\sqrt{2}$. We chose this to match the CFD value $Q = 0.525\,\text{mN}\cdot\text{m}$. Therefore, it is assumed that $\omega_0(10\,\text{m/s}) = \sqrt{2} \times 366.52\,\text{rad/s}$ and $Q_{\max}(10\,\text{m/s}) = (4\sqrt{2}/3\sqrt{3}) \times 0.525\,\text{mN}\cdot\text{m}$. From Equation (4), we determine that

$$c_1 = \frac{Q_{\max}(V_i)}{V_i^2}$$
$$= \frac{4\sqrt{2} \times 0.525\,\mathrm{mN \cdot m}}{3\sqrt{3} \times (10\,\mathrm{m/s})^2}$$
$$= 5.72 \times 10^{-6}\,\mathrm{kg}. \tag{24}$$

Additionally, from Equation (5), we obtain

$$c_2 = \frac{\sqrt{2} \times 366.52\,\mathrm{rad/s}}{10\,\mathrm{m/s}}$$
$$= 51.8\,\frac{\mathrm{rad}}{\mathrm{m}}. \tag{25}$$

Furthermore, in the CFD simulation, the angular velocity becomes stationary when the load torque is chosen to be 95% of the optimum rotor torque. We assume the same and thus obtain

$$c_4 = \frac{L_i}{\omega_i^2}$$
$$= \frac{0.95 \times 0.525\,\mathrm{mN \cdot m}}{(366.52\,\mathrm{rad/s})^2}$$
$$= 3.71 \times 10^{-9}\,\frac{\mathrm{kg \cdot m^2}}{\mathrm{rad^2}}. \tag{26}$$

We will set the value of c_3 in Section 3.3.

By using these parameters, we obtain the torque curves shown in Figure 4. The rotor torque is plotted for $V = 6, 8, 10$ and $12\,\mathrm{m/s}$ without considering the modulation by $F_{n_i}(\psi)$ and the induced velocities. Note that the rotor torque characteristics at angular velocities lower than the maximum torque for each V are assumed to be different from those of the CFD and the experimental wind turbines to simplify the analysis. The intersection of the rotor torque curve for a given flow velocity and the load torque curve is the a stable steady state. For example, $\omega \simeq 370\,\mathrm{rad/s}$ is the angular velocity at the steady state when $V = 10\,\mathrm{m/s}$. Note that the angular velocity at the steady state for a given flow velocity does not change when the rotor torque Q or $Q_{\max}(V)$ is multiplied by a constant factor, since the load torque L was chosen to also be multiplied by the same factor in the present study.

Note that the adopted angular velocity values $\omega = 366.52\,\mathrm{rad/s}$ (3500 rpm) and the torque $Q = 0.525\,\mathrm{mN \cdot m}$ at the maximum power are slightly different from the values at the steady state $\omega_{\mathrm{SI}} = 366.1\,\mathrm{rad/s}$ (3496 rpm) and $Q_{\mathrm{SI}} = 0.485\,\mathrm{mN \cdot m}$ obtained in the CFD simulation based on the DFBI for a single rotor [14]. "SI" stands for single. We can use these values of ω_{SI} and Q_{SI} instead of those used above to determine $Q_{\max}(V)$, $\omega_0(V)$ and c_4 and then c_1 and c_2 via Equations (4) and (5) by assuming the following three conditions: (i) the rotor torque balances the load torque at $\omega = \omega_{\mathrm{SI}}$, (ii) the balanced torque is $Q = Q_{\mathrm{SI}}$, and (iii) Q_{SI} is 95% of the torque at the maximum power. We can immediately obtain $Q_{\max}(V)$ or c_1 from the second condition and c_4 from the third condition, respectively. We can also obtain $\omega_0(V)$ from the first one, which becomes a cubic equation for $\omega_0(V)$ for a given ω_{SI}. The resulting values are $c_1 = 5.56 \times 10^{-6}\,\mathrm{kg}$, $c_2 = 49.5\,\mathrm{rad/m}$, and $c_4 = 3.62 \times 10^{-9}\,\mathrm{kg \cdot m^2/rad^2}$, of which relative differences from the values obtained in Equations (24)–(26) and used in the simulations presented in Sections 3.2–3.4 are within 5%. Note that $\omega_0(V)$ was calculated from the cubic equation by a perturbation technique based on a trivial approximate solution $\omega_0(V)/\sqrt{2} \simeq \omega_{\mathrm{SI}}$ that gives the maximum power in our model.

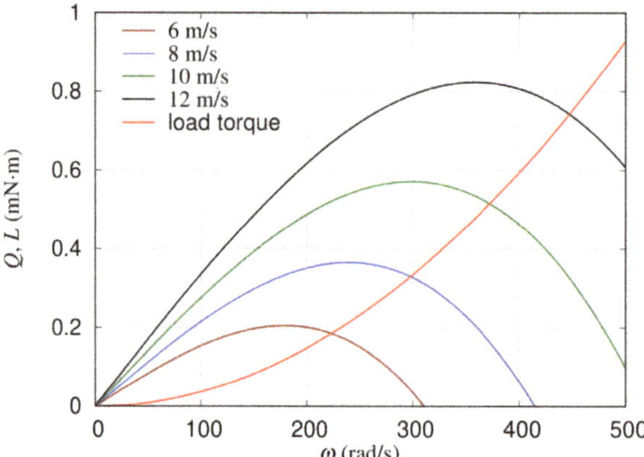

Figure 4. The rotor torques for $V = 6, 8, 10$, and $12\,\text{m/s}$ and load torques are plotted against the angular velocity of the rotor.

3.2. Phase Synchronization

In this Section 3.2, we demonstrate that phase synchronization occurs due to the interaction torque generated by the pressure fluctuation. In order to focus on the effect of the interaction torque, we set $a_i = 0$ and $c_3 = 0$, so that $V_i = V$, i.e., the effective flow velocity is the same as the upstream flow velocity. The upstream velocity used in this section is $V = 10\,\text{m/s}$. Furthermore, we set $F_{n_i}(\psi_i) \equiv 1$ to exclude the effects of torque modulation due to the azimuthal angle of the blade.

We solve the evolution Equations (1) and (2) for $i = 1, 2$ by the fourth-order Runge–Kutta method. The step size is $(2\pi/n\omega_{\text{av}}(0))/30$, where $\omega_{\text{av}}(0) = (|\omega_1(0)| + |\omega_2(0)|)/2$.

Figure 5 shows the time evolution of ω_i for $\alpha = 0, 0.05, 0.1$, and 0.2 in the CD layout. The gap between the rotors is $g_{12} = 10\,\text{mm}$. The initial conditions are $\omega_1(0) = 420\,\text{rad/s}$, $\omega_2(0) = -320\,\text{rad/s}$, $\psi_1(0) = \pi$, and $\psi_2(0) = -0.5$. The initial angular velocities have different magnitudes. One of the blades of rotor 1 is at the narrowest position in the gap initially and that of rotor 2 is slightly ahead.

When $\alpha = 0$, the rotors just reach their individual steady states since there is no interaction between the rotors. The rotors have identical characteristics, and thus they rotate at the same angular velocity at the steady state.

When α is finite, the angular velocities oscillate. This simulates the phase synchronization observed in the experiments as well as in the CFD simulations. Such phase synchronization and oscillation of phase difference in the wind tunnel experiments of a pair of two-bladed H-type Darrieus turbines were reported in Section 6.2 of Ref. [4] and in Section 4.4 of Ref. [5]. The phase synchronization and the oscillation of angular velocities in the two-dimensional CFD were reported in Section 3.5 and Figures 17–19 of Ref. [14]. The oscillation period becomes shorter as α increases since the interaction between the rotors becomes stronger. The oscillation period of the angular velocities is about $0.5\,\text{s}$ for $\alpha = 0.05$ at $t \simeq 4\,\text{s}$, while it is about $0.25\,\text{s}$ for $\alpha = 0.2$. Moreover, the oscillation period becomes shorter over time.

The oscillation amplitude of the angular velocity becomes larger as α is increased. This is also because the interaction becomes stronger as α is increased.

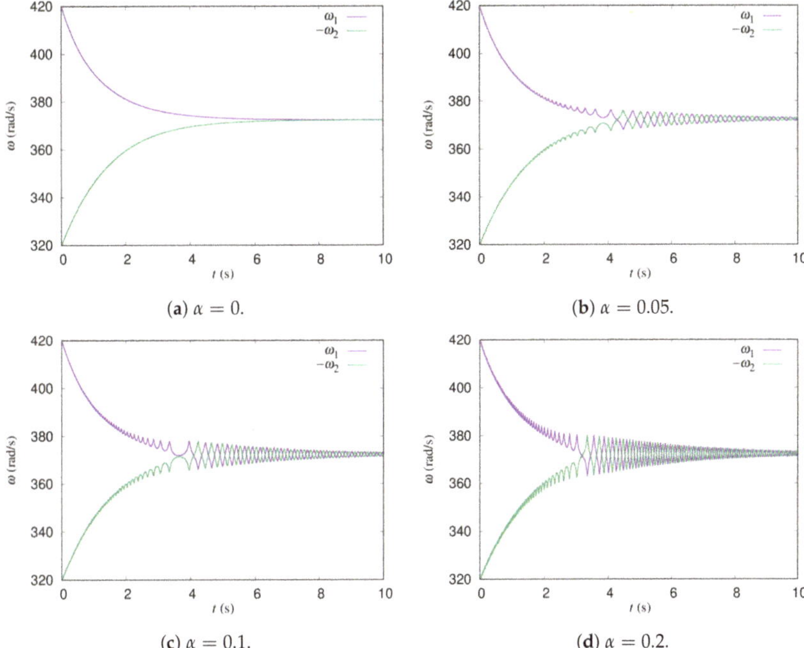

Figure 5. Time evolution of ω_i for $\alpha = 0, 0.05, 0.1$, and 0.2 in the CD layout with a 10 mm gap.

Figure 6 shows the time evolution of the phase differences $\delta\psi_k = (\psi_{11} + \psi_{2k}) \bmod 2\pi$ when $\alpha = 0.05$ and $\alpha = 0.2$. The index k takes a value of $1, 2$, or 3. Before the phase synchronization, $\delta\psi_k$ runs over the whole range of azimuthal angles. During phase synchronization, on the other hand, $\delta\psi_k$ oscillates around a fixed angle. One of the three blades of rotor 2 has a phase difference of π with blade 1 of rotor 1, which means that those blades meet in the gap region every rotation. We see that the oscillation period of the angular velocity during phase synchronization is shorter for larger α values, as shown in Figure 6. Moreover, the phase synchronization starts earlier for larger α values.

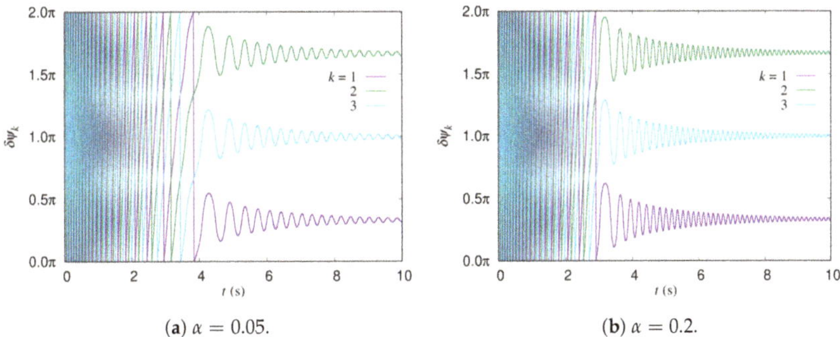

Figure 6. Time evolution of phase differences $\delta\psi_k = (\psi_{11} + \psi_{2k}) \bmod 2\pi$ for $\alpha = 0.05$ and 0.2 in the CD layout with 10 mm gap.

Phase synchronization occurs also in the CU layout. Since the induced velocity is not taken into account in the simulations presented in this section, the oscillation period and amplitude are the same as those in the CD layout.

Surprisingly, phase synchronization also occurs in the CO layout. The time evolution of ω for $\alpha = 0.05$ and 0.2 is shown in Figure 7. Compared with the data presented in

Figure 5 for the CD layout, the oscillation period of the angular velocities of the CO layout is longer than those of the CD layout with the same α values. We can also see that the oscillation amplitude of the angular velocities is smaller in the CO layout than that in the CD layout. This is of course because the interaction in the CO layout is much weaker than in the CD and CU layouts. In this simulation, the characteristics of the rotors are identical, and thus, the angular velocity is the same at the steady state without the interaction. In the beginning, the magnitudes of ω_1 and ω_2 are largely different. Thus, they have almost no interaction. However, as the magnitudes of ω_1 and ω_2 become closer to their steady-state values, the time period that the blades of the rotors stay in the narrow gap region together becomes longer, although it must still be much shorter than that for the CD and CU layouts. This makes the interaction effect visible, even in the CO layout.

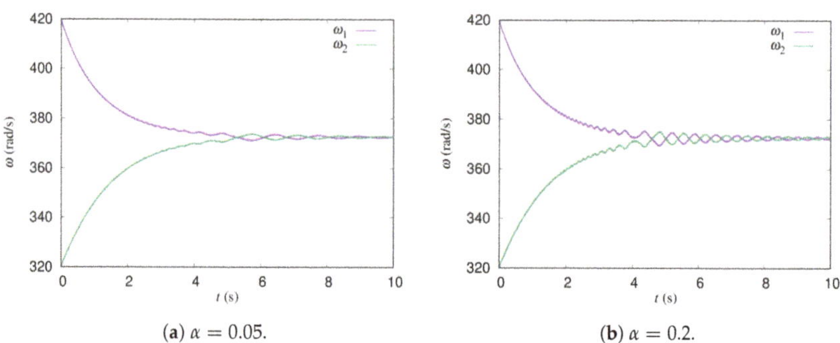

(**a**) $\alpha = 0.05$.   (**b**) $\alpha = 0.2$.

Figure 7. Time evolution of ω_i for $\alpha = 0.05$ and 0.2 in the CO layout with 10 mm gap. Induced velocities are not taken into account.

In fact, it becomes difficult for phase synchronization to occur if the angular velocities of the rotors at steady state without interaction are different from each other. We set the rotor torque of rotor 2 as 95% of that of rotor 1 while keeping the load torque unchanged. The other parameters are the same as those used in Figure 7. This makes the angular velocity of rotor 2 about 366 rad/s at its steady state without interaction, which is about 1.7% smaller than that of rotor 1. The time evolution of ω for $\alpha = 0.1$ is plotted in Figure 8a. Phase synchronization does not occur in this case. The angular velocities oscillate around each steady-state value. Note that phase synchronization occurs when $\alpha \geq 0.5$, even in this case.

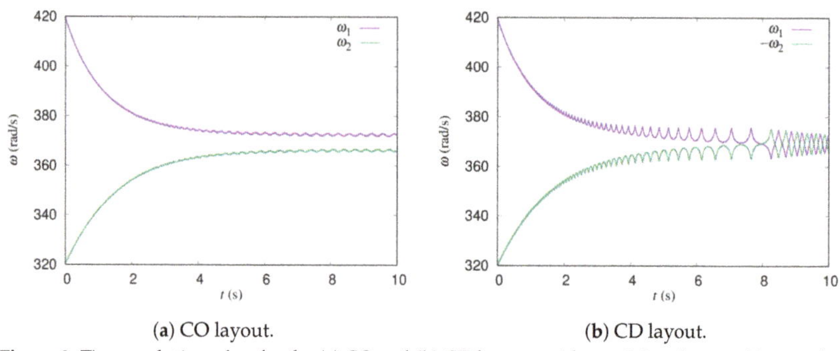

(**a**) CO layout.   (**b**) CD layout.

Figure 8. Time evolution of ω_i for the (**a**) CO and (**b**) CD layouts with $\alpha = 0.1$ and $g_{12} = 10$ mm when the rotor torque of rotor 2 is 95% that of rotor 1. The induced velocities are not taken into account.

On the other hand, phase synchronization occurs for the CD layout, even at $\alpha = 0.1$, which is shown in Figure 8b. The angular velocity of the rotors during phase synchronization is about 369 rad/s, which is about an average of the natural values of the two rotors without interaction. Phase synchronization occurs for smaller interactions because the relative speed of the blades of the two rotors is smaller for the CD layout than for the CO layout, leading to a longer interaction duration.

It may be worth pointing out that α can be used to qualitatively reproduce the parameter dependence of the experimental and CFD results on aspects such as the solidity and upstream flow velocity by assuming that α is dependent on these parameters. This is left as a future issue.

3.3. Dependence of the Synchronized Angular Velocity on the Gap

In this Section 3.3, we focus on the dependence of the synchronized angular velocity on the gap, especially when we take into account the mutually induced velocities generated by each component. Therefore, we set $|c_3| = 0.0427$ m according to the CFD simulation results [13,14]. The sign of c_3 is positive when $\omega > 0$ and negative when $\omega < 0$. On the contrary, we set $a_i = 0$, since the self-induced velocity by the rotor itself should be the same for identical rotors. When $a_1 = a_2$, the effect is only a change in the upstream velocity by a factor of $a_1 = a_2$ for both rotors. Note that the torque modulation $F_{n_i}(\psi_i)$ in Equation (3) with $n_i = 3$ is taken into account in the results presented in this section.

The angular velocity at steady state is plotted against the gap width in Figure 9. Each angular velocity is obtained using the Fourier transform of the time series data at the steady state. We show the results for $\alpha = 0.05$, although they are the same for different α values. Additionally, $F_{n_i}(\psi_i)$ just introduces modulation of the angular velocity, of which the period is one-third of the rotation period of the rotor and does not affect the phase-synchronized angular velocity.

The magnitudes of the angular velocities ω_1 and ω_2 agree well in both the CD and CU layouts. The dashed line at $\omega \simeq 372$ rad/s is the angular velocity without the mutually induced velocity. The angular velocities at each g_{12}/D are larger (smaller) than this value in the CD (CU) layout. This is due to the induced velocity. The effective velocity is larger (smaller) than the upstream velocity in the CD (CU) layout, as found in the CFD simulation [13,14]. Furthermore, the angular velocity increases (decreases) as the gap is narrowed in the CD (CU) layout because the magnitude of the induced velocity becomes larger for smaller gaps. This dependence can be clearly observed for the CD layout in the experiments [7,8] and the CFD simulations, except for the small gap distances [13,14]. For the CU layout, on the other hand, further analysis is required for comparison with the experiments and the CFD simulations.

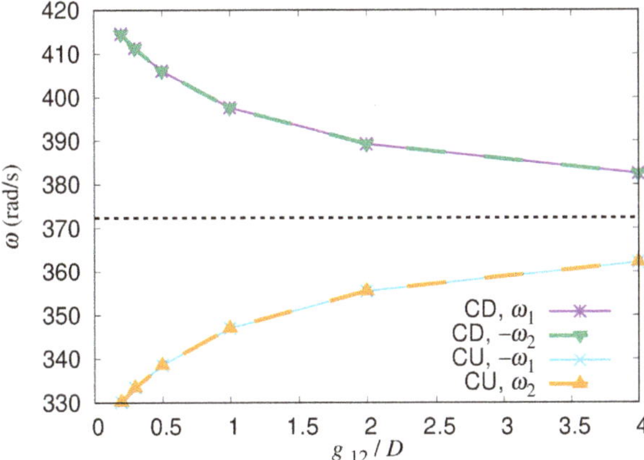

Figure 9. The angular velocity in the phase-synchronized steady state is plotted against the gap g_{12}/D. The dashed line around $\omega \simeq 372$ rad/s shows the angular velocity at the steady state without the mutually induced velocity.

3.4. Oscillation Period of the Difference in Angular Velocities

In this Section 3.4, we analyze the oscillation period of the difference of angular velocities $\Delta\omega := |\omega_1| - |\omega_2|$. The parameters used are the same as those in Section 3.3. Figure 10 shows the time evolution of $\Delta\omega$ for the CD and CU layouts with $g_{12}/D = 0.2$ and 0.3. The parameter $\alpha = 0.05$ was chosen to control the interaction strength for which the oscillation period of $\Delta\omega$ has a comparable order of magnitude with the experiments [4,5] and the CFD simulations [13,14]. The difference $\Delta\omega$ oscillates around 0 during phase synchronization and decays gradually. The oscillation period becomes shorter as time proceeds in all cases, as shown in Figure 10.

First, we found that the oscillation period is longer for the CU layout than the CD layout if the gap distance is the same. For example, the oscillation period is about 0.36 s for the CD layout with $g_{12}/D = 0.2$ at around $t = 5$ s, while it is about 0.55 s for the CU layout.

Second, we found that the oscillation period is longer for a larger gap distance for a given layout. For example, the oscillation period is about 0.36 s when $g_{12}/D = 0.2$ for the CD layout at around $t = 5$ s, while it is about 0.62 s when $g_{12}/D = 0.3$.

The oscillation period of $\Delta\omega$ seems to be related to the mean angular velocity during phase synchronization; $\Delta\omega$ is longer for smaller mean angular velocities. The mean angular velocity is smaller in the CU layout than in the CD layout because of the mutual induced velocity (see Figure 9). Additionaly, the interaction becomes weaker if the gap distance is larger, thereby the oscillation period becomes longer.

Note that it is difficult to find such trends for the amplitude of $\Delta\omega$, since it decays over time. However, we found that the decay is slower for the CU layout than the CD layout if the gap distance is the same. The damping rate seems to be smaller for longer oscillation periods. The oscillation amplitude for the CD layout with $g_{12}/D = 0.2$ is about 7 rad/s at around $t = 5$ s. The relative magnitude for the mean angular velocity $\omega \simeq 415$ rad/s is about 1.7%. The relative magnitude is a bit larger for the CU layout, since the mean angular velocity is smaller while the oscillation amplitude is comparable if the gap distance is the same.

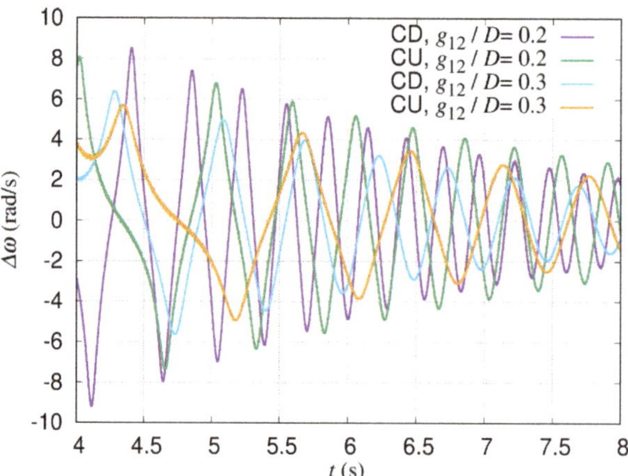

Figure 10. Time evolution of difference of angular velocities $\Delta\omega := |\omega_1| - |\omega_2|$ is shown. The oscillation period is longer for larger gap distances, as well as for the CU layout compared with the CD layout.

4. Conclusions

We developed an analytical model of the interaction torque between two vertical-axis wind turbines through pressure fluctuation. In this model, the pressure fluctuation is obtained according to Bernoulli's law, taking into account the temporal change in distance between the blades. Although rather crude assumptions were made in the development of the model, our simulations successfully demonstrated the phase synchronization as well as the oscillation of angular velocities around the mean value observed in the experiments and CFD simulations.

Our simulation results show that the angular velocities of the rotors oscillate in time during phase synchronization. When an artificial parameter is changed to strengthen the interaction, the oscillation period becomes shorter and the amplitude becomes larger. This is reasonable physically.

It was also found that phase synchronization occurs even for a pair of rotors with slightly different torque characteristics. This is important because the characteristics of the rotors cannot be identical in experiments.

Our model includes the induced velocities, which change the effective wind speed at each rotor. The simulation results also show that the mutually induced velocity can explain the qualitative dependence of the phase-synchronized angular velocities on the rotational direction of the rotors and the gap distance between them.

The oscillation period of the difference in angular velocities was found to be longer in the CU layout than in the CD layout. Additionally, the oscillation period was found to be longer for larger gap distances. This dependence seems to be related to the mean angular velocity of the rotors during phase synchronization. The mutually induced velocity changed the mean angular velocity in our simulations. The weaker interaction for larger gap distance made the oscillation period longer.

Author Contributions: Conceptualization, M.F., Y.H. and Y.J.; methodology, M.F.; software, M.F.; validation, Y.H. and Y.J.; formal analysis, M.F.; investigation, M.F.; resources, M.F.; data curation, M.F.; writing—original draft preparation, M.F.; writing—review and editing, Y.H. and Y.J.; visualization, M.F.; supervision, Y.H.; project administration, Y.H. and Y.J.; funding acquisition, Y.H. and Y.J. All authors have read and agreed to the published version of the manuscript.

Funding: This work was supported by JSPS KAKENHI Grant Number JP18K05013. The second author was supported in part by the International Platform for Dryland Research and Education (IPDRE), Tottori University.

Institutional Review Board Statement: Not applicable

Informed Consent Statement: Not applicable

Data Availability Statement: The data presented in this study are available from the corresponding author upon reasonable request.

Conflicts of Interest: The authors declare no conflict of interest.

Nomenclature

D_i	Diameter of rotor i
R_i	Radius of rotor i
g_{ij}	Gap between rotors i and j
n_i	Number of blades on rotor i
x	Coordinate in streamwise direction
y	Coordinate in spanwise direction
x_{ik}	x coordinate of blade k of rotor i
y_{ik}	y coordinate of blade k of rotor i
ψ_{ik}	Azimuthal angle of blade k of rotor i
c_{ik}	Chord length of blade k of rotor i
S_{ik}	Projected area of blade k of rotor i along rotation direction
$W_{ik,j\ell}$	Distance between blade k of rotor i and blade ℓ of rotor j
ϕ_{ij}	Angle of rotor j observed from rotor i
V	Upstream flow speed
V_i	Effective flow velocity at rotor i
I_i	Moment of inertia of rotor i
ω_i	Angular velocity of rotor i
Q_i	Rotor torque on rotor i

L_i	Load torque on rotor i
Q_{pi}	Torque due to pressure fluctuation on rotor i
ψ_i	Azimuthal angle of representative blade of rotor i
$Q_{max}(V_i)$	Maximum rotor torque for given V_i
$\omega_0(V_i)$	No-load angular velocity for given V_i
$F_n(\psi)$	Torque-modulation function for n-blades rotor
c_1	Parameter for $Q_{max}(V_i)$
c_2	Parameter for $\omega_0(V_i)$
λ_i	Tip-speed ratio of rotor i
ψ	Azimuthal angle fixed in space
σ_i	Solidity of rotor i
$F_1(\psi)$	Torque-modulation function for single-blade rotor
a_i	Parameter for self-induced velocity
Γ_j	Circulation of rotor j
c_3	Parameter for Γ_j
c_4	Parameter for L_i
V_{av}	Average flow velocity $(V_1 + V_2)/2$
δV	Change in flow velocity
δp	Pressure fluctuation
p	Air pressure
α	Parameter controlling strength of interaction between rotors
ω_{SI}	Steady-state angular velocity of single rotor by CFD using DFBI model
Q_{SI}	Steady-state rotor torque of single rotor by CFD using DFBI model
$\omega_{av}(0)$	Average of initial angular velocities $(\lvert\omega_1(0)\rvert + \lvert\omega_2(0)\rvert)/2$
$\delta\psi_k$	Phase differences between representative blade of rotor 1 and blade k of rotor 2
$\Delta\omega$	Difference of angular velocities $\lvert\omega_1\rvert - \lvert\omega_2\rvert$
C_{qi}	Torque coefficient of rotor i
$\hat{Q}_{max}$	Normalized maximum rotor torque
λ_0	No-load tip-speed ratio
ρ	Mass density of air
A_i	Swept area of rotor i
$\hat{\Gamma}_j$	Normalized circulation of rotor j
C_{Li}	Normalized load torque for rotor i
c_L	Parameter for C_{Li}

Appendix A. Verification of Parameter Dependence

In this Appendix A, CFD and experimental data to determine dependence of Q_{max} on flow velocity V in Equation (4), dependence of ω_0 on V in Equation (5), CFD data to determine expression of the torque-modulation function in Equation (6), and dependence of Γ on V in Equation (8) are shown.

First, let us show the dependence of Q_{max} on flow velocity V read from Figure 2 of Ref. [14] and Figure 3 of Ref. [8]. Figure A1 shows the CFD and experimental data, as well as their fitting curves taken to be quadratic in V. The fitting curves agree well with the CFD and experimental data.

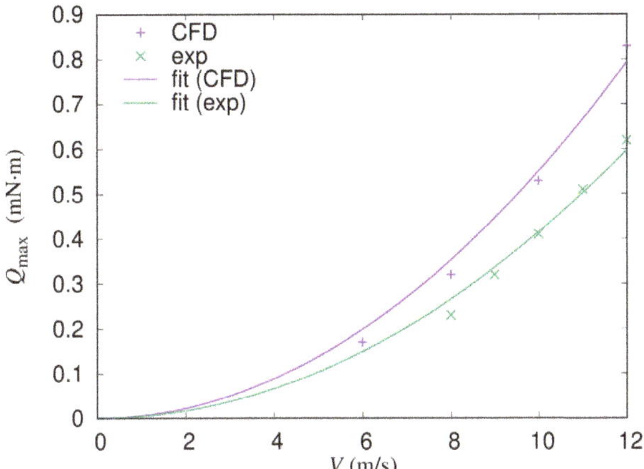

Figure A1. $Q_{\max}$ from Figure 2 of Ref. [14], Figure 3 of Ref. [8], and their fitting curves are shown.

Second, let us show the dependence of ω_0 on flow velocity V, also read from Figure 2 of Ref. [14] and Figure 3 of Ref. [8]. Figure A2 shows the CFD and experimental data, as well as their fitting curves taken to be linear in V. Although the number of data points is not enough, the fitting curves appear to agree well with the CFD and experimental data.

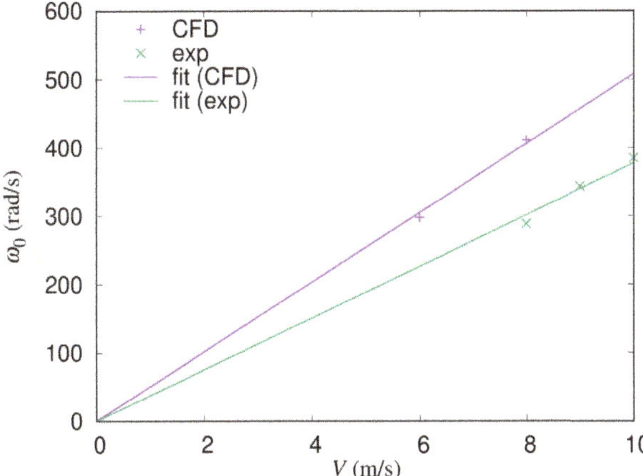

Figure A2. ω_0 from Figure 2 of Ref. [14], Figure 3 of Ref. [8], and their fitting curves are shown.

Third, the torque dependence on the azimuthal angle obtained by the CFD is shown. A rotor with three blades, of which dimensions are the same as described in Section 3.1, was placed in the flow field of which upstream flow velocity was 10 m/s. The modulation of the torque, expressed by Equation (6), tries to simulate this dependence. Figure A3 shows the averaged torques over the 16th–20th rotations of the CFD data, which was not published previously. For $n = 3$, the torque is the summation of torques on all blades of the rotor. The torque for a single blade has its maximum at $\psi \simeq \pi/2$. In the downstream half, the torque on a single blade becomes negative in the CFD, although it is set to be zero in Equation (6) for simplicity.

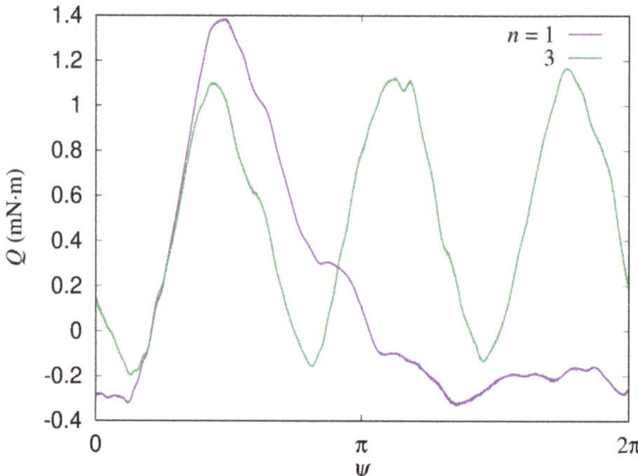

Figure A3. Torque dependence on the azimuthal angle obtained by CFD. Equation (6) tries to simulate this dependence.

Finally, let us show the dependence of Γ on flow velocity. Figure A4 shows Γ obtained by CFD simulations, and their fitting curves taken to be linear in flow velocity. The CFD data were obtained by the time average of the simulation results. In the figure, "inf" means a plot of asymptotic values of Γ at infinity against upstream flow speeds, and the "ep", or evaluation point, means a plot of Γ evaluated on a circle with a 36.1 mm radius centered at the rotor against flow speeds on the upstream side at 36.1 mm from the rotor center. Precisely, the flow speeds of the "ep" case are obtained by averaging over a 50 mm range in the spanwise direction. Note that the horizontal axis is not the upstream velocity for the "ep" case, although it is labeled by V. The fitting curves agree well with the CFD data.

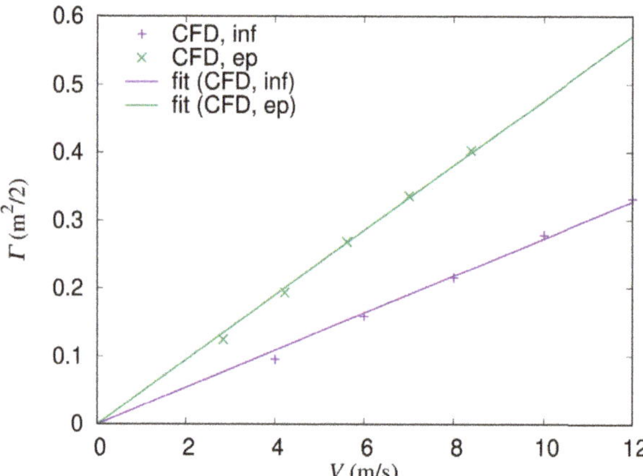

Figure A4. Γ obtained by CFD results and the fitting curves are shown.

Appendix B. Normalized Expression

In this Appendix B, normalized expressions of our model are summarized. First, let us normalize $Q_{\max}(V_i)$ and $\omega_0(V_i)$ given in Equations (4) and (5), respectively, as follows:

$$
\begin{aligned}
\hat{Q}_{\max} &:= \frac{Q_{\max}(V_i)}{\rho V_i^2 A_i R_i / 2} \\
&= \frac{2c_1}{\rho A_i R_i},
\end{aligned}
\tag{A1}
$$

$$
\begin{aligned}
\lambda_0 &:= \frac{\omega_0(V_i)}{V_i / R_i} \\
&= c_2 R_i,
\end{aligned}
\tag{A2}
$$

where ρ is the mass density of air, and A_i is the swept area of rotor i. Note that $\hat{Q}_{\max}$ and λ_0 are independent of V_i. By using Equations (A1) and (A2), the torque coefficient C_{qi}, or normalized rotor torque Q_i given in Equation (3) with $F_{n_i}(\psi_i) \equiv 1$, can be expressed as

$$
\begin{aligned}
C_{qi} &:= \frac{Q_i}{\rho V_i^2 A_i R_i / 2} \\
&= -\frac{3\sqrt{3}\hat{Q}_{\max}}{2\lambda_0^3} \lambda_i (\lambda_i + \lambda_0)(\lambda_i - \lambda_0),
\end{aligned}
\tag{A3}
$$

where $\lambda_i := R_i \omega_i / V_i$ is the tip–speed ratio of rotor i. The torque coefficient C_{qi} is expressed only by λ_i, and the dimensionless parameters $\hat{Q}_{\max}$ and λ_0 only. The rotor torque becomes zero when λ becomes equal to the no-load tip–speed ratio λ_0.

Next, the circulation in Equation (8) is normalized as

$$
\begin{aligned}
\hat{\Gamma}_j &:= \frac{\Gamma_j}{R_i V_i} \\
&= \frac{c_3}{R_j}.
\end{aligned}
\tag{A4}
$$

Note that $\hat{\Gamma}_j$ is independent of V_i.

Lastly, let us normalize the load torque given in Equation (9) as

$$
\begin{aligned}
C_{Li} &:= \frac{L_i}{\rho V_i^2 A_i R_i / 2} \\
&= c_L \lambda_i^2,
\end{aligned}
\tag{A5}
$$

where

$$
c_L := \frac{2c_4}{\rho A_i R_i^3}.
\tag{A6}
$$

Again, the parameter c_L is independent of V_i.

By using these expressions, numerical values of $\hat{Q}_{\max}$, λ_0, $\hat{\Gamma}$, and c_L corresponding to the parameters used in Section 3 are obtained as follows:

$$
\hat{Q}_{\max} = 0.176,
\tag{A7}
$$

$$
\lambda_0 = 1.30,
\tag{A8}
$$

$$
\hat{\Gamma} = 1.71,
\tag{A9}
$$

$$
c_L = 0.182.
\tag{A10}
$$

Note that the mass density of air is assumed to be $\rho = 1.20\,\text{kg/m}^3$. The rotor radius is $R = 50\,\text{mm}$ and the swept area is $A = 50\,\text{mm} \times 43.4\,\text{mm} = 2.17 \times 10^{-3}\,\text{m}^2$.

Let us point out that the simulation results shown in Section 3 can be interpreted, if ω_i is expressed by the tip–speed ratio λ_i, as results with a different set of a flow velocity V_i, a rotor radius R_i, and a swept are A_i that give the same dimensionless parameters $\hat{Q}_{\max}$, λ_0, $\hat{\Gamma}$, and c_L.

References

1. Thomas, R.N. Coupled Vortex Vertical Axis Wind Turbine. US Patent US 6,784,566 B2, 31 August 2004.
2. Dabiri, J.O.; Greer, J.R.; Koseff, J.R.; Moin, P.; Peng, J. A new approach to wind energy: Opportunities and challenges. *AIP Conf. Proc.* **2015**, *1652*, 51–57. [CrossRef]
3. Ahmadi-Baloutaki, M.; Carriveau, R.; Ting, D.S.K. A wind tunnel study on the aerodynamic interaction of vertical axis wind turbines in array configurations. *Renew. Energy* **2016**, *96*, 904–913. [CrossRef]
4. Vergaerde, A.; De Troyer, T.; Kluczewska-Bordier, J.; Parneix, N.; Silvert, F.; Runacres, M.C. Wind tunnel experiments of a pair of interacting vertical-axis wind turbines. *J. Phys. Conf. Ser.* **2018**, *1037*, 072049.[CrossRef]
5. Vergaerde, A.; De Troyer, T.; Standaert, L.; Kluczewska-Bordier, J.; Pitance, D.; Immas, A.; Silvert, F.; Runacres, M.C. Experimental validation of the power enhancement of a pair of vertical-axis wind turbines. *Renew. Energy* **2020**, *146*, 181–187. [CrossRef]
6. Jiang, Y.; Zhao, P.; Stoesser, T.; Wang, K.; Zou, L. Experimental and numerical investigation of twin vertical axis wind turbines with a deflector. *Energy Convers. Manag.* **2020**, *209*, 112588. [CrossRef]
7. Jodai, Y.; Hara, Y.; Sogo, Y.; Marusasa, K.; Okinaga, T. Wind tunnel experiments on interaction between two closely spaced vertical axis wind turbines. In Proceedings of the 23rd Chu-Shikoku-Kyushu Branch Meeting, The Japan Society of Fluid Mechanics, Yamaguchi, Japan, 1–2 June 2019; pp. 11-1–11-2.
8. Jodai, Y.; Hara, Y. Wind Tunnel Experiments on Interaction between Two Closely Spaced Vertical-Axis Wind Turbines in Side-by-Side Arrangement. *Energies* **2021**, *14*, 7874. [CrossRef]
9. Zanforlin, S.; Nishino, T. Fluid dynamic mechanisms of enhanced power generation by closely spaced vertical axis wind turbines. *Renew. Energy* **2016**, *99*, 1213–1226. [CrossRef]
10. Chen, W.H.; Chen, C.Y.; Huang, C.Y.; Hwang, C.J. Power output analysis and optimization of two straight-bladed vertical-axis wind turbines. *Appl. Energy* **2017**, *185*, 223–232. [CrossRef]
11. De Tavernier, D.; Ferreira, C.; Li, A.; Paulsen, U.S.; Madsen, H.A. Towards the understanding of vertical-axis wind turbines in double-rotor configuration. *J. Phys. Conf. Ser.* **2018**, *1037*, 022015. [CrossRef]
12. Ma, Y.; Hu, C.; Li, Y.; Li, L.; Deng, R.; Jiang, D. Hydrodynamic Performance Analysis of the Vertical Axis Twin-Rotor Tidal Current Turbine. *Water* **2018**, *10*, 1694. [CrossRef]
13. Hara, Y.; Jodai, Y.; Okinaga, T.; Sogo, Y.; Kitoro, T.; Marusasa, K. A synchronization phenomenon of two closely spaced vertical axis wind turbines. In Proceedings of the 25th Chu-Shikoku-Kyushu Branch Meeting, The Japan Society of Fluid Mechanics, Takamatsu, Japan, 31 May 2020; pp. 3-1–3-2.
14. Hara, Y.; Jodai, Y.; Okinaga, T.; Furukawa, M. Numerical Analysis of the Dynamic Interaction between Two Closely Spaced Vertical-Axis Wind Turbines. *Energies* **2021**, *14*, 2286. [CrossRef]
15. Jin, G.; Zong, Z.; Jiang, Y.; Zou, L. Aerodynamic analysis of side-by-side placed twin vertical-axis wind turbines. *Ocean Eng.* **2020**, *209*, 107296. [CrossRef]
16. Whittlesey, R.W.; Liska, S.; Dabiri, J.O. Fish schooling as a basis for vertical axis wind turbine farm design. *Bioinspir. Biomimetics* **2010**, *5*, 035005. [CrossRef] [PubMed]
17. Novak, P.; Ekelund, T.; Jovik, I.; Schmidtbauer, B. Modeling and control of variable-speed wind-turbine drive-system dynamics. *IEEE Control Syst. Mag.* **1995**, *15*, 28–38. [CrossRef]

 *energies*

Article

Method to Predict Outputs of Two-Dimensional VAWT Rotors by Using Wake Model Mimicking the CFD-Created Flow Field

Jirarote Buranarote [1,*], Yutaka Hara [2], Masaru Furukawa [2] and Yoshifumi Jodai [3]

[1] Department of Mechanical and Aerospace Engineering, Tottori University, 4-101 Koyama-Minami, Tottori 680-8552, Japan

[2] Advanced Mechanical and Electronic System Research Center, Faculty of Engineering, Tottori University, 4-101 Koyama-Minami, Tottori 680-8552, Japan; hara@tottori-u.ac.jp (Y.H.); furukawa@tottori-u.ac.jp (M.F.)

[3] Department of Mechanical Engineering, Kagawa National Institute of Technology (KOSEN), Kagawa College, 355 Chokushi, Takamatsu 761-8058, Japan; jodai@t.kagawa-nct.ac.jp

* Correspondence: jirarote@eng.src.ku.ac.th; Tel.: +81-070-7490-6660

Abstract: Recently, wind farms consisting of clusters of closely spaced vertical-axis wind turbines (VAWTs) have attracted the interest of many people. In this study, a method using a wake model to predict the flow field and the output power of each rotor in a VAWT cluster is proposed. The method uses the information obtained by the preliminary computational fluid dynamics (CFD) targeting an isolated single two-dimensional (2D) VAWT rotor and a few layouts of the paired 2D rotors. In the method, the resultant rotor and flow conditions are determined so as to satisfy the momentum balance in the main wind direction. The pressure loss of the control volume (CV) is given by an interaction model which modifies the prepared information on a single rotor case and assumes the dependence on the inter-rotor distance and the induced velocity. The interaction model consists of four equations depending on the typical four-type layouts of selected two rotors. To obtain the appropriate circulation of each rotor, the searching range of the circulation is limited according to the distribution of other rotors around the rotor at issue. The method can predict the rotor powers in a 2D-VAWT cluster including a few rotors in an incomparably shorter time than the CFD analysis using a dynamic model.

Keywords: vertical-axis wind turbine; wake model; computational fluid dynamics; closely spaced arrangement; rotor cluster; interaction effect; momentum conservation; control volume

Citation: Buranarote, J.; Hara, Y.; Furukawa, M.; Jodai, Y. Method to Predict Outputs of Two-Dimensional VAWT Rotors by Using Wake Model Mimicking the CFD-Created Flow Field. *Energies* **2022**, *15*, 5200. https://doi.org/10.3390/en15145200

Academic Editor: Charalampos Baniotopoulos

Received: 19 June 2022
Accepted: 16 July 2022
Published: 18 July 2022

Publisher's Note: MDPI stays neutral with regard to jurisdictional claims in published maps and institutional affiliations.

1. Introduction

Wind power generation is one of the alternatives to fossil energy. The obvious advantage of wind power is almost no pollution to the environment. To realize carbon neutrality, a large amount of renewable energy is expected to be introduced [1,2]. Therefore, wind power generators are becoming increasingly larger in size, and their application to offshore generation is increasing [3,4]. However, the efficiency of a wind farm can be reduced owing to the wake effects [5–7]. One of the challenges in the sector of wind power is to exactly evaluate the influence of the wakes, so as to improve the efficiency of a wind farm [8–14]. Additionally, to maximize the power output from a wind farm, it is necessary to deploy wind turbines with the optimal layout for the wind condition of the planned site [15–19]. The wake control methods using the blade's pitch [20] or the yaw [21,22] of the propeller-typed wind turbines were studied to optimize the wind farm.

According to the studies by Whittlesey et al. [23] and Dabiri [24], the output per unit land area of a wind farm consisting of small-size vertical-axis wind turbines (VAWTs) with a high aspect ratio (rotor height/diameter), which are closely arranged using a unit of counter-rotating paired rotors, can be much greater than that of a conventional wind farm consisting of large-size horizontal-axis wind turbines (HAWTs), which are deployed with the inter-rotor intervals of several multiples of the rotor diameter (in general, the interval in the dominant wind direction is about 10 times as long as the diameter). Since then, many researchers have been interested in the closely arranged VAWT wind farms and, especially, the interaction effects between two VAWT rotors. For example, Zanforlin and Nishino [25] performed a two-dimensional (2D) computational fluid dynamics (CFD) analysis of a pair of inversely rotating VAWTs to show the greater averaged output than the output of an isolated single VAWT. De Tavernier et al. [26] carried out the 2D-CFD based on the panel vortex method of a closely arranged VAWT pair, each of which had a 10 m rotor radius, to show the effects of the load and rotor spacing on the paired rotor performance. Bangga et al. [27] proposed two layouts of a VAWT array based on their CFD study of rotor pairs arranged side by side. Sahebzadeh et al. [28] numerically analyzed the output performance of a co-rotating rotor pair by widely changing the rotor spacing and the relative angle to the main stream. Peng et al. [29] investigated the effects of configuration parameters such as airfoil section, solidity, pitch angle, rotational direction, and turbine spacing on twin VAWTs by CFD analysis. The effects of a three-rotor cluster of VAWTs were also numerically studied by Hezaveh et al. [30] and Silva and Danao [31].

As examples of experimental studies of paired VAWTs, Vergaerde et al. [32] conducted the wind tunnel test using two H-type Darrieus rotors (rotor diameter: 0.5 m, rotor height: 0.8 m [33]). Their turbines were placed side-by-side against the main flow and were adequately controlled by DC motors. They observed the power increase up to 16% for the counter-rotating VAWTs and reported the stable synchronized operation of twin rotors. Jodai and Hara [34] studied the interaction between two closely spaced VAWTs by using miniature 3D printed rotors (diameter: 50 mm, low aspect ratio of 0.87, and high solidity of 0.382) arranged side-by-side. Their experimental results showed a maximum 15% increase in power in the case of the counter-down layout when the inter-rotor space became the shortest (gap space: 10% of the rotor diameter).

Hara et al. [35] applied the dynamic fluid body interaction (DFBI) model to the CFD analysis to simulate the closely arranged paired 2D rotors corresponding to the equator-cross section of the experimental model used in Ref. [34]. Their CFD analysis considering the time-varying rotor speed, for the first time, simulated the synchronization operation of twin rotors and showed the alternation in the angular velocities of two rotors. Furukawa et al. [36] developed an analytical model considering the pressure fluctuation (or increase in flow velocity) observed in the gap region between twin rotors in the above experiments and CFD analyses. The model successfully demonstrated the alternation in the angular velocities of two rotors and showed that the period of the variation in rotor speed depended on the strength of the interaction between the two rotors.

Although the increase in the averaged power of a closely spaced side-by-side VAWT pair is clear, the effects of the distribution of wind direction on the VAWT cluster consisting of many rotors must be investigated more extensively to search for the optimal layout of VAWTs. The CFD analysis, especially the simulation using the DFBI model, can give reliable results, but it needs a long calculation time. Although the experiments can also give useful information, the cost is high and the time for preparation is long. If the number of rotors in a target wind farm increases, the simulation by CFD or the experiment using a lot of rotors is non-realistic. Therefore, a method that can simulate precisely and in a short time the flow field of a wind farm including a large number of VAWTs is necessary.

Buranarote et al. [37] proposed a wake model of a 2D-VAWT rotor, which was named the ultra-super-Gaussian function because the model improved the super-Gaussian function proposed by Shapiro et al. [38] which could express the transformation of the wake profile from top-hat shape to Gaussian. The ultra-super-Gaussian function includes a correction function to express the acceleration regions and the deflection of a VAWT wake. The method proposed by Buranarote et al. [37] was based on the potential flow and included the velocity deficit artificially, like the method by Whittlesey et al. [23]. Buranarote et al. [39] improved their method by including the modification of the y-component (cross-flow) of the flow velocity to mimic the CFD results. Moreover, they introduced the Biot-Savart law to consider the effects of the interaction between the rotors [40] on the wake shift and width. However, in the previous method of Buranarote et al. [39,40], the circulation around each rotor, which was used to estimate the power output, was newly calculated with the modified flow field after adding the velocity deficits; the calculated circulation was sensitive to the flow condition resulted in failure in the reliable prediction of the power outputs of rotors.

This study proposes a new method to predict outputs of two-dimensional VAWT rotors by using a wake model mimicking the CFD-created flow field. The new method is based on the previous method but the circulation that is used to estimate the output of each rotor is the same as the input value to calculate the potential flow in a wind farm. The decision on the appropriate flow and rotor conditions is conducted by evaluating the momentum balance, which is calculated using the momentum transports and the pressure forces at the boundaries of the control volume (CV) and thrust forces of rotors. The necessary CFD data, or the available and reliable experimental data, are the power performance of an isolated single VAWT and the averaged flow velocity distribution in the CV, and the pressure distributions at the boundaries under several wind speed conditions. In addition, the power output data of closely spaced paired VAWTs in typical four layouts in the case of a specific inter-rotor distance are necessary. The interaction effects are considered by modifying the given pressure loss of the isolated single rotor, according to the relative layout of the selected two rotors and considering the distance and the induced velocity.

The method will be validated with the CFD results for two or three rotors studied by Hara et al. [35] and Okinaga et al. [41], in which the rotor height was considered so as to correspond to the experimental rotor used in the experimental study by Jodai et al. [34]. The 2D-CFD analysis of an isolated single rotor is outlined in Appendix B. The same 2D-CFD rotor model is used for the CFD analysis of four-rotor arrays conducted in this study. The present method does not include the three-dimensional effects caused by the finite rotor height because our target in the future is a wind farm consisting of small-scale VAWTs of 14 m diameter with a low aspect ratio.

The final goal of our project is to provide a cost-effective and relatively short-time method to optimize the layout of VAWTs in an arbitrary wind farm. In this paper, at the early stage of the project, we show the possibility to predict a reasonable condition of a VAWT cluster. Therefore, the round robin, which needs a long calculation time when the number of rotors is large, is utilized in the search for adequate conditions. The maximum number of rotors in a VAWT cluster considered in this study is four due to the problem of calculation time. However, if some advanced optimization method like the genetic algorithm (GA) is adopted, the problem of the computation time will be mitigated.

The detail of the new method is described in the next section. The application of the method to an isolated single rotor, paired rotors, three-rotor clusters, and four-rotor layouts is discussed in Section 3. Finally, Section 4 concludes the discussion.

2. Model

2.1. Method

Figure 1 shows a schematic image of a VAWT wind farm, where a VAWT rotor is shown by a circle and our method does not need detailed information on the configuration of a turbine such as the number of blades and the cross-section. In our method, the VAWTs are dealt with as 2D rotors with each having a diameter, D. Let us assume the wind farm consists of N VAWTs. In Figure 1, the coordinate axis x is defined as parallel to the upstream wind speed U_∞; the coordinate axis y is perpendicular to the dominant wind direction. The center position (rotational axis) of the k-th rotor is expressed as (x_k, y_k). Our method assumes that the circulation around an isolated single rotor (Γ) is given as a linear function of wind speed (U_∞) in advance. Therefore, the blockage effect (dipole μ), thrust force (Th), and power output (P) are given as functions of the circulation, respectively, for each rotor.

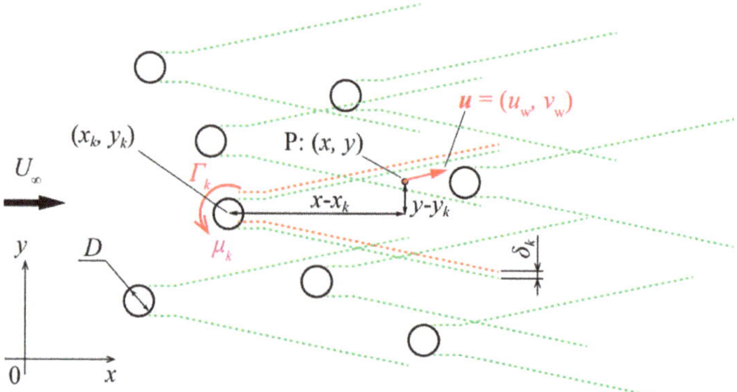

Figure 1. Schematic image of a wind farm of vertical-axis wind turbines (VAWTs).

If an arbitrary set of the conditions of circulation (Γ_k) and blockage effect (μ_k) is given for a wind farm, the complex velocity potential $W(z)$, where $z = x + iy$ is expressed by Equation (1) [23].

$$W(z) = U_\infty z + \sum_{k=1}^{N} \left\{ -i\frac{\Gamma_k}{2\pi} \ln(z - z_k) + \mu_k (z - z_k)^{-1} \right\} \tag{1}$$

Using the above potential $W(z)$, the potential flow ($u_{\mathrm{p}}, v_{\mathrm{p}}$) at an arbitrary position P:(x, y) in the wind farm can be calculated as follows:

$$u_{\mathrm{p}}(x, y) = U_\infty - \sum_{k=1}^{N} \left[\frac{\Gamma_k}{2\pi} \frac{(y - y_k)}{(x - x_k)^2 + (y - y_k)^2} + \mu_k \frac{(x - x_k)^2 - (y - y_k)^2}{\left\{ (x - x_k)^2 + (y - y_k)^2 \right\}^2} \right] \tag{2}$$

$$v_{\mathrm{p}}(x, y) = \sum_{k=1}^{N} \left[\frac{\Gamma_k}{2\pi} \frac{(x - x_k)}{(x - x_k)^2 + (y - y_k)^2} - \mu_k \frac{2(x - x_k)(y - y_k)}{\left\{ (x - x_k)^2 + (y - y_k)^2 \right\}^2} \right] \tag{3}$$

However, the potential flow is an ideal flow and cannot express the actual velocity deficit or wake generated by each rotor. Therefore, as Whittlesey et al. introduced [23], the component in the x-direction of the potential flow is modified by a wake function du_k showing the velocity deficit of each rotor; the resultant flow $u_{\mathrm{w}}(x, y)$ is expressed by Equation (4). The subscript "w" means the flow field obtained by applying the wake function (or wake model) in this study.

$$u_{\mathrm{w}}(x, y) = u_{\mathrm{p}}(x, y) \left\{ 1 - \sum_{k=1}^{N} du_k(x, y) \right\} \tag{4}$$

In our method, the wake function du_k is given by Equation (5) including the ultra-super-Gaussian function $f_{\mathrm{USG\_}k}$ defined in our previous study [37]. The function $f_{\mathrm{USG\_}k}$ modifies the super-Gaussian function $f_{\mathrm{SG\_}k}$ proposed by Shapiro et al. [38] expressing the wake profile transformation from top-hat shape to Gaussian by adding the correction function $f_{\mathrm{COR\_}k}$ expressing the acceleration regions and the deflection δ_k of the wake. μ_{ref} is a reference value of the blockage effect and $C_{\mathrm{W\_}k}$ is a fitting parameter to the prepared flow field of an isolated single rotor.

$$du_k(x, y) = C_{\mathrm{W\_}k} \frac{\mu_k}{\mu_{\mathrm{ref}}} f_{\mathrm{USG\_}k} = C_{w\_k} \frac{\mu_k}{\mu_{\mathrm{ref}}} \left\{ f_{\mathrm{SG\_}k} - f_{\mathrm{COR\_}k} \right\} \tag{5}$$

The present method modifies the wake deflection δ_k and the wake width d_W, which are used in the ultra-super-Gaussian function f_{USG_k}, by using the induced velocity given by the Biot–Savart law. The details are described in Appendix A.

The component in the y-direction of the potential flow is also modified by Equation (6) in our method [40].

$$v_{\text{w}}(x,y) = v_{\text{p}}(x,y) + U_\infty \sum_{k=1}^{N} \frac{|\Gamma_k|}{|\Gamma_{\text{SI}}|} dv_k(x,y) \tag{6}$$

where dv_k shows the difference in y-component velocity between the potential flow and the prepared flow (by CFD or experiments) around an isolated single rotor, the circulation of which is defined as Γ_{SI} and depends on U_∞ (see Equation (A17) in Appendix B). The dv_k is approximated using the sum of four Gaussian-type functions f_{G_i} and four resonant-type functions f_{R_i} as shown in Equation (7). The details of functions f_{G_i} and functions f_{R_i} are described in Appendix A.

$$dv_k(x,y) = \sum_{i=1}^{4} f_{\text{G}\_i} + \sum_{i=1}^{4} f_{\text{R}\_i} \tag{7}$$

If an arbitrary set of circulations (Γ_k) of N rotors is given, a flow field and a set of rotor outputs are calculated by using Equations (2)–(7) and the prepared relation between the circulation and rotor power. However, the result is not always the actual correct condition. Therefore, we have to find out the set of circulations that give the condition satisfying the conservation of momentum expressed by Newton's law of motion.

Figure 2 shows a schematic image of a control volume (CV) used for the calculation of the flow field around an isolated single rotor. The prepared data of an isolated single rotor, regardless of CFD data or experimental data, must satisfy Equation (8). The left-hand side of Equation (8) is the total force acting on the fluid in the x-direction (main flow direction) and the right-hand side shows the variation in the momentum in the x-direction per unit time. The variation in the momentum $\Delta(\dot{m}u_x)$ is calculated from the flow velocity at the boundaries of CV by Equation (9), in which the mass flow rate $\dot{m}_x$ and $\dot{m}_y$ are expressed by $\rho u dy$ and $\rho v dx$, respectively. Here, ρ is the air density; dx and dy are small boundary elements. The total force F_{total_x} is given by Equation (10), in which the pressure forces Fp_{in} and Fp_{out} acting on the inlet and outlet boundaries are considered and are calculated by the integration along each boundary. The Fp in Equation (10) shows the total pressure force in the x-direction and means the pressure loss. The forces ($F_{\_\tau\_\text{top}}$, $F_{\_\tau\_\text{bottom}}$) caused by shear stress on the top and bottom boundaries are neglected in this study. From Equations (8) and (10), the relation of Equation (11) is obtained for an isolated single rotor. In our method, the pressure loss Fp has to be prepared as a function of the upstream speed or the corresponding circulation of a single rotor for an applied CV. Note that the pressure loss Fp cannot be calculated from the model flow field obtained by using Equations (4) and (6).

$$F_{\text{total}\_x} = \Delta(\dot{m}u_x) \tag{8}$$

$$\Delta(\dot{m}u_x) = -\int \dot{m}_x u_{\text{in}} + \int \dot{m}_x u_{\text{out}} - \int \dot{m}_y u_{\text{bottom}} + \int \dot{m}_y u_{\text{top}} \tag{9}$$

$$F_{\text{total}\_x} = -Th + \int p_{\text{in}} dy - \int p_{\text{out}} dy = -Th + Fp_{\text{in}} - Fp_{\text{out}} = -Th + Fp \tag{10}$$

$$\Delta(\dot{m}u_x) = -Th + Fp \tag{11}$$

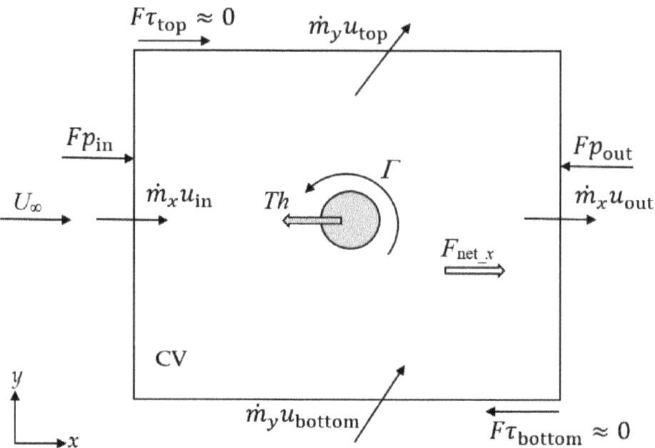

Figure 2. Schematic image showing the momentum transports and relevant forces in the x-direction of the control volume (CV) in the case of an isolated single rotor.

To apply the momentum balance given by Equation (11) to a wind farm, an evaluation function *Err* is defined by Equation (12).

$$Err = \Delta(\dot{m}u_x) + \sum_k Th_k - \sum_k fp_k Fp_k \tag{12}$$

where Th_k and Fp_k are the thrust force and pressure loss caused by the k-th rotor under the condition of an isolated single rotor, respectively. Th_k and Fp_k are calculated using the input value of circulation (Γ_k) based on the prepared information for an isolated single rotor. fp_k is the correction function introduced to express the interaction between the k-th rotor and other rotors and is defined by Equation (13), in which I_j is a function expressing the interaction effects from the j-th rotor to the k-th rotor.

$$fp_k = 1 - \sum_{j \neq k} I_j \tag{13}$$

The layouts between selected two 2D-VAWT rotors can be roughly categorized into four kinds according to the relative rotation condition against the main flow U_∞ as shown in Figure 3; i.e., co-rotating (CO), counter-down (CD), counter-up (CU), and tandem (TD).

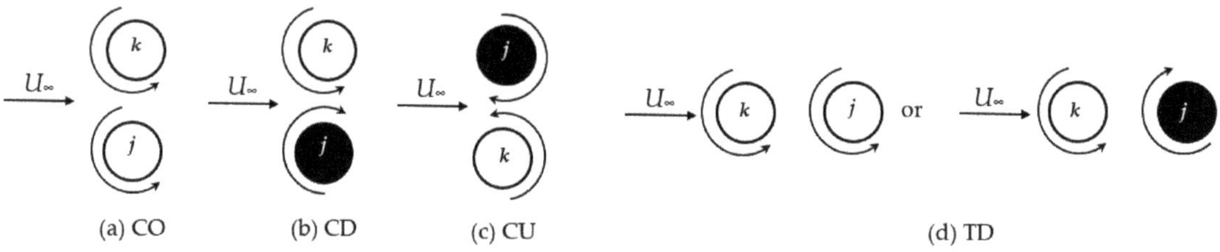

Figure 3. Four categories of the layouts between the selected two rotors: (**a**) co-rotating (CO); (**b**) counter-down (CD); (**c**) counter-up (CU); (**d**) tandem (TD).

In our method, the interaction function I_j is defined separately for each category of the selected two-rotor layout as shown in Equations (14)–(17).

$$I_j = \alpha_1 \frac{|\Gamma_j|}{\Gamma_{\text{SI}}} |\sin \varphi_j| \left(\frac{D}{r_{jk}}\right)^2 \quad ; \text{for CO} - \text{like} \tag{14}$$

$$I_j = \alpha_2 \frac{\Gamma_j}{\Gamma_{SI}} \sin \varphi_j \left(\frac{D}{r_{jk}}\right)^2 \quad ; \text{for CD} - \text{like} \tag{15}$$

$$I_j = \alpha_3 \frac{\Gamma_j}{\Gamma_{SI}} \sin \varphi_j \left(\frac{D}{r_{jk}}\right)^2 \quad ; \text{for CU} - \text{like} \tag{16}$$

$$I_j = \alpha_4 \frac{|\Gamma_j|}{\Gamma_{SI}} \cos \varphi_j \left(\frac{D}{r_{jk}}\right)^2 \quad ; \text{for TD} - \text{like} \tag{17}$$

φ_j is defined as the angle between the direction seen from the k-th rotor to the j-th rotor and the upwind direction as shown in Figure 4. r_{jk} is the distance between the centers of the two rotors. A constant angle γ_1 divides the CV into two zones, i.e., the wake zone (in gray) and the out-of-wake zone (in white). If the j-th rotor (counterpart to the k-th rotor) exists in the wake zone, Equation (17) is used for the calculation of the interaction function I_j as the TD-like layout. When the counterpart rotor exists in the out-of-wake zone, the equation used for the calculation of the interaction function I_j is determined as one of three equations (Equations (14)–(16)) according to the corresponding layout category of the two rotors. The fitting parameters of α_1, α_2, α_3, and α_4 are introduced so as to obtain a sufficient correspondence between the results of rotor power prediction using the model and the CFD analysis, in the specific four conditions of paired rotors, i.e., CO, CD, CU, and TD layouts.

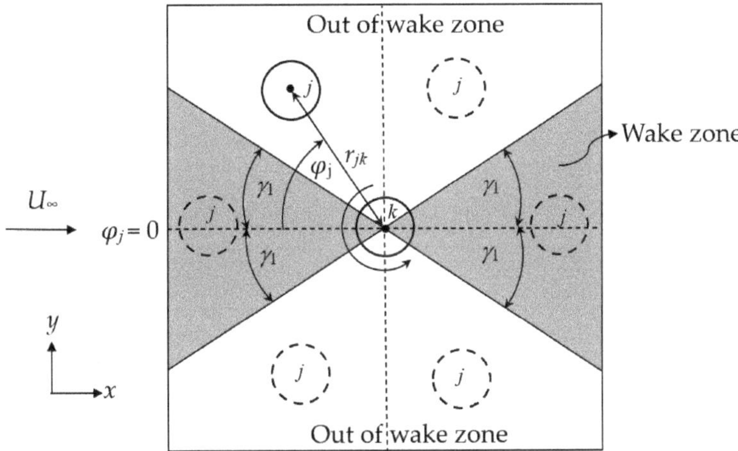

Figure 4. Definitions of the angle φ_j of the selected two rotors and the wake zone.

The flow chart of the method is shown in Figure 5. Before starting the calculation, the relations, such as between the circulation and rotor output, and the fitting parameters must be prepared from the information of an isolated single rotor and four specific conditions of two rotors. The first step of the actual model calculation is to decide the calculation conditions such as the upstream wind speed (U_∞), the positions of the rotors, and the rotational directions. The input parameters are the fitting parameters obtained by comparison of the velocity distribution (x- and y-components) of the model with that of the CFD in the case of an isolated single rotor. For example, the fitting parameters include C_{N0}, C_{N1}, C_{N2}, C_{N3}, C_{P0}, C_{P1}, C_{P2}, and C_{P3}, which are used to define the correction function f_{COR} of the ultra-super-Gaussian function and are defined in Equations (A4)–(A9) in Appendix A. The fitting parameters must be determined at several positions in the flow field according to the complexity of the local velocity profiles. The present study uses 35 parameters. All the fitting parameters are shown in Tables A1–A6 of Appendix A. After the initial condition of the rotors and a set of parameters are input, the process of searching for a reasonable combination of circulations is initiated starting from the lowest value of the circulation in a given searching range for each rotor (determining the searching ranges will be described in Section 2.2). For a given combination of circulations, the potential flow is calculated first and it is modified by our wake model to include the effects of the velocity deficits. With the flow field (u_w, v_w) obtained, the momentum changes in the x-direction per unit of time $\Delta(\dot{m}u_x)$ is calculated from the four boundary conditions. The thrust force Th_k and the (isolated single rotor) pressure loss Fp_k are calculated by Equations (A19) and (A20) in Appendix B by using the given circulation value for each rotor. Using the interaction function I_j, the correction function fp_k which expresses the effects of the inter-rotor interaction is obtained for each rotor by Equation (13). Then, the momentum balance Err is evaluated by Equation (12). The *"error"* of momentum balance is defined in Equation (18) in the present study. The denominator in Equation (18) is the momentum per unit time entering into the CV from the inlet boundary.

$$error = \frac{\sqrt{Err^2}}{\int \dot{m}_x u_{\text{in}}} \tag{18}$$

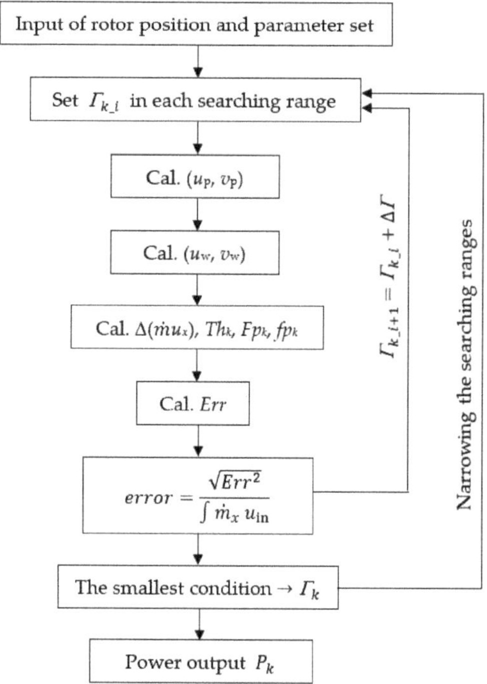

Figure 5. The flow chart of the method.

The above calculation is repeated by changing the combination of circulations with the interval of $\Delta\Gamma$ throughout all the searching ranges. After the round robin, a combination of circulation Γ_k which gives the smallest *"error"* is obtained. In the present study, the similar searching process (sub-searching) is repeated twice by narrowing the searching ranges as $(\Gamma_k - \Delta\Gamma) \leq \Gamma_{k\_i} \leq (\Gamma_k + \Delta\Gamma)$. Finally, for the smallest *"error"* condition, the power output of each rotor can be obtained by using Equation (A21) in Appendix B.

2.2. Specific Calculation and Searching Range of Circulation

In this section, we consider a specific case assuming small rotors in order to explain the selection of searching range to obtain adequate circulation. The target rotors are the same as those investigated in the previous study [35] and the rotor configuration is depicted in Figure A4 in Appendix B. The necessary relations on the performance of the single rotor are given in Equation (A17)–(A20). The values of α_1, α_2, α_3, and α_4 were determined as 0.2304, 0.19, -0.2853, and 1.07, respectively, from the preliminary calculations of the specific four layouts of paired rotors. The CV was defined as $20D \times 20D \times 0.868D$. The thickness of the CV given by $0.868D$ corresponds to the rotor height of the target rotor; by considering the rotor height, our 2D method can be applied to the prediction of the power output of an actual 3D-VAWT rotor.

Figure 6a,d show the averaged distributions of the x- and y-component velocities calculated by the CFD analysis around two VAWT rotors rotating CCW direction in the CO-like layout. The commercial code STAR-CCM+ was utilized for the CFD simulation. The circulation of the upper rotor (rotor-k) is 0.334 m^2/s, which was obtained by the integration of the flow field along a circular path at $r/D = 0.75$. The circulation of the lower rotor (rotor-j) is 0.345 m^2/s. The details of the rotor shape are not shown but are replaced by circles in white in Figure 6a. The black solid lines around the rotors are path lines.

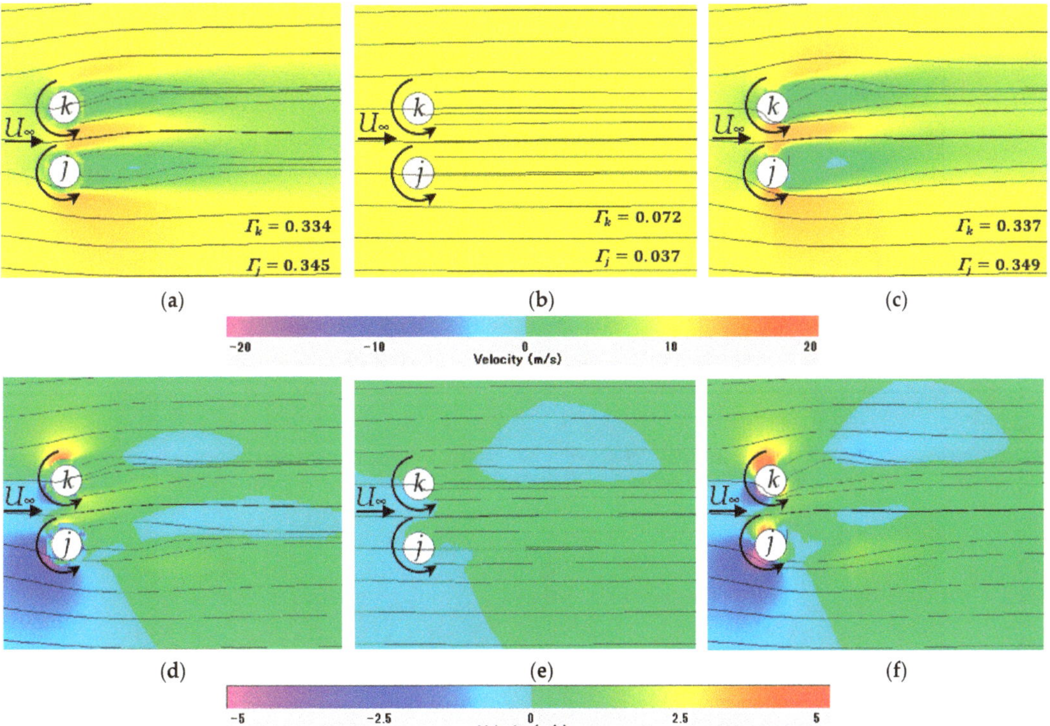

Figure 6. The distribution of the x-component velocity around two rotors in the CO-like layout: (**a**) CFD; (**b**) model with a wide searching range; (**c**) model with a narrow searching range. The distribution of the y-component velocity around two rotors in the CO-like layout: (**d**) CFD; (**e**) model with a wide searching range; (**f**) model with a narrow searching range.

Figure 6b,e show the resultant flow field obtained by the proposed method using an in-house code, in which the circulation of each rotor was changed step by step in a round robin over a wide searching range from $0.1\Gamma_{SI}$ to $1.1\Gamma_{SI}$. Here, Γ_{SI} is the circulation of an isolated single rotor obtained from the CFD analysis and the value is 0.326 m^2/s in the case of $U_\infty = 10$ m/s. The obtained circulations of two rotors in Figure 6b are $\Gamma_k = 0.072$ m^2/s and $\Gamma_j = 0.037$ m^2/s, respectively, and they are different from the CFD results in Figure 6a.

To demonstrate the variation in the error of momentum balance graphically, we define a new indicator $(1 - error)$ that approaches 1 when the error becomes small. The distribution of the values of $(1 - error)$ obtained in the searching process of the smallest error in the case of Figure 6b is shown in Figure 7, which includes 625 results corresponding to the combinations of circulation values (Γ_j, Γ_k) with the interval of $(1.1 - 0.1)\Gamma_{SI}/25$. Figure 7 shows that there are a lot of combinations of circulation which might give a small error.

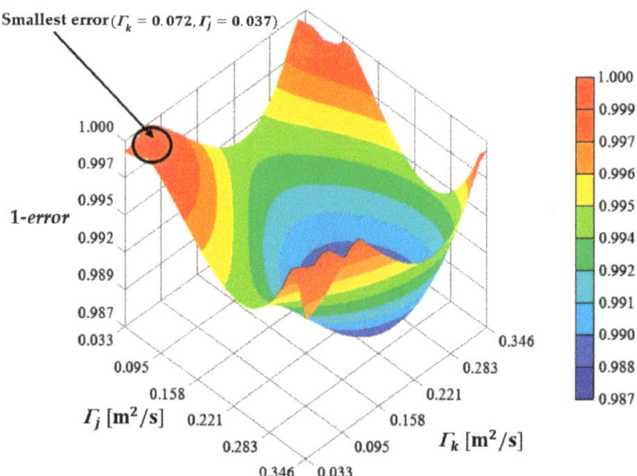

Figure 7. Three-dimensional chart showing the distribution of the value of $1 - error$ in the model calculation of two rotors in the CO-like layout with a wide searching range from $0.1\Gamma_{SI}$ to $1.1\Gamma_{SI}$.

In the parallel layouts, it is expected that two rotors have large circulations close to the value of Γ_{SI}. Therefore, the search range can be narrowed. Figure 6c,f are is the flow field obtained using a searching range limited from $0.95\Gamma_{SI}$ to $1.1\Gamma_{SI}$ for each rotor in the CO layout. The circulation combination giving the smallest error, shown in Figure 6c, is $\Gamma_k = 0.337$ m^2/s and $\Gamma_j = 0.349$ m^2/s, respectively. Figure 8 is the distribution of the 625 values of $(1 - error)$ obtained in the searching process with a limited range from $0.95\Gamma_{SI}$ to $1.1\Gamma_{SI}$ for each rotor. The results of our method with appropriately limited searching ranges can approach the CFD results.

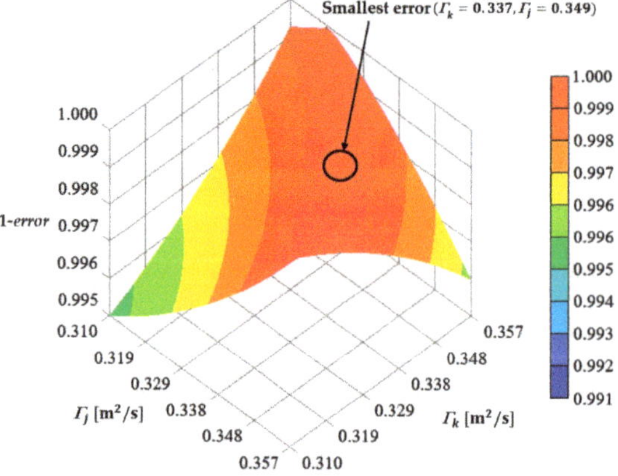

Figure 8. Three-dimensional chart showing the distribution of the value of $1 - error$ in the model calculation of two rotors in the CO-like layout with a narrow searching range from $0.95\Gamma_{SI}$ to $1.1\Gamma_{SI}$.

Incorrect results are also obtained when the rotor at issue exists in the wake of other rotors if the wide searching range of circulation is applied. In our method, we propose a procedure to search for the appropriate result close to the CFD result, in which the searching range of circulation of each rotor in a VAWT wind farm is limited in the manner described below.

Let us define a circular region of the radius R_{int} around a rotor (k) at issue, and assume there are N_{int} rotors in the region. We consider that only the rotors (j) included in the interaction region defined by R_{int} are used to decide the searching range of circulation of the rotor (k).

First of all, the angle φ_j defined in Figure 4 is checked for all rotors in the interaction region. If the absolute value of one of the angles is less than a constant angle γ_2, the searching range expressed by Equation (19) is applied. Note that the angle γ_2 is different from the angle γ_1, and γ_2 is less than γ_1. The interaction region is determined to be a circle of $R_{\mathrm{int}} = 10D$ in this specific calculation; therefore, the maximum of the inter-rotor distance (between rotor centers) r_{jk} is $10D$ in Equation (19) in this study.

$$0.3\Gamma_{\mathrm{SI}} \le \Gamma_k < \left(0.075\frac{r_{jk}}{D} + 0.35\right)\Gamma_{\mathrm{SI}} \quad ; \quad \text{for} \quad |\varphi_j| < \gamma_2 \tag{19}$$

In the second step of the procedure, the average of the distance d_j between the center of the j-th rotor and the stream-wise center line through the k-th rotor at issue is calculated with all the counterpart rotors of the k-th rotor in the interaction region using Equation (20).

$$\overline{d_j} = \frac{\sum_{j \ne k}|d_j|}{N_{\mathrm{int}} - 1} \tag{20}$$

In the third step of the procedure, the searching range of the circulation Γ_k of the k-th rotor is determined as one of Equations (21)–(23) according to the averaged lateral distance $\overline{d_j}$. Figure 9 schematically shows the definition of the angle γ_2 and three zones determining the searching range of circulation of the k-th rotor. The delimiting lateral distances are determined as $d_1 = 0.9D$ and $d_2 = 1.5D$ in this study.

$$0.85\Gamma_{\mathrm{SI}} \le \Gamma_k < 1.0\Gamma_{\mathrm{SI}} \quad ; \quad \text{for} \quad 0 \le \overline{d_j} < d_1 \tag{21}$$

$$0.9\Gamma_{\mathrm{SI}} \le \Gamma_k < 1.0\Gamma_{\mathrm{SI}} \quad ; \quad \text{for } d_1 \le \overline{d_j} < d_2 \tag{22}$$

$$0.95\Gamma_{\mathrm{SI}} \le \Gamma_k < 1.1\Gamma_{\mathrm{SI}} \quad ; \quad \text{for } \overline{d_j} \ge d_2 \tag{23}$$

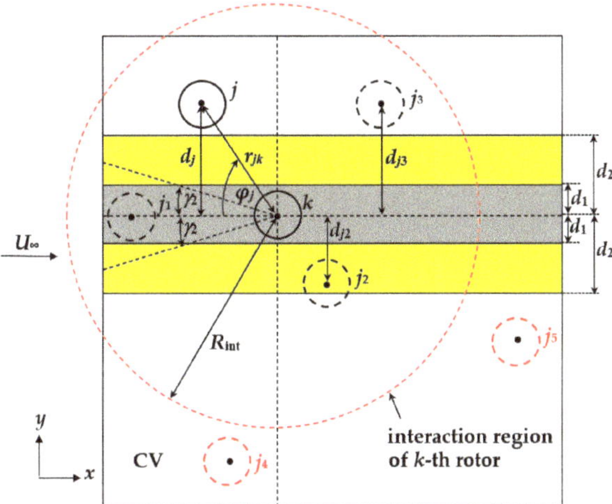

Figure 9. Definition of the angle γ_2 and three zones (in gray, yellow, and white) determining the searching range of circulation of the k-th rotor.

3. Results and Discussion

The method described in Section 2 is validated in this section by applying it to several layouts consisting of the same small 2D rotors as used in the previous section (Section 2.2). Firstly, in Section 3.1, the power dependency of the isolated single rotor on the wind speed is confirmed. Then, in Section 3.2, the power dependency of the paired rotors on the 16 wind directions is compared between the present method and the CFD analysis obtained in the previous study. As for the more complicated layouts, the power dependency of the three-rotor cluster on the 12 wind directions is investigated in Section 3.3. Finally, to show the applicability of the proposed method to VAWT wind farms, the power prediction in the four-rotor layouts arranged in a line is tried in Section 3.4.

3.1. An Isolated Single Rotor

Table 1 shows the comparison between the CFD analysis [35] and the model simulation using the present method based on the momentum balance in the cases of an isolated single rotor in the five different upstream wind speeds (U_∞). Although the original CFD analysis was conducted using a wide calculation domain of $40\,D \times 50\,D$, the square region of $20\,D \times 20\,D$ enclosing the single rotor is extracted as the CV, which is the same size as that of the model simulation, in order to acquire the dependence of the pressure loss Fp on the circulation of the isolated single rotor. The output power shown in the unit of "mW" is equivalent to that of the small rotor, the size of which is 50 mm in diameter × 43.4 mm in height. The parameters necessary to rebuild the flow field calculated by the CFD were obtained at the reference wind speed of $U_\infty = 10$ m/s. Therefore, the smallest percentage difference (% *Error*) between the model and CFD results is obtained in the case of $U_\infty = 10$ m/s. The output power predicted for a single rotor in other wind speeds agrees well with the CFD result; this means the method using momentum conservation works well.

Table 1. Predicted power output of an isolated single rotor by the present method and the comparison with the CFD results (reproduced with permission from Hara et al. [35]).

U_∞ (m/s)	P_Model (mW)	P_CFD (mW)	% *Error*
4	8.14	7.99	1.88
6	32.87	31.88	3.11
8	85.50	83.76	2.08
10	177.62	177.62	0.0
12	314.23	322.17	−2.46

3.2. Paired Rotors

In this section, the prediction of the powers of paired rotors is investigated. Figure 10 shows the definition of relative wind direction (θ) to the rotor pair which is categorized into two configurations in terms of the relative rotational direction of two rotors. The configuration in Figure 10a is defined as the co-rotation (CO), which includes the CO and TD layouts where two rotors rotate in the same direction (see Figure 3). On the other hand, the configuration in Figure 10b is defined as the inverse rotation (IR), which includes the CD, CU, and TD layouts where two rotors rotate mutually in opposite directions. The dotted lines in red in Figure 10 correspond to the boundaries of the wake zone defined by the angle γ_1 in Figure 4. The angle γ_1 is set at $30°$ in this study.

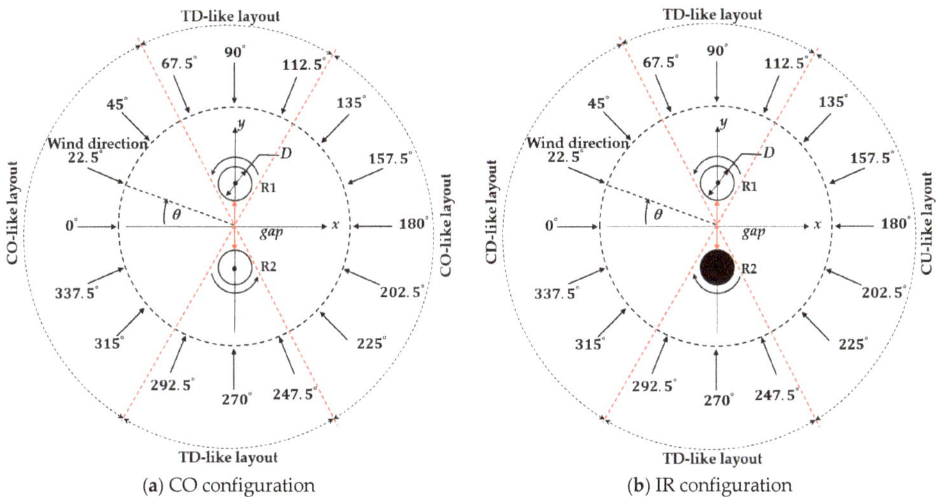

(**a**) CO configuration

(**b**) IR configuration

Figure 10. Definition of 16 wind directions in the two configurations in the case of paired rotors: (**a**) co-rotation (CO) configuration; (**b**) inverse rotation (IR) configuration.

The power outputs of the upper (Rotor 1: R1) and lower (Rotor 2: R2) rotors and the averaged power of both rotors were predicted in each of the 16 wind directions by the present method under the condition of the upstream wind speed U_∞ of 10 m/s. The results in the case the inter-rotor space (*gap*) is equal to the rotor diameter are shown in Figure 11a–c for CO and Figure 12a–c for the IR configurations. Each of the predicted rotor power is normalized by the power of the isolated single rotor (SI) and compared with the normalized CFD results in Figures 11 and 12. The conditions of $\theta = 0°$ and $180°$ in Figure 11c are the same CO layout and their output powers were adjusted to agree with the CFD results to get the adequate value of the parameter α_1. The conditions of $\theta = 0°$ and $180°$ in Figure 12c correspond to the CD and CU layouts, respectively, and the values of the parameters α_2 and α_3 were adjusted so as to agree with the CFD results. Similarly, the conditions of $\theta = 90°$ and $270°$ in Figures 11c and 12c correspond to either of the two TD layouts shown in Figure 3d, and the parameter α_4 can be determined by fitting the resultant circulations obtained by the model simulation to those of the CFD analysis in one of the four TD conditions. The actual values of α_1, α_2, α_3, and α_4 were already shown in Section 2.2. Although the difference between the model simulation and the CFD analysis is somewhat large in a few wind directions as shown in Figures 11 and 12, the method can predict the power of a rotor pair with the *gap* = 1.0*D* very well.

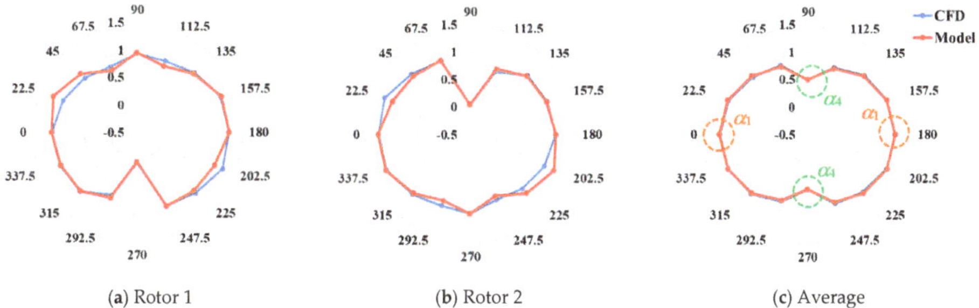

(**a**) Rotor 1

(**b**) Rotor 2

(**c**) Average

Figure 11. Distributions of normalized rotor powers in the CO-configuration in the case of *gap* = 1.0*D*: (**a**) Rotor 1; (**b**) Rotor 2; (**c**) average.

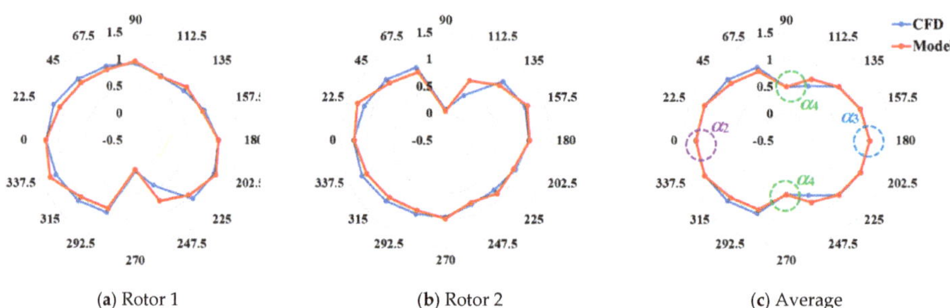

(a) Rotor 1 (b) Rotor 2 (c) Average

Figure 12. Distributions of normalized rotor powers in the IR−configuration in the case of *gap* = 1.0*D*: (a) Rotor 1; (b) Rotor 2; (c) average.

Figures 13 and 14 are the results of the rotor power distribution of the paired rotors with the *gap* = 0.5*D*. The model simulation seems to underestimate the rotor powers compared with the CFD analysis in many wind directions.

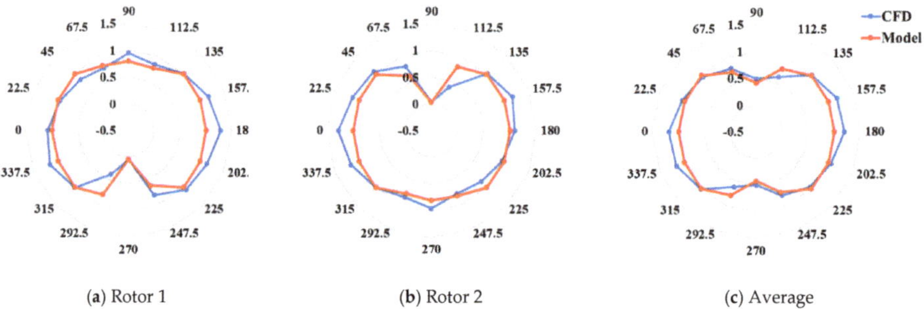

(a) Rotor 1 (b) Rotor 2 (c) Average

Figure 13. Distributions of normalized rotor powers in the CO-configuration in the case of *gap* = 0.5*D*: (a) Rotor 1; (b) Rotor 2; (c) average.

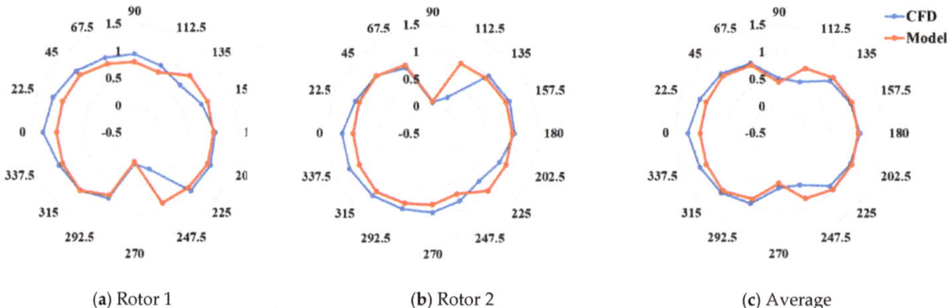

(a) Rotor 1 (b) Rotor 2 (c) Average

Figure 14. Distributions of normalized rotor powers in the IR−configuration in the case of *gap* = 0.5*D*: (a) Rotor 1; (b) Rotor 2; (c) average.

Figures 15 and 16 show the results in the case of *gap* = 2.0*D*. Except for a few directions, the model and CFD predictions agree well in the long inter-rotor distance case. This fact may suggest the necessity to modify the interaction function I_j defined in Equations (14)–(17) in the case of short inter-rotor distance.

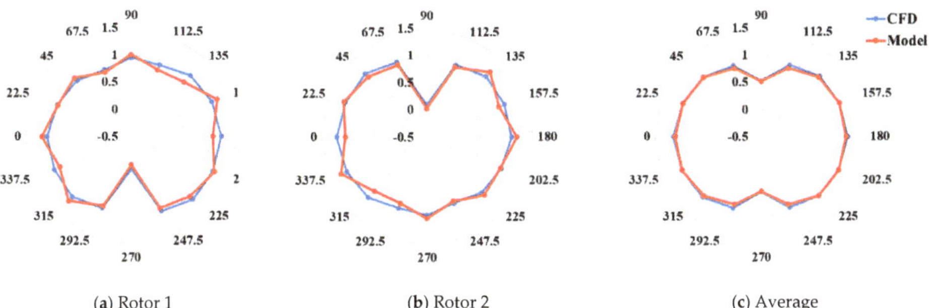

(a) Rotor 1   (b) Rotor 2   (c) Average

Figure 15. Distributions of normalized rotor powers in the CO-configuration in the case of *gap* = 2.0*D*: (**a**) Rotor 1; (**b**) Rotor 2; (**c**) average.

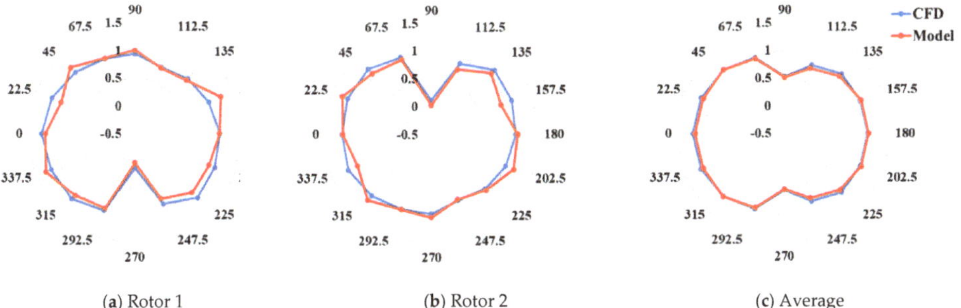

(a) Rotor 1   (b) Rotor 2   (c) Average

Figure 16. Distributions of normalized rotor powers in the IR−configuration in the case of *gap* = 2.0*D*: (**a**) Rotor 1; (**b**) Rotor 2; (**c**) average.

Although further improvements may be needed in the present method, it is worth noting that the prediction of the power of paired rotors in a specific wind direction can be performed by the method in about 40 min, which is 500 times shorter than the calculation time with the CFD analysis using the DFBI model (about 2 weeks).

3.3. Three-Rotor Cluster

The three-rotor cluster is also categorized into the CO and IR configurations. Figure 17 shows the definition of the 12 wind directions for the two configurations in the cases of three-rotor clusters which are arranged like a triangular shape with an inter-rotor space of *gap* = 1.0*D*. Under the condition of the upstream wind speed U_∞ of 10 m/s, the prediction of the averaged output power of the rotor clusters was carried out. All the parameters and necessary relations such as the pressure loss function are the same as those used in the case of paired rotors except for the division of the searching range of circulation. The number of divisions of the searching range was set as 10 in the three-rotor case instead of the 25 used in paired rotors case to save the calculation time. Therefore, the error in the prediction is anticipated to be a little worse.

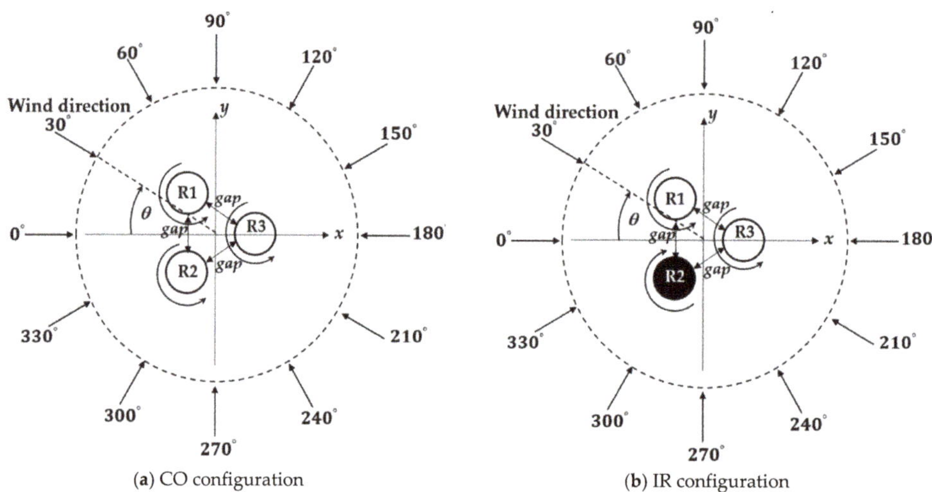

Figure 17. Definition of 12 wind directions in the two configurations in the cases of three-rotor clusters: (**a**) co-rotation (CO) configuration; (**b**) inverse rotation (IR) configuration.

Figure 18 is the prediction results of the averaged power of the three rotors. The model calculation underestimates the averaged power of the rotor clusters. However, the dependence on the wind direction is well simulated and the model simulation gives the same trend as the CFD analysis [41].

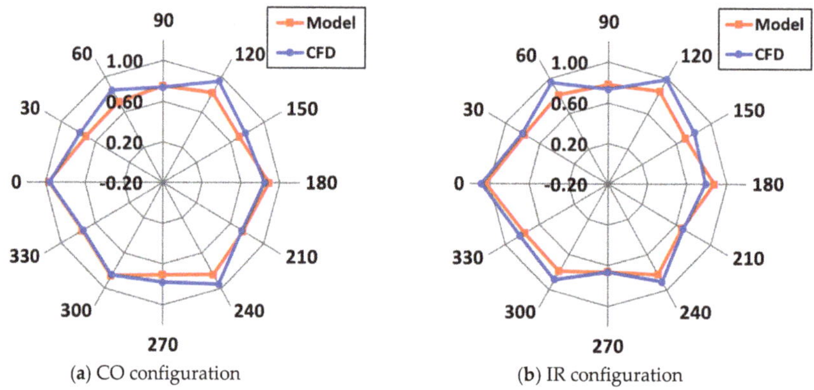

Figure 18. Distributions of averaged rotor power of three rotors in the configurations: (**a**) CO-rotation (CO); (**b**) inverse rotation (IR). The power is normalized by the single rotor power. Red symbols show the results of model simulation and blue symbols show CFD analysis (reproduced with permission from Okinaga et al. [41]). The inter-rotor space (*gap*) is 1.0D.

3.4. Four-Rotor Layout

Finally, to show the applicability of the proposed method to VAWT wind farms, four-rotor layouts arranged in line with an inter-rotor space of 3.0D were selected as the targets of the output power prediction. To cover a large number of rotors, the CV was set to be a large size of $40D \times 40D \times 0.868D$. As the size of the CV was changed, the function expressing the pressure loss Fp of the CV was newly prepared from the CFD flow field of the isolated single rotor; i.e., the reference value Fp_{ref} in Equation (A20) in Appendix B was changed from 229.61 to 347.95 mN. The upstream wind speed was assumed to be 10 m/s.

The number of divisions of the searching range of the circulation of each rotor was set as 8. In this case, the calculation time is about 50 times as long as the calculation time in a two-rotor case with the 25 division because the subdivision process is executed twice in the actual calculation using the in-house code.

Figure 19a–d show the results of the power prediction and the distributions of x- and y-direction velocity components in two kinds of four-rotor layouts, i.e., parallel and tandem, respectively. The fitting parameters α_1, α_2, α_3, and α_4 are set to be the same values as those used in the paired rotors (see Section 2.2). The model predicts the decrease in the rotor power in the order of R1, R2, R3, and R4 in the parallel array of four rotors. Similarly, in the tandem array of four rotors, the output power decreases in the same order.

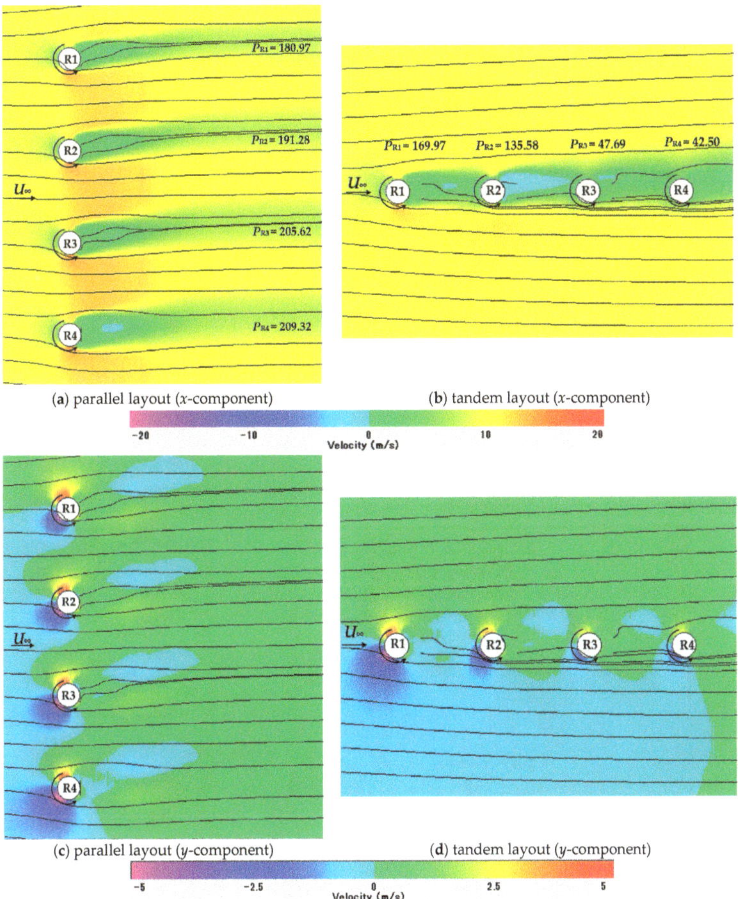

(a) parallel layout (*x*-component)   (b) tandem layout (*x*-component)

(c) parallel layout (*y*-component)   (d) tandem layout (*y*-component)

Figure 19. The upper figures show the prediction of the distributions of x-direction velocity components around four-rotor layouts: (**a**) parallel; (**b**) tandem. The lower figures show that of the y-direction velocity component: (**c**) parallel; (**d**) tandem. The upstream wind speed is 10 m/s and the inter-rotor space (*gap*) is 3.0*D*.

In this study, the CFD analysis of the two kinds of four-rotor layouts was carried out by using the DFBI model. The mesh size and whole domain size of each simulation of the four-rotor layouts are the same as those used in the previous CFD analysis of one-, two-, and three-rotor arrangements. The details of the meshes created for the four-rotor layouts in the present study are shown in Appendix C. The upstream wind speed U_∞ is set at 10 m/s in the CFD analysis of four-rotor layouts.

Figure 20a,b show the distributions of x-component of unsteady velocity obtained by the CFD for the parallel and tandem layouts, respectively. Figure 20c,d illustrate the unsteady flow field shown by the y-component velocity for the two layouts of four rotors. The condition shown in Figure 20 corresponds to the state at 4 s from the beginning of the simulation. The time history of the angular velocity of each rotor is shown in Appendix C. The CFD analysis was conducted by a high-performance PC with 28 cores; it took about 1 week for each calculation of two layouts.

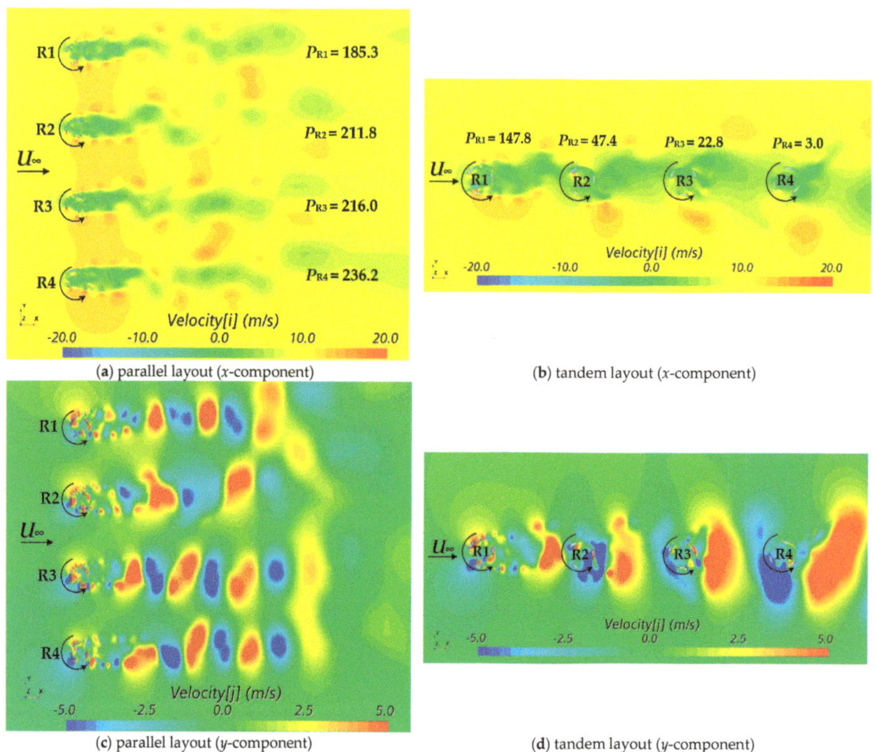

Figure 20. CFD results using the DFBI model of the two kinds of four-rotor layouts; (**a**) x-component distribution in the parallel array; (**b**) x-component distribution in the tandem array; (**c**) y-component distribution in the parallel array; (**d**) y-component distribution in the tandem array. The distributions show the unsteady velocity field at the time of 4 s (see Appendix C).

Table 2 shows the comparison of each rotor power in the parallel and tandem layouts between the present model and CFD analysis. The percentage error of the rotor power (*Err.P*) is defined by Equation (24) in this study. P_{Model} is the rotor power predicted by our model using Equation (A21) and the P_{CFD} is the CFD result of the rotor power which is obtained by multiplying the averaged angular velocity and the averaged rotor torque during the last 0.5 s. The P_{ref} is the output power of an isolated single rotor.

$$Err.P = \frac{P_{\mathrm{Model}} - P_{\mathrm{CFD}}}{P_{\mathrm{ref}}} \times 100 \tag{24}$$

Table 2. Comparison of each rotor power in the parallel and tandem layouts between the present model and the CFD analysis.

Layout	Parallel				Tandem			
Rotor	R1	R2	R3	R4	R1	R2	R3	R4
Power (model) (mW)	181.0	191.3	205.6	209.3	170.0	135.6	47.7	42.5
Power (CFD) (mW)	185.3	211.8	216.0	236.2	147.8	47.4	22.8	3.0
Err.P (%)	−2.4	−11.5	−5.8	−15.1	12.5	49.5	14.0	22.2

The comparison shown in Table 2 does not give satisfactory results. In particular, the difference between the model and CFD is large in the tandem layout of four rotors. One of the reasons for the disagreement is the small number of initial divisions of the searching range used for the model prediction. An advanced method like the genetic algorithm (GA) should be used to give the possible combinations of circulations at random. Further improvement in the model of the interaction among rotors also is necessary. Nevertheless, the present model can predict the same trend in the rotor power in the parallel and tandem layouts of four rotors as that of the CFD analysis.

4. Conclusions

We developed a method to predict the output powers of the vertical-axis wind turbine (VAWT) rotors of an arbitrary layout; the calculation time can be incomparably shorter than computational fluid dynamics (CFD) using the dynamic fluid/body dynamics (DFBI) model, which enables the time variation in rotor speed. However, our method needs to know information, such as the flow velocity distributions and pressure loss and circulation, and so on, around an isolated single rotor in several wind speed conditions. In addition, four fitting parameters must be determined in advance to express the interaction effects between two rotors in the CO, CD, CU, and TD layouts having a specific inter-rotor space (*gap*). By applying the law of the momentum conservation with including the interaction effects through the modification of the pressure loss of the control volume (CV), the method could predict the power dependence of the paired rotor with the rotor gap of $1.0D$ on the 16 wind directions, which agreed with the CFD analysis well. However, the dependence on the inter-rotor distance (between rotor centers) should be improved. The application of the method to the three-rotor cluster showed almost the same trend as the CFD in the averaged power distribution over 12 wind directions. The method was applied to the four-rotor layouts arranged in line of parallel or tandem to the main flow to show the applicability to VAWT wind farms. The important advantages of our method include the very small number of the calculation grid (actually 400×400 used) and the random positioning of rotors as well as the significantly short calculation time compared with the CFD using the DFBI model. Unlike the conventional wake models, our wake model can express the acceleration regions existing on both sides of the velocity deficit. Once the necessary parameters are prepared, the averaged flow field around an arbitrary rotor layout of the VAWT cluster can be reproduced with fidelity as high as the CFD analysis. To apply our method to the problem of finding the optimal layout of rotors in a VAWT wind farm, it is necessary to increase the accuracy of the prediction and to introduce other optimization methods such as the genetic algorithm (GA) instead of the round robin.

Author Contributions: Conceptualization, Y.H.; methodology, J.B. and Y.H.; software, J.B. and Y.H.; validation, Y.H.; formal analysis, J.B. and Y.H.; investigation, J.B.; resources, J.B. and Y.H.; data curation, J.B.; writing—original draft preparation, J.B.; writing—review and editing, Y.H., M.F. and Y.J.; visualization, J.B.; supervision, Y.H.; project administration, Y.H.; funding acquisition, Y.H., M.F. and Y.J. All authors have read and agreed to the published version of the manuscript.

Funding: This research was supported by JSPS KAKENHI Grant Number JP 18K05013, the International Platform for Dryland Research and Education (IPDRE), Tottori University, and the 2021 Special Joint Project of the Faculty of Engineering, Tottori University.

Data Availability Statement: The data that support the findings of this study are available from the corresponding author, J.B., upon reasonable request.

Acknowledgments: The first author thanks Tomoyuki Okinaga at Tottori University for helping with the CFD simulation. The authors thank the Advanced Mechanical and Electronic System Research Center (AMES) for supporting the article processing charge.

Conflicts of Interest: The authors declare no conflict of interest.

Appendix A

Shapiro et al. [38] proposed the super-Gaussian function f_{SG} which simulates the transformation of the wake profile (x-component of velocity deficit) of a horizontal-axis wind turbine (HAWT) from a top-hat shape to a Gaussian shape. We utilize the super-Gaussian function to simulate the wake profile of VAWT. However, the function cannot express the acceleration regions existing on both sides of the velocity deficit. Therefore, in our model, the correction function f_{COR} is introduced to express the acceleration regions [37,40]. In addition, the deflection of the wake of VAWT is considered.

Figure A1 compares the four profiles of the x-component velocity in the wake of an isolated rotor at $x_n = 2.0$ [37]. As shown in Figure A1, the ultra-super-Gaussian function f_{USG} which subtracts the correction function (see Figure A2) from the super-Gaussian function f_{SG} can reproduce the profile obtained by the CFD. Note that the super-Gaussian function f_{SG} and the ultra-super-Gaussian function f_{USG} are defined as the profile of the velocity deficit.

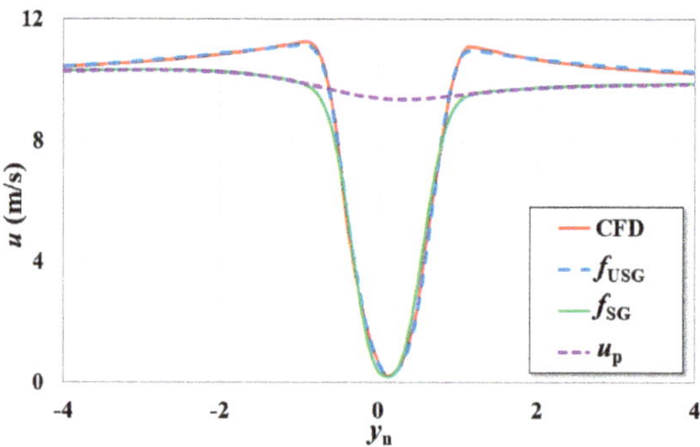

Figure A1. Comparison of the profiles of the x-component velocity of the wake of an isolated rotor at $x_n = 2.0$ (reproduced with permission from Buranarote et al. [37]).

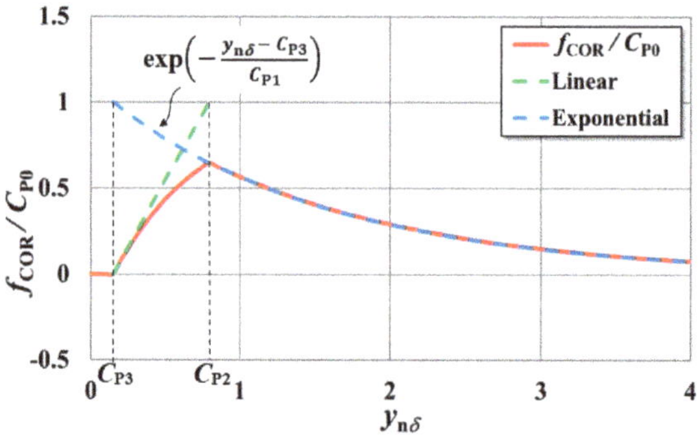

Figure A2. Correction function for positive range in $y_{n\delta}$. (reproduced with permission from Buranarote et al. [37]).

The super-Gaussian function is expressed by the following Equation (A1):

$$f_{SG} = \exp\left[-\frac{D^2}{8\sigma_0{}^2}\left\{ \frac{2|y_{n\delta}|}{Dd_w(x_n)} \right\}^{p(x_n)} \right] \tag{A1}$$

where $\sigma_0 = D/4$. In our model [37], the function $p(x_n)$ is defined as the following Equation (A2):

$$p(x_n) = 2\left(1 + f_p/x_n\right) \tag{A2}$$

Here, x_n is the non-dimensional coordinate defined by $x_n = (x - x_k)/D$. f_p is one of the fitting parameters, which is not included in the original super-Gaussian function. The normalized coordinate $y_{n\delta}$, which includes the wake shift δ, in Equation (A1) is defined as Equation (A3)

$$y_{n\delta} = \frac{y - y_k - \delta_k}{D} \tag{A3}$$

The correction function f_{COR} for $y_{n\delta} \geq 0$ is defined as follows:

$$f_{COR} = 0 \qquad \text{for } \{0 \leq y_{n\delta} < C_{P3}\} \tag{A4}$$

$$f_{COR} = C_{P0}\exp\left(-\frac{y_{n\delta} - C_{P3}}{C_{P1}}\right)\left[\frac{1}{C_{P2} - C_{P3}}(y_{n\delta} - C_{P3})\right] \qquad \text{for } \{C_{P3} \leq y_{n\delta} \leq C_{P2}\} \tag{A5}$$

$$f_{COR} = C_{P0}\exp\left(-\frac{y_{n\delta} - C_{P3}}{C_{P1}}\right) \qquad \text{for } \{y_{n\delta} > C_{P2}\} \tag{A6}$$

In the region $y_{n\delta} < 0$, the correction function f_{COR} is given by Equations (A7)–(A9):

$$f_{COR} = 0 \qquad \text{for } \{-C_{N3} < y_{n\delta} < 0\} \tag{A7}$$

$$f_{COR} = C_{N0}\exp\left(\frac{y_{n\delta} + C_{N3}}{C_{N1}}\right)\left[\frac{-1}{C_{N2} - C_{N3}}(y_{n\delta} + C_{N3})\right] \qquad \text{for } \{-C_{N2} \leq y_{n\delta} \leq -C_{N3}\} \tag{A8}$$

$$f_{COR} = C_{N0}\exp\left(\frac{y_{n\delta} + C_{N3}}{C_{N1}}\right) \qquad \text{for } \{y_{n\delta} < -C_{N2}\} \tag{A9}$$

In the present method, the wake deflection δ_k in Equation (A3) is modified by Equation (A10) to consider the interaction between rotors with the induced velocities u_{ind} (Equation (A11)) and v_{ind} (Equation (A12)) given by the Biot–Savart law. The distance r_j and angle ϕ_j are defined in Figure A3. Factors β_1 and β_2 are the correction constants. The sign of the second term in parentheses in Equation (A10) depends on the rotational direction of the selected two rotors and is positive when the rotors rotate in the same direction. The sign in front of the induced velocity v_{ind} in the index of the exponential function in Equation (A10) is positive when the k-th rotor rotates counterclockwise.

The wake width d_W, which is used in the super-Gaussian function f_{SG_k} [38], is defined by Equation (A13) in our method. k_W is one of the fitting parameters in our method. The function α including a correction constant β_3 is defined by Equation (A14), in which the summation of the induced velocity does not include the effect from the k-th rotor.

$$\delta_k = \frac{\Gamma_k}{\Gamma_{SI}}\delta_{SI}\left[1 \pm (1 - \beta_1 u_{ind})|v_{ind}|\left(1 - e^{-(x - x_k)/\{\beta_2 D(1 \pm v_{ind})\}}\right)\right] \tag{A10}$$

$$u_{ind} = \sum_j \frac{\Gamma_j}{r_j}\sin\phi_j \tag{A11}$$

$$v_{ind} = \sum_j \frac{\Gamma_j}{r_j}\cos\phi_j \tag{A12}$$

$$d_{\mathrm{w}}(x_{\mathrm{n}}) = 1 + k_{\mathrm{w}} \ln\left(1 + e^{2(1+\alpha)x_{\mathrm{n}}}\right) \qquad (A13)$$

$$\alpha = \beta_3 \sum_{j \neq k} \frac{\Gamma_j}{r_j} \sin\phi_j \qquad (A14)$$

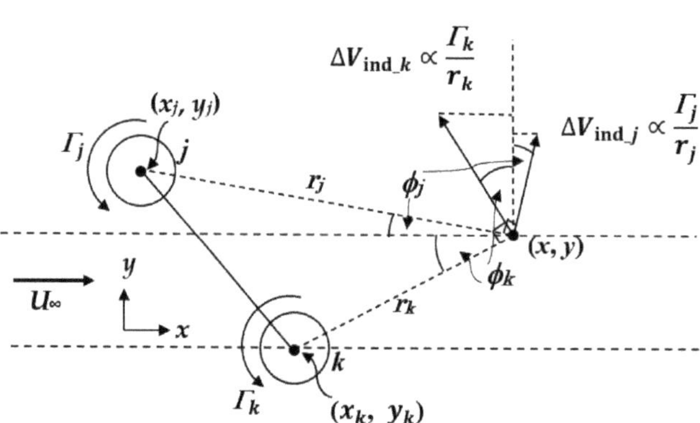

Figure A3. Schematic image of a pair of rotors and the induced velocity. Angle ϕ_j is defined at an arbitrary point P:(x, y) when seeing the direction of the center position of a rotor on the basis of the upstream direction.

The non-dimensional function dv_k for the y-component velocity correction is expressed by the superposition of four Gaussian-type functions f_{G_i} and four resonance-type functions f_{R_i} as shown in Equation (7). The Gaussian-type function and resonance-type function are defined by Equations (A15) and (A16), respectively.

$$f_{\mathrm{G}\_i} = C1_{\mathrm{fg}\_i} \exp\left(-\frac{\left(y_{\mathrm{n}\delta} - C3_{\mathrm{fg}\_i}\right)^2}{C2_{\mathrm{fg}\_i}}\right) \quad ; \ (i = 1 \ \mathrm{to} \ 4) \qquad (A15)$$

$$f_{\mathrm{R}\_i} = \frac{C1_{\mathrm{fr}\_i}}{1 + C2_{\mathrm{fr}\_i}(y_{\mathrm{n}\delta} - C3_{\mathrm{fr}\_i})^2} \quad ; \ (i = 1 \ \mathrm{to} \ 4) \qquad (A16)$$

There are 11 fitting parameters (C_w, k_w, f_{p}, C_{N0}, C_{N1}, C_{N2}, C_{N3}, C_{P0}, C_{P1}, C_{P2}, and C_{P3}) in f_{USG}, the values are shown in Tables A1 and A2. There are 24 fitting parameters ($C1_{\mathrm{fg}\_i}$, $C2_{\mathrm{fg}\_i}$, $C3_{\mathrm{fg}\_i}$, $C1_{\mathrm{fr}\_i}$, $C2_{\mathrm{fr}\_i}$, and $C3_{\mathrm{fr}\_i}$; $[i = 1 \ \mathrm{to} \ 4]$) in function dv_k, the values are shown in Tables A3 and A4 for the functions f_{G_i} and Tables A5 and A6 for the functions f_{R_i}. The fitting parameters are determined at 46 positions between $x_{\mathrm{n}} = -10$ and $x_{\mathrm{n}} = 10$ in the x-direction by comparison between the velocity profiles of the model and the CFD result. At an arbitrary x-position in the flow field, the parameters are used for interpolation.

Table A1. The fitting parameters in the ultra-super-Gaussian function, f_{USG}, ($x_n < 0$).

x_n	C_w	k_w	f_p	C_{N0}	C_{N1}	C_{N2}	C_{N3}	C_{P0}	C_{P1}	C_{P2}	C_{P3}
−10	0.0091	0.6582	0.7010	0.2542	0.9057	0.2054	0.1845	0.4361	2.1068	0.2156	0.1880
−8	0.0114	0.6905	0.0993	−0.4745	4.7472	2.6759	1.1221	0.2326	0.7857	0.4652	0.1646
−6	0.0168	0.7932	1.3848	−0.5536	5.5377	3.1215	1.3090	0.1482	0.9012	0.5873	0.0326
−4	0.0274	0.9092	0.8591	−0.3956	7.7506	1.4080	0.5688	−0.0644	1.6639	3.3820	0.4386
−2	0.0768	0.7449	0.3266	−0.6307	1.8931	2.0142	0.7956	0.1582	15.5537	5.7160	2.4223
−1	0.1826	0.7898	0.0532	−0.2961	1.0724	1.4556	0.8866	0.1070	13.8469	3.3443	1.4025
−0.75	0.2772	0.5971	0.0423	−0.3472	0.7244	1.4133	0.7344	0.0878	9.8289	2.8971	1.5263
−0.7	0.3000	0.6001	0.0161	−0.4380	0.5724	1.6683	0.7654	0.0849	12.3316	3.0527	0.4652
−0.6	0.2882	0.6330	0.0552	−0.2869	0.6256	1.3519	0.7741	0.1008	9.7876	2.6759	0.5537
−0.5	0.3282	0.5557	0.0002	−0.3087	0.5715	1.2986	0.6040	0.1042	7.4638	2.5409	0.7335
−0.4	0.2887	0.5019	0.0501	−0.3079	0.4466	1.1460	0.6543	0.1325	6.5829	1.9413	0.8771
−0.3	0.4235	0.2310	0.0501	−0.2739	0.3718	0.9942	0.5110	0.1035	5.9861	1.6361	0.7088
−0.2	0.6754	0.2062	0.4530	−0.2698	0.3480	0.1711	0.1711	0.0797	5.2713	1.2941	0.3692
−0.1	0.6232	0.0334	0.1954	−0.0935	0.2087	0.5196	0.5196	0.1074	4.5032	0.2434	0.2433

Table A2. The fitting parameters in the ultra-super-Gaussian function, f_{USG}, ($x_n \geq 0$).

x_n	C_w	k_w	f_p	C_{N0}	C_{N1}	C_{N2}	C_{N3}	C_{P0}	C_{P1}	C_{P2}	C_{P3}
0	0.5428	0.0002	0.0503	0.0376	3.2977	1.8589	0.4196	0.1575	3.6227	0.2587	0.2586
0.1	0.6577	0.0002	0.3936	0.0636	2.3979	1.7094	0.0005	0.1713	3.0411	0.2162	0.2161
0.2	0.7026	0.0002	0.8119	0.0569	2.7857	0.9608	0.0870	0.1869	2.7598	0.2382	0.2343
0.3	0.5986	0.2500	2.4680	0.0966	2.3364	0.1319	0.0618	0.2673	2.5117	0.4382	0.1426
0.4	0.7250	0.2348	2.5906	0.0879	2.0929	0.7942	0.4227	0.2282	2.4431	0.7059	0.2423
0.5	0.9100	0.2414	2.0028	0.0762	1.8844	0.7625	0.7146	0.2028	2.2853	0.5807	0.2810
0.6	0.9017	0.2699	1.4722	0.1082	1.9000	0.2629	0.2554	0.2149	2.0951	0.6432	0.4851
0.7	0.8632	0.2434	2.0263	0.1524	1.7128	0.4202	0.1218	0.2324	2.1085	0.6836	0.4465
0.75	0.8382	0.2401	2.2507	0.1566	1.7334	0.5260	0.1615	0.2393	2.1085	0.6402	0.4687
0.8	0.8504	0.2306	2.3917	0.1651	1.7208	0.5245	0.1327	0.2456	2.0862	0.6898	0.4399
0.9	0.8672	0.2166	2.4366	0.1739	1.7442	0.5439	0.0802	0.2452	2.1060	0.6962	0.4270
1	0.8762	0.2012	2.2059	0.2108	1.5943	0.8172	0.0553	0.2418	2.1336	0.7414	0.4284
1.1	0.8914	0.1931	2.1107	0.2157	1.6410	0.8202	0.0080	0.2379	2.1930	0.7759	0.3961
1.2	0.8879	0.1777	1.9914	0.2077	1.6346	0.8721	0.1508	0.2282	2.2008	0.8100	0.5021
1.3	0.9051	0.1636	1.8485	0.2067	1.7130	0.9247	0.0650	0.2064	2.2741	0.7501	0.6179
1.4	0.9128	0.1570	1.7436	0.2088	1.7008	0.9034	0.0864	0.2033	2.2947	0.7559	0.6180
1.5	0.9219	0.1498	1.6507	0.2090	1.7150	0.8972	0.0837	0.2025	2.3173	0.8285	0.5823
1.6	0.9303	0.1494	1.3965	0.2043	1.7353	0.8768	0.0966	0.1889	2.4271	0.7589	0.6133
1.7	0.9296	0.1336	1.4457	0.1977	1.7899	0.9275	0.1035	0.1875	2.4322	0.8566	0.6182
1.8	0.9378	0.1281	1.2945	0.1940	1.8314	0.9296	0.0805	0.1801	2.4907	0.8648	0.6216
1.9	0.9436	0.1284	1.1287	0.1920	1.8235	0.8183	0.0833	0.1748	2.5667	0.9122	0.5759
2	0.9740	0.1333	1.3191	0.1849	1.8677	0.3759	0.0424	0.1971	2.6559	0.9214	0.0583
2.1	0.9750	0.1277	1.1529	0.1757	1.9256	0.3570	0.0216	0.1867	2.7499	0.9503	0.0442
2.2	0.9441	0.1243	1.1448	0.1661	2.0006	0.4031	0.0826	0.1743	2.8841	0.9444	0.1459
2.3	0.9400	0.1194	1.0655	0.1572	2.0721	0.3583	0.0821	0.1666	2.9870	0.9797	0.1179
2.4	0.9266	0.1147	0.9121	0.1495	2.1548	0.3858	0.0739	0.1531	3.0981	0.9936	0.2387
2.5	0.9105	0.1129	0.8607	0.1441	2.2145	0.3849	0.0788	0.1480	3.2481	1.0166	0.1923
3	0.7977	0.1058	0.6963	0.1077	2.8396	0.3861	0.0758	0.1184	4.2106	1.1886	0.1689
4	0.5657	0.1063	0.4007	0.3131	2.1568	6.6251	0.1685	0.3866	2.6141	6.8336	0.3954
6	0.3571	0.1143	0.3388	0.1253	3.5691	7.9674	1.5521	0.2249	3.9554	7.7057	1.1499
8	0.3064	0.1051	0.0357	0.0941	4.8910	8.3139	0.8055	0.1531	5.3229	7.8979	1.2282
10	0.2920	0.0905	0.3596	0.1307	3.5955	8.6259	1.2900	0.1461	5.2149	7.7429	1.0662

Table A3. The fitting parameters in Gaussian-type functions, f_{G_i}, ($x_n < 0$).

x_n	$C1_{fg1}$	$C2_{fg1}$	$C2_{fg1}$	$C1_{fg2}$	$C2_{fg2}$	$C2_{fg2}$	$C1_{fg3}$	$C2_{fg3}$	$C2_{fg3}$	$C1_{fg4}$	$C2_{fg4}$	$C3_{fg4}$
−10	0.0235	0.1686	−0.3653	−0.0231	0.1752	−0.3619	−0.0597	0.0938	1.5479	0.0597	0.0952	1.5486
−8	0.0108	0.1888	−0.2683	−0.0081	0.1372	−0.3694	−0.0103	0.1976	0.1644	0.0078	0.1417	0.2737
−6	0.0135	0.1554	−0.3257	−0.0159	0.1723	−0.2669	−0.0427	0.1721	0.1944	0.0448	0.1779	0.1726
−4	0.0229	0.4217	−1.3449	−0.0224	0.4184	−1.3278	−0.0531	0.0965	0.4337	0.0540	0.0973	0.4337
−2	0.0082	0.3241	−1.1351	−0.0044	0.3085	−1.0243	0.0142	0.2679	0.3599	−0.0101	0.7884	0.4118
−1	0.0250	0.4050	−1.2649	−0.0250	0.3950	−1.1774	0.0156	0.4366	0.2983	−0.0092	1.1056	0.5103
−0.75	−0.0212	0.2538	−0.5622	0.0362	0.1196	−0.2382	0.0133	0.1835	0.9382	−0.0112	0.4931	0.8562
−0.7	−0.0198	0.1009	−0.9601	−0.0194	0.0485	−0.5937	−0.0272	0.2134	0.5007	0.0221	0.7050	0.1507
−0.6	−0.0421	0.5509	−0.2054	0.0235	0.0507	−0.5007	0.0202	0.2292	0.7502	−0.0123	1.0304	0.4118
−0.5	−0.1213	1.1116	−0.4966	0.0780	1.1743	−0.7618	0.1627	0.1053	0.8753	−0.1438	0.1041	0.8753
−0.4	−0.1678	0.9429	−0.7003	0.1329	1.0688	−0.8001	0.0942	0.1765	0.3735	−0.1902	0.1649	0.2601
−0.3	−0.0451	0.6440	−0.7994	0.0202	1.1743	−1.0633	−0.0198	0.1237	0.5260	−0.0968	0.0549	0.4159
−0.2	−0.0870	0.3531	−0.4945	0.0488	0.1115	−0.6442	−0.0277	0.0174	0.4979	−0.1414	0.1458	0.2983
−0.1	−0.0111	0.1642	−1.1036	0.0371	0.0441	−0.5663	0.0461	0.2134	0.8890	−0.1499	0.5382	0.2635

Table A4. The fitting parameters in Gaussian-type functions, f_{G_i}, ($x_n \geq 0$).

x_n	$C1_{fg1}$	$C2_{fg1}$	$C2_{fg1}$	$C1_{fg2}$	$C2_{fg2}$	$C2_{fg2}$	$C1_{fg3}$	$C2_{fg3}$	$C2_{fg3}$	$C1_{fg4}$	$C2_{fg4}$	$C3_{fg4}$
0	−0.0282	0.2644	−0.7399	0.0321	0.2067	−0.5096	−0.0188	0.1606	0.6415	−0.1988	0.1055	0.3195
0.1	−0.0101	0.1185	−1.0749	0.0648	0.0471	−0.5048	−0.0347	0.0912	0.5759	−0.1326	0.1690	0.1507
0.2	−0.0086	0.1905	−0.8931	0.0829	0.0559	−0.4788	0.1798	0.0244	0.5212	−0.1983	0.0467	0.4911
0.3	−0.0207	0.4419	−0.3236	0.0560	0.1071	−0.4590	0.0998	0.1044	0.3243	−0.0432	0.0118	0.7140
0.4	−0.0301	0.2881	−0.8644	0.0458	0.5590	−0.4720	0.0930	0.0640	0.4118	−0.0321	0.0200	0.7194
0.5	−0.0782	0.4384	−0.6955	0.1074	0.5869	−0.4583	0.0617	0.1149	0.4528	−0.0458	0.0221	0.7659
0.6	−0.0075	0.1009	−0.8507	0.0620	0.1972	−0.3284	0.0980	0.0508	0.4549	−0.0153	0.1649	0.3134
0.7	−0.0116	0.2116	−0.5554	0.0341	0.0302	−0.4679	0.0823	0.6071	0.2744	−0.0290	0.8212	0.6627
0.75	−0.0559	0.4797	−0.5663	0.0822	0.6792	−0.2601	0.0606	0.8796	0.3729	−0.0480	1.1384	0.5431
0.8	−0.0641	0.4331	−0.5472	0.0928	0.5810	−0.2874	0.0702	0.8866	0.4494	−0.0569	1.0591	0.6319
0.9	−0.0324	0.2019	−0.6852	0.0464	0.5345	−0.4195	0.2196	0.8986	0.2774	−0.1847	1.1113	0.2855
1	−0.0195	0.0280	−0.6645	0.0231	0.0927	−0.3909	0.1242	0.8241	0.7064	−0.1016	0.6778	0.9532
1.1	−0.0107	0.2134	−0.5882	0.0321	0.0485	−0.3264	0.1567	0.9455	0.5062	−0.1289	1.0222	0.6326
1.2	−0.0219	0.1079	−0.4754	0.0590	0.0661	−0.3271	0.1216	0.9341	0.6162	−0.1017	0.9743	0.7481
1.3	−0.0558	0.0745	−0.5868	0.0794	0.1665	−0.3756	0.0442	0.1009	0.5868	−0.0063	1.0618	1.4181
1.4	−0.0680	0.0872	−0.5884	0.0808	0.1703	−0.4052	0.0379	0.0807	0.5984	−0.0082	0.8341	1.3434
1.5	−0.0694	0.0877	−0.5999	0.0810	0.1966	−0.4046	0.0406	0.0807	0.6152	−0.0074	1.1495	1.6060
1.6	−0.0722	0.0915	−0.5991	0.1011	0.2151	−0.3756	0.0540	0.1458	0.6705	−0.0273	0.7772	0.5111
1.7	−0.0551	0.0708	−0.5765	0.0867	0.1519	−0.3620	0.0340	0.0817	0.6651	−0.0060	1.3904	1.0672
1.8	−0.0155	0.0552	−0.5759	0.0634	0.0961	−0.2594	0.1763	0.0736	0.6538	−0.1443	0.0795	0.6524
1.9	−0.0372	0.0745	−0.5528	0.0779	0.1463	−0.2945	0.0737	0.2138	0.7138	−0.0492	0.4508	0.6401
2	−0.0160	0.0745	−0.5143	0.0693	0.0894	−0.2747	0.1489	0.0990	0.7382	−0.1203	0.1036	0.7424
2.1	−0.0602	0.0877	−0.5690	0.0880	0.1811	−0.3981	0.0455	0.3259	0.6873	−0.0301	0.8226	0.5003
2.2	−0.0496	0.0991	−0.5868	0.0715	0.2031	−0.3920	0.0619	0.1132	0.6155	−0.0466	0.1882	0.5007
2.3	−0.0718	0.1378	−0.5964	0.0937	0.2646	−0.4255	0.0416	0.1352	0.6401	−0.0276	0.5382	0.2819
2.4	−0.0469	0.1202	−0.5868	0.0655	0.2646	−0.3790	0.0355	0.1202	0.6436	−0.0234	0.4507	0.3257
2.5	−0.0363	0.1149	−0.6060	0.0529	0.2998	−0.3735	0.0284	0.1158	0.6360	−0.0172	0.5382	0.2382
3	−0.0111	1.4491	−1.3743	0.0219	0.8227	−0.7618	0.0533	0.3804	0.3011	−0.0405	0.7993	0.2054
4	0.0096	0.1062	−1.1569	−0.0037	0.2368	−1.0284	0.0125	0.1835	0.3284	−0.0099	0.2470	1.3101
6	0.0224	0.6458	−0.8438	−0.0217	0.7993	−0.8329	−0.0067	0.0921	0.2259	0.0139	0.1854	0.1944
8	0.0231	0.4278	−0.6121	−0.0225	0.3979	−0.6374	−0.0258	0.4929	0.5649	0.0301	0.4274	0.5417
10	0.0247	0.4050	−1.0838	−0.0246	0.3745	−1.0626	−0.0270	0.4929	0.4884	0.0312	0.4425	0.4774

Table A5. The fitting parameters in resonance-type functions, f_{R_i}, $(x_n < 0)$.

x_n	$C1_{fr1}$	$C2_{fr1}$	$C3_{fr1}$	$C1_{fr2}$	$C2_{fr2}$	$C3_{fr2}$	$C1_{fr3}$	$C2_{fr3}$	$C3_{fr3}$	$C1_{fr3}$	$C2_{fr3}$	$C3_{fr3}$
−10	0.2350	0.0101	−0.0006	−0.2396	0.0101	0.3111	1.6824	1.4410	−0.0001	−1.6824	1.4410	0.0001
−8	0.2405	0.0087	−0.2206	−0.2456	0.0087	0.2860	1.0335	1.3209	−0.0001	−1.0335	1.3209	0.0001
−6	0.2470	0.0161	−0.0988	−0.2538	0.0163	0.4355	1.2991	0.6408	−0.0004	−1.2991	0.6408	0.0004
−4	0.2460	0.0237	−0.4757	−0.2534	0.0237	0.2733	1.7798	0.8008	0.0007	−1.7798	0.8008	−0.0007
−2	0.3042	0.0648	−0.4639	−0.3168	0.0652	0.3341	2.6455	1.2809	0.0024	−2.6455	1.2809	−0.0024
−1	0.2864	0.0647	−0.6180	−0.3014	0.0660	0.1831	4.6927	0.6408	0.0159	−4.6927	0.6408	−0.0159
−0.75	0.2980	0.0607	−0.6671	−0.3132	0.0622	0.0819	5.0415	0.8016	0.0214	−5.0415	0.8016	−0.0214
−0.7	0.2742	0.0568	−0.2993	−0.2964	0.0632	0.4784	5.5835	0.8906	0.0209	−5.5835	0.8906	−0.0209
−0.6	0.2550	0.0575	−0.4743	−0.2768	0.0636	0.3882	6.0181	1.0445	0.0213	−6.0181	1.0445	−0.0213
−0.5	0.2636	0.0556	−0.5181	−0.2821	0.0600	0.3116	6.0327	1.1109	0.0240	−6.0327	1.1109	−0.0240
−0.4	0.2591	0.0565	−0.5181	−0.2791	0.0619	0.3062	6.0938	1.1992	0.0259	−6.0938	1.1992	−0.0259
−0.3	0.2593	0.0562	−0.4743	−0.2776	0.0613	0.3062	6.0913	1.0992	0.0268	−6.0913	1.0992	−0.0268
−0.2	0.2771	0.0544	−0.2200	−0.2958	0.0603	0.4757	6.1938	1.0992	0.0275	−6.1938	1.0992	−0.0275
−0.1	0.2784	0.0538	−0.1899	−0.2959	0.0594	0.4757	6.1182	0.9992	0.0273	−6.1182	0.9992	−0.0273

Table A6. The fitting parameters in resonance-type functions, f_{R_i}, $(x_n \geq 0)$.

x_n	$C1_{fr1}$	$C2_{fr1}$	$C3_{fr1}$	$C1_{fr2}$	$C2_{fr2}$	$C3_{fr2}$	$C1_{fr3}$	$C2_{fr3}$	$C3_{fr3}$	$C1_{fr3}$	$C2_{fr3}$	$C3_{fr3}$
0	0.2802	0.0552	−0.1681	−0.2976	0.0613	0.4873	6.1572	0.9992	0.0267	−6.1572	0.9992	−0.0267
0.1	0.2839	0.0538	−0.3431	−0.2976	0.0577	0.3075	6.0498	0.9992	0.0267	−6.0498	0.9992	−0.0267
0.2	0.2344	0.0547	−0.4962	−0.2447	0.0578	0.3116	5.8423	1.0453	0.0269	−5.8423	1.0453	−0.0269
0.3	0.2285	0.0538	−0.5071	−0.2355	0.0557	0.3021	5.7690	0.9570	0.0257	−5.7690	0.9570	−0.0257
0.4	0.2608	0.0557	−0.2966	−0.2665	0.0575	0.3896	5.6689	0.9008	0.0244	−5.6689	0.9008	−0.0244
0.5	0.2589	0.0510	−0.2316	−0.2619	0.0519	0.4347	5.3271	0.8008	0.0244	−5.3271	0.8008	−0.0244
0.6	0.2679	0.0500	−0.1260	−0.2689	0.0505	0.5311	5.3979	0.8008	0.0214	−5.3979	0.8008	−0.0214
0.7	0.2399	0.0501	−0.0382	−0.2391	0.0505	0.7026	4.7998	0.8008	0.0217	−4.7998	0.8008	−0.0217
0.75	0.2698	0.0542	0.0493	−0.2668	0.0539	0.7136	4.1235	0.8039	0.0238	−4.1235	0.8039	−0.0238
0.8	0.2462	0.0580	0.0493	−0.2421	0.0575	0.7806	4.5215	0.8227	0.0196	−4.5215	0.8227	−0.0196
0.9	0.2860	0.0580	0.0518	−0.2794	0.0566	0.6763	4.8775	0.7918	0.0158	−4.8775	0.7918	−0.0158
1	0.2571	0.0519	0.0542	−0.2494	0.0501	0.7363	5.2356	0.7008	0.0131	−5.2356	0.7008	−0.0131
1.1	0.2719	0.0569	0.0542	−0.2620	0.0544	0.7061	4.3723	0.7208	0.0117	−4.3723	0.7208	−0.0117
1.2	0.1681	0.0445	0.0710	−0.1605	0.0426	1.0612	2.2775	0.5074	0.0210	−2.2775	0.5074	−0.0210
1.3	0.3056	0.0543	−0.2436	−0.2901	0.0497	0.3166	2.4834	0.6488	0.0115	−2.4834	0.6488	−0.0115
1.4	0.3011	0.0707	0.0688	−0.2866	0.0673	0.6694	3.0307	1.0250	0.0035	−3.0307	1.0250	−0.0035
1.5	0.3249	0.0551	−0.2494	−0.3071	0.0503	0.2907	2.3580	1.6010	0.0019	−2.3580	1.6010	−0.0019
1.6	0.2896	0.0539	−0.2406	−0.2722	0.0488	0.3316	2.5045	1.6788	−0.0028	−2.5045	1.6788	0.0028
1.7	0.2894	0.0482	−0.1825	−0.2739	0.0442	0.3709	2.3087	1.5857	−0.0061	−2.3087	1.5857	0.0061
1.8	0.2839	0.0425	−0.3007	−0.2681	0.0386	0.2556	1.9482	1.1992	−0.0108	−1.9482	1.1992	0.0108
1.9	0.3027	0.0386	−0.0691	−0.2893	0.0361	0.4328	2.6024	1.3038	−0.0101	−2.6024	1.3038	0.0101
2	0.2838	0.0284	−0.1701	−0.2721	0.0264	0.3567	2.1167	1.6008	−0.0142	−2.1167	1.6008	0.0142
2.1	0.2867	0.0319	−0.0874	−0.2739	0.0297	0.4251	4.2651	1.0992	−0.0097	−4.2651	1.0992	0.0097
2.2	0.2430	0.0313	−0.1366	−0.2300	0.0285	0.4306	4.7534	1.2008	−0.0100	−4.7534	1.2008	0.0100
2.3	0.2664	0.0291	−0.0942	−0.2542	0.0270	0.4251	5.3882	1.0398	−0.0098	−5.3882	1.0398	0.0098
2.4	0.2539	0.0294	−0.0901	−0.2412	0.0271	0.4306	5.8081	1.0281	−0.0100	−5.8081	1.0281	0.0100
2.5	0.2307	0.0250	−0.1487	−0.2185	0.0228	0.4444	5.9830	0.9945	−0.0104	−5.9830	0.9945	0.0104
3	0.2566	0.0226	−0.1334	−0.2425	0.0203	0.3124	4.3969	0.8562	−0.0176	−4.3969	0.8562	0.0176
4	0.2470	0.0164	0.0075	−0.2347	0.0147	0.3545	5.3616	0.5986	−0.0135	−5.3616	0.5986	0.0135
6	0.2499	0.0098	0.0062	−0.2410	0.0089	0.1392	2.5100	0.4103	−0.0108	−2.5100	0.4103	0.0108
8	0.2223	0.0098	0.0395	−0.2160	0.0094	0.1740	2.9275	0.4103	−0.0014	−2.9275	0.4103	0.0014
10	0.2540	0.0253	0.0462	−0.2489	0.0254	0.1833	3.1876	0.3155	0.0001	−3.1876	0.3155	−0.0001

Appendix B

Figure A4 shows the 2D-VAWT rotor [35] as the target of test calculation in this study. The rotor has three blades (cross-section: NACA 0018) of a chord length of $c = 20$ mm. The diameter D is 50 mm. The rotor height is assumed to be 43.4 mm which is equivalent to the experimental model used in the wind tunnel experiments [34]. The CFD analysis [35] utilizes the commercial application software STAR-CCM+ as the numerical solver. The equation of continuity and two-dimensional unsteady incompressible Reynolds averaged Navier–Stokes (RANS) equations are solved by applying the SST k–ω turbulence model. The calculation domain has the same size as that of the four-rotor array cases shown in Appendix C. The constant wind speed (10 m/s) is set at the inlet boundary (left side) and the constant gage pressure (0 Pa) is kept at the outlet boundary (right side). The top and bottom boundaries are defined as slip walls. The CFD analysis also applies the DFBI (dynamic fluid/body interaction) model [35] to simulate the change in the rotational speed of each rotor. In the 2D-CFD calculation, the moment of inertia of each rotor with unit height was utilized for the DFBI model which solves the equation of motion of each rotor. The output power obtained by the CFD of each rotor was converted to that of the equivalent rotor to the experimental one with a height of 43.4 mm. From the CFD analysis using the DFBI model for an isolated single rotor in different upstream wind speeds U_∞, the linear relation between the circulation Γ_{SI} and the wind speed is obtained as shown in Equation (A17) (see Figure 4 in [40]). The subscripts SI and ref show "Single Rotor" and "reference". We assume that the performance of a rotor in a rotor cluster can be given by the same function of the circulation as the relation obtained from the CFD of the isolated single rotor. That is, the blockage effect μ (m^3/s) and the thrust force Th (mN) of the 2D rotor in this study are given by Equations (A18) and (A19), respectively, as the functions of circulation Γ (m^2/s). The x-direction pressure loss Fp (mN) of the isolated single rotor in the CV ($20D \times 20D \times 0.868D$) is expressed by Equation (A20). The power output P (mW) is calculated by Equation (A21). The angular velocity ω (rad/s) is calculated by Equation (A22). The reference values, which correspond to the values at the reference wind speed U_{∞_ref} = 10 m/s, are μ_{ref} = 0.0015 m^3/s, Γ_{ref} = 0.3264 m^2/s, Th_{ref} = 141.37 mN, Fp_{ref} = 229.61 mN, P_{ref} = 177.62 mW, and ω_{ref} = 363.60 rad/s, respectively.

$$\Gamma_{SI} = \Gamma_{ref}\frac{U_\infty}{U_{\infty\_ref}} \tag{A17}$$

$$\mu = \mu_{ref}\left(\frac{\Gamma}{\Gamma_{ref}}\right)^2 \tag{A18}$$

$$Th = Th_{ref}\left(\frac{\Gamma}{\Gamma_{ref}}\right)^2 \tag{A19}$$

$$Fp = Fp_{ref}\left(\frac{\Gamma}{\Gamma_{ref}}\right)^3 \tag{A20}$$

$$P = P_{ref}\left(\frac{\Gamma}{\Gamma_{ref}}\right)^3 \tag{A21}$$

$$\omega = \omega_{ref}\frac{\Gamma}{\Gamma_{ref}} \tag{A22}$$

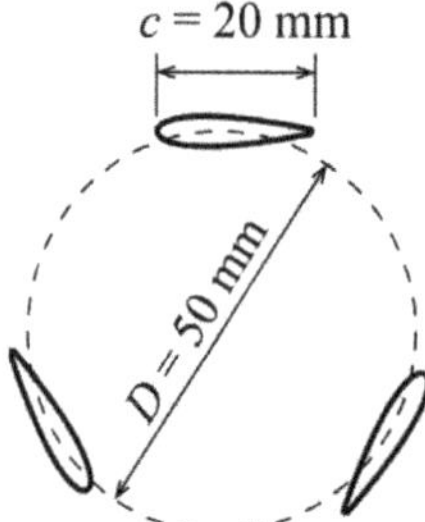

Figure A4. 2D-VAWT rotor as the target of the CFD calculation.

Appendix C

The computation meshes used in the CFD analysis in this study for the two kinds of four-rotor layouts are shown in Figure A5. Figure A5a,b show the whole domains of the parallel and tandem layouts, respectively. The size of the whole domain is $80D \times 100D$. The center of each four-rotor array is located at $40D$ from the inlet boundary. Figure A5c,d show the mesh around the rotors of the parallel and tandem layouts, respectively. The distance between the centers of the adjacent rotors is equal to $4D$ (i.e., inter-rotor *gap* = 3D). Figure A5e,f show the details of the mesh created around a rotor and a blade, respectively. These mesh sizes are the same as that used for the CFD analysis of one-, two-, and three-rotor arrangements. The total number of cells is 593,880 in the case of the four-rotor parallel layout and 473,725 in the case of the four-rotor tandem layout.

Figure A6 shows the CFD results of the time history of the angular velocity of each rotor in the two kinds of four-rotor layouts. The calculation using the DFBI model was conducted until 4 s for each layout when the convergence was almost attained. In the last 0.5 s, the angular velocity was averaged to be used to evaluate the power output of each rotor. In the calculations shown in Figure A6, the initial angular velocity is 360 rad/s for all rotors in the parallel layout. On the other hand, in the tandem case, the initial values are set at 366, 250, 200, and 180 rad/s for R1, R2, R3, and R4, respectively.

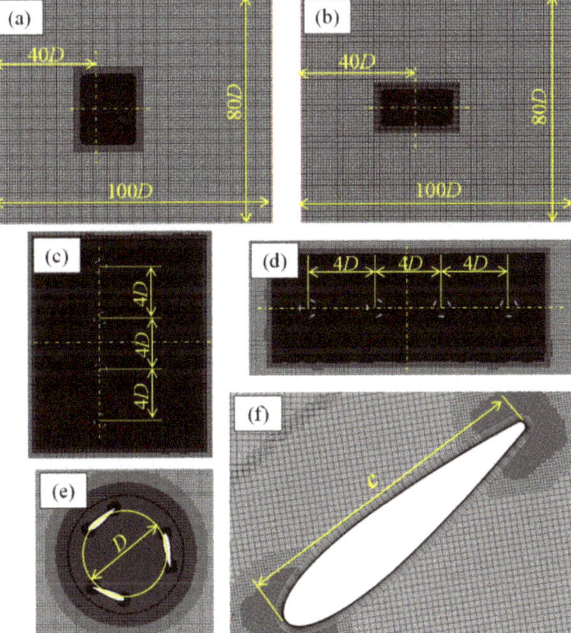

Figure A5. Computation meshes for the CFD analysis of four-rotor layouts: (**a**) whole domain of the parallel layout; (**b**) whole domain of the tandem layout; (**c**) the mesh around the 4 rotors in the parallel array; (**d**) the mesh around the 4 rotors in the tandem array; (**e**) the mesh around a rotor; (**f**) the mesh around a blade.

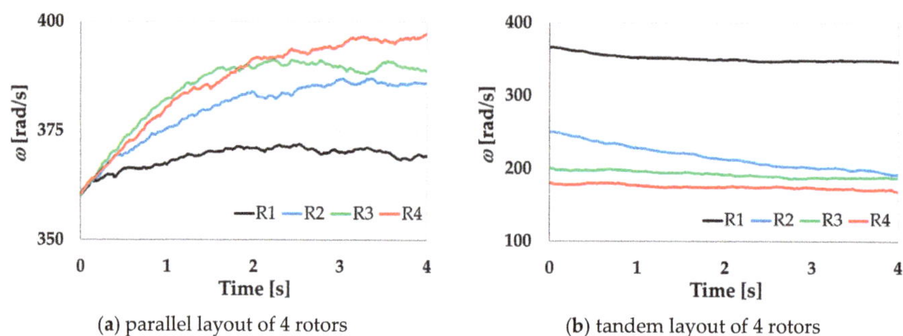

Figure A6. CFD results of the time history of the angular velocity of each rotor in (**a**) parallel layout and (**b**) tandem layouts of 4 rotors, respectively.

References

1. Barthelmie, R.J.; Pryor, S.C. Climate change mitigation potential of wind energy. *Climate* **2021**, *9*, 136. [CrossRef]
2. McKenna, R.; Pfenninger, S.; Heinrichs, H.; Schmidt, J.; Staffell, I.; Bauer, C.; Gruber, K.; Hahmann, N.A.; Jansen, M.; Klingler, M.; et al. High-resolution large-scale onshore wind energy assessments: A review of potential definitions, methodologies and future research needs. *Renew. Energy* **2022**, *182*, 659–684. [CrossRef]
3. Kou, L.; Li, Y.; Zhang, F.; Gong, X.; Hu, Y.; Quande, Y.; Ke, W. Review on monitoring, operation and maintenance of smart offshore wind farms. *Sensors* **2022**, *22*, 2822. [CrossRef] [PubMed]
4. Chen, J.; Kim, M.-H. Review of recent offshore wind turbine research and optimization methodologies in their design. *J. Mar. Sci. Eng.* **2022**, *10*, 28. [CrossRef]
5. Vermeer, L.J.; SΦrensen, J.N.; Crespo, A. Wind turbine wake aerodynamics. *Prog. Aerosp. Sci.* **2003**, *39*, 467–510. [CrossRef]
6. Fleming, P.; Sinner, M.; Young, T.; Lannic, M.; King, J.; Simley, E.; Doekemeijer, B. Experimental results of wake steering using fixed angles. *Wind Energy Sci.* **2021**, *6*, 1521–1531. [CrossRef]
7. Porté-Agel, F.; Wu, Y.-T.; Chen, C.-H. A numerical study of the effects of wind direction on turbine wakes and power losses in a large wind farm. *Energies* **2013**, *6*, 5297–5313. [CrossRef]
8. Jensen, N.O. *A Note on Wind Generator Interaction*; Risø National Laboratory: Roskilde, Denmark, 1983.
9. Niayifar, A.; Porté-Agel, F.; Diaz, A.P. Analytical modeling of wind farms: A new approach for power prediction. *Energies* **2016**, *9*, 741. [CrossRef]
10. Göçmen, T.; Laan, P.; Réthoré, P.E.; Diaz, A.P. Wind turbine wake models developed at the technical university of Denmark: A review. *J. Renew. Sustain. Energy* **2016**, *60*, 752–769. [CrossRef]
11. Zhang, Z.; Huang, P.; Sun, H. A novel analytical wake model with a cosine-shaped velocity deficit. *Energies* **2020**, *13*, 335. [CrossRef]
12. Gao, X.; Li, Y.; Zhao, F.; Sun, H. Comparisons of the accuracy of different wake models in wind farm layout optimization. *Energy Explor. Exploit.* **2020**, *38*, 1725–1741. [CrossRef]
13. Porté-Agel, F.; Bastankhah, M.; Shamsoddin, S. Wind-turbine and wind-farm flows: A review. *Bound.-Layer Meteorol.* **2020**, *174*, 1–59. [CrossRef]
14. Vahidi, D.; Porté-Agel, F. A physics-based model for wind turbine wake expansion in the atmospheric boundary layer. *J. Fluid Mech.* **2022**, *943*, A49. [CrossRef]
15. Mosetti, G.; Poloni, C.; Diviacco, B. Optimization of wind turbine positioning in large windfarms by means of a genetic algorithm. *J. Wind. Eng. Ind. Aerodyn.* **1994**, *51*, 105–116. [CrossRef]
16. Feng, J.; Shen, W.Z. Solving the wind farm layout optimization problem using random search algorithm. *Renew. Energy* **2015**, *78*, 182–192. [CrossRef]
17. Kirchner-Bossi, N.; Porté-Agel, F. Realistic Wind Farm Layout Optimization through Genetic Algorithms Using a Gaussian Wake Model. *Energies* **2018**, *11*, 3268. [CrossRef]
18. Rinker, J.M.; Soto Sagredo, E.; Bergami, L. The Importance of wake meandering on wind turbine fatigue loads in wake. *Energies* **2021**, *14*, 7313. [CrossRef]
19. Liang, Z.; Liu, H. Layout optimization of a modular floating wind farm based on the full-field wake model. *Energies* **2022**, *15*, 809. [CrossRef]
20. Serrano González, J.; López, B.; Draper, M. Optimal pitch angle strategy for energy maximization in offshore wind farms considering Gaussian wake model. *Energies* **2021**, *14*, 938. [CrossRef]
21. Munters, W.; Meyers, J. Dynamic strategies for yaw and induction control of wind farms based on large-eddy simulation and optimization. *Energies* **2018**, *11*, 177. [CrossRef]
22. Qian, G.-W.; Ishihara, T. A new analytical wake model for yawed wind turbines. *Energies* **2018**, *11*, 665. [CrossRef]

23. Whittlesey, R.W.; Liska, S.; Dabiri, J.O. Fish schooling as a basis for vertical axis wind turbine farm design. *Bioinspiration Biomim.* **2010**, *5*, 035005. [CrossRef]
24. Dabiri, J.O. Potential order-of-magnitude enhancement of wind farm power density via counter-rotating vertical-axis wind turbine arrays. *J. Renew. Sustain. Energy* **2011**, *3*, 043104. [CrossRef]
25. Zanforlin, S.; Nishino, T. Fluid dynamic mechanisms of enhanced power generation by closely spaced vertical axis wind turbines. *Renew. Energy* **2016**, *99*, 1213–1226. [CrossRef]
26. De Tavernier, D.; Ferreira, C.; Li, A.; Paulsen, U.S.; Madsen, H.A. Towards the understanding of vertical-axis wind turbines in double-rotor configuration. *J. Phys. Conf. Ser.* **2018**, *1037*, 022015. [CrossRef]
27. Bangga, G.; Lutz, T.; Krämer, E. Energy assessment of two vertical axis wind turbines in side-by-side arrangement. *J. Renew. Sustain. Energy* **2018**, *10*, 033303. [CrossRef]
28. Sahebzadeh, S.; Rezaeiha, A.; Montazeri, H. Impact of relative spacing of two adjacent vertical axis wind turbines on their aerodynamics. *J. Phys.* **2020**, *1618*, 042002. [CrossRef]
29. Peng, H.Y.; Han, Z.D.; Liu, H.J.; Lin, K.; Lam, H.F. Assessment and optimization of the power performance of twin vertical axis wind turbines via numerical simulations. *Renew. Energy* **2020**, *147*, 43–54. [CrossRef]
30. Hezaveh, S.H.; Bou-Zeid, E.; Dabiri, J.; Kinzel, M.; Cortina, G.; Martinelli, L. Increasing the Power Production of Vertical-Axis Wind-Turbine Farms Using Synergistic Clustering. *Bound.-Layer Meteorol* **2018**, *169*, 275–296. [CrossRef]
31. Silva, J.E.; Danao, L.A.M. Varying VAWT Cluster Configuration and the Effect on Individual Rotor and Overall Cluster Performance. *Energies* **2021**, *14*, 1567. [CrossRef]
32. Vergaerde, A.; De Troyer, T.; Standaert, L.; Kluczewska-Bordier, J.; Pitance, D.; Immas, A.; Silvert, F.; Runacres, M.C. Experimental validation of the power enhancement of a pair of vertical-axis wind turbines. *Renew. Energy* **2020**, *146*, 181–187. [CrossRef]
33. Vergaerde, A.; De Troyer, T.; Molina, A.C.; Standaert, L.; Runacres, M.C. Design, manufacturing and validation of a vertical-axis wind turbine setup for wind tunnel tests. *J. Wind Eng. Ind. Aerodyn.* **2019**, *193*, 103949. [CrossRef]
34. Jodai, Y.; Hara, Y. Wind tunnel experiments on interaction between two closely spaced vertical-axis wind turbines in side-by-side arrangement. *Energies* **2021**, *14*, 7874. [CrossRef]
35. Hara, Y.; Jodai, Y.; Okinaga, T.; Furukawa, M. Numerical analysis of the dynamic interaction between two closely spaced vertical-axis wind turbines. *Energies* **2021**, *14*, 2286. [CrossRef]
36. Furukawa, M.; Hara, Y.; Jodai, Y. Analytical model for phase synchronization of a pair of vertical-axis wind turbines. *Energies* **2022**, *15*, 4130. [CrossRef]
37. Buranarote, J.; Hara, Y.; Jodai, Y. Proposal of a model simulating the velocity profile of the wake of a two-dimensional vertical axis wind turbine. In Proceedings of the JSFM Annual General Meeting, Yamaguchi, Japan, 18–20 September 2020.
38. Shapiro, C.R.; Starke, G.M.; Meneveau, C.; Gayme, D.F. A wake modeling paradigm for wind farm design and control. *Energies* **2019**, *12*, 2956. [CrossRef]
39. Buranarote, J.; Hara, Y.; Jodai, Y. Construction of a model simulating exactly the velocity profile around a two-dimensional vertical axis wind turbine. In Proceedings of the JSME 59th Chugoku-Shikoku Branch Meeting, Okayama, Japan, 5 March 2021.
40. Buranarote, J.; Hara, Y.; Jodai, Y.; Furukawa, M. A wake model simulating the velocity profile of a two-dimensional vertical axis wind turbine. In Proceedings of the 7th International Conference on Jets, Wakes and Separated Flows (ICJWSF-2022), Tokyo, Japan, 15–17 March 2022.
41. Okinaga, T.; Hara, Y.; Yoshino, K.; Jodai, Y. Numerical simulation considering the variation in rotational speed of three closely spaced vertical-axis wind turbines. In Proceedings of the JWEA 43rd Wind Energy Utilization Symposium, Tokyo, Japan, 18–19 November 2021.

Article

A Study on a Casing Consisting of Three Flow Deflectors for Performance Improvement of Cross-Flow Wind Turbine

Tadakazu Tanino [1,*], Ryo Yoshihara [2] and Takeshi Miyaguni [3]

[1] Department of Mechanical Engineering, National Institute of Technology, Kurume College, 1-1-1 Komorino, Kurume 830-8555, Japan
[2] Interdisciplinary Graduate School of Engineering Sciences Kyushu University, 6-1 Kasuga-koen, Kasuga 816-8580, Japan
[3] Department of Mechanical Systems Engineering, University of Kitakyushu, 1-1 Hibikino, Wakamatsu-ku, Kitakyushu 808-0135, Japan
* Correspondence: ttanino@kurume-nct.ac.jp; Tel.: +81-942-35-9366

Abstract: We investigated the effective use of cross-flow wind turbines for small-scale wind power generation to increase the output power by using a casing, which is a kind of wind-collecting device, composed of three flow deflector plates having the shape of a circular-arc airfoil. Drag-type vertical-axis wind turbines have an undesirable part of about half of the swept area where the inflow of wind results in low output performance. To solve this problem, we devised a casing consisting of three flow deflector plates, two of which were to block the unwanted inflow of wind and the remaining flow deflector plate having an angle of attack with respect to the wind direction to increase the flow toward the rotor. In this study, output performance experiments using a wind tunnel and numerical fluid analysis were conducted on a cross-flow wind turbine with three flow deflector plates to evaluate the effectiveness of the casing on output performance improvement. As a result, it was confirmed that the casing could improve the output performance of the cross-flow wind turbine by approximately 60% at the maximum performance point and could also maintain the output performance about 50% higher compared to the bare cross-flow wind turbine without the casing within a deviation angle of ±10 degrees, even when the casing direction was inclined against the wind direction due to changes in wind direction.

Keywords: drag-type wind turbine; vertical axis wind turbine; cross-flow wind turbine; wind-collecting device; flow deflector; output performance improvement

Citation: Tanino, T.; Yoshihara, R.; Miyaguni, T. A Study on a Casing Consisting of Three Flow Deflectors for Performance Improvement of Cross-Flow Wind Turbine. *Energies* **2022**, *15*, 6093. https://doi.org/10.3390/en15166093

Academic Editors: Yutaka Hara and Yoshifumi Jodai

Received: 8 July 2022
Accepted: 15 August 2022
Published: 22 August 2022

Publisher's Note: MDPI stays neutral with regard to jurisdictional claims in published maps and institutional affiliations.

1. Introduction

For the type of wind turbine [1], wind turbines are classified by the difference in fluid force acting on the wind turbine blade into lift-type wind turbines and drag-type wind turbines. The propeller-type wind turbines, which can be seen everywhere, are representative of lift-type wind turbines. Other lift-type wind turbines include the Darrieus wind turbine. On the other hand, Savonius wind turbines and the cross-flow wind turbines are representative of drag-type wind turbines. Furthermore, wind turbines can also be classified by the relationship between the direction of wind and the direction of rotor axis into horizontal-axis wind turbine (HAWTs) and vertical-axis wind turbine (VAWTs). The propeller-type wind turbine is a kind of HAWT and Darrieus wind turbine, and the Savonius wind turbine and cross-flow wind turbine are classified into VAWT. Lift-type wind turbines have the advantages of high rotating speed and high output so these are widely used, while drag-type wind turbines have not been widely used because of the disadvantages of low speed and low output. However, VAWTs such as the Savonius wind turbine and the cross-flow wind turbine do not require yaw control, and therefore have a simple structure and are less affected by changes in wind direction.

In addition, the usage of wind turbines for power generation began in Denmark. In the 1890s, Poul La Cour incorporated aerodynamic design principles in the blade design of the wind turbine [2]. For a while after that, wind power generation was far from being a successful business. After the oil crisis of the 1970s, the development and introduction of wind turbines were vigorously promoted mainly in Europe and United States. Now, the movement toward decarbonization due to global warming is accelerating the use of wind energy.

Despite such circumstances, Japan is an island nation with many mountainous regions, so the land areas where large-scale wind power generation is feasible are limited in Japan. Therefore, offshore wind power generation is expected as a promising technology, and various studies are being conducted to introduce it [3]. In addition, the Agency for Natural Resources and Energy has proposed a vigorous and strategic target for the introduction of the wind power generation of 30% of the total domestic power generation capacity requiring approximately 130 GW of the installed capacity [4]. In order to achieve the goal, the spread of medium- and small-scale wind power generation is also needed. In particular, medium- and small-scale wind power generation is expected to be used in urban and mountain areas where wind conditions, such as wind direction and speed, are highly variable, so it is expected that not only horizontal propeller-type wind turbines but also vertical Darrius, Savonius, and the cross-flow wind turbine will be used. However, as mentioned above, Savonius and the cross-flow wind turbine have a disadvantage of low power output in spite of the advantage of low starting torque, so the output performance of these wind turbines needs to be improved. For these reasons, we began our study on the output performance improvement for the cross-flow wind turbines.

As shown in Figures 1 and 2, a cross-flow wind turbine consists of many small blades which are arranged in a circle, and the structure is simple. In Japan, relatively many studies on the cross-flow wind turbine have been conducted. Ushiyama et al. [5] and Tan et al. [6] studied the relations between the design factors, such as the number of blades and blade pitch angle, and the wind turbine performance. Additionally, we conducted similar studies using a cross-flow wind turbine with a diameter of 114 mm; as a result, it was shown that the optimal solidity and blade pitch angle are $\sigma = 0.76$ and $\beta = 45°$, respectively [7]. Moreover, we studied upscaling of cross-flow wind turbines using scale-up models with a 1.5 times larger diameter [8]. Two types of the cross-flow wind turbine of which the solidity and the blade pitch angle were same but the size of the blade airfoil and the number of blades were different were investigated, and it was confirmed that the output performance characteristics of three types of cross-flow wind turbines including the wind turbine with a diameter of 114 mm were almost equivalent. In any case, all of the studies for cross-flow wind turbines described here [5–8] have shown that the maximum power coefficient is as low as about 0.1, and these results indicate the necessity to devise some way to achieve higher output performance.

Cross-flow wind turbines have a characteristic that the wind turbine can rotate by the wind from any direction. The half swept area catching the wind is useful to rotate the wind turbine itself, but for the rest half swept area, the entering flow into the area prevents from the rotor rotation. Therefore, the key to improving output performance is to prevent undesirable wind from entering the rotor and to allow some of the prevented wind to enter the effective half swept area. With this in mind, we studied a more effective use of cross-flow wind turbines, which are expected to be used for small-scale wind power generation, by improving the output performance of cross-flow wind turbines. In other words, we studied methods of improving the flow of cross-flow wind turbines to achieve high output, for example, a method of using the flow at the edge of a structure such as a building, a method of using a wind-collecting device as a casing composed of two deflector plates, and so on. For the effective use of cross-flow wind turbines, not only us but also various unique studies have been conducted on suitable installation locations for cross-flow wind turbines and additional devices that improve the ambient wind flow suitable for cross-flow wind turbines. These studies have focused on improving the output performance of the cross-flow wind turbine because of its low output performance. Shimizu et al. developed a ring diffuser with guide vanes for higher performance of a cross-flow wind turbine [9]. The ring diffuser consisted of multiple guide vanes, and these guide vanes were arranged radially and with an inclination angle, which guided the entering wind flow to the rotor favorably around the wind turbine rotor. They showed that the ring diffuser with a diameter of 1150 mm, 18 vanes, and vane inclination angles of 30° to 60° was applied to a cross-flow wind turbine with a diameter of 350 mm and an axial length of 360 mm, and the ring diffuser could improve the output performance of the cross-flow wind turbine by about 1.5 to 2 times. Kiwata et al. studied a cross-flow wind turbine installed at the top of a windbreak fence [10]. In their research, a cross-flow wind turbine with a diameter of 80 mm installed above a windbreak fence having a height of 500 mm and a geometric shielding rate of 60% and 100% was examined, and the relations among the output performance of the cross-flow wind turbine and the rotating direction of the rotor, the clearance between the rotor and the top edge of the fence, and the geometric shielding rate were investigated. In the case of the geometric shielding rate of 100%, the flow above the fence was clearly increased so the maximum power coefficient reached about 0.6, whereas when the wind turbine was located above the fence having a geometric shielding rate of 60%, the increase in the flow above the fence decreased and the maximum power coefficient was reduced to about 0.3. However, it was still higher than that of the bare cross-flow wind turbine. Their results indicated that installing cross-flow wind turbines above the windbreak fence is one of the effective uses of cross-flow wind turbines. Mohamed et al. proposed the wind concentrator for cross-flow wind turbines to improve the output performance [11]. The wind concentrator consists of an arc-shaped windshield device and a wind augmentation device, which is a type of wind lens [12] as mentioned later. As mentioned above, cross-flow wind turbines have an undesirable swept area where the blades are moving in the upstream direction, and the wind flow entering this area prevents the rotor rotation. The arc-shaped windshield device was used to block this undesirable flow. In addition, the wind augmentation device was used to increase the flow rate by the wind-lens effect. In their research, using a cross-flow wind turbine with a diameter of 80 mm, an arc-shaped windshield device of an inner radius of 43.5 mm which covered 1/4 of the rotor where the blades were moving to the upstream side and a wind augmentation device of a wind lens type which was two parallel plates with flanges having a height of 50 mm and covered the rear side of the rotor were added as a wind concentrator to the cross-flow wind turbine. As a result, it was shown that the addition of the wind augmentation device to the arc-shaped windshield device could improve the power coefficient of the wind turbine by 88% and more, and the maximum power coefficient became higher by about 108%. Shigemitsu et al. focused on the two-directional prevailing winds generated by the land breeze and the sea breeze such as in coastal areas and investigated the use of cross-flow wind turbines in a prevailing wind environment. In order to improve the performance of cross-flow wind turbines, symmetrical casings combining a nozzle and a diffuser with the same shape, which could use the prevailing winds effectively, have been proposed [13,14]. A nozzle was arranged on the side where the blades were rotating along with the direction of the wind to make a flow path like a spiral casing, and a diffuser, the shape of which is the same as the nozzle, was arranged symmetrically to the nozzle, making an outlet flow path. The symmetrical casing with the nozzle and the diffuser was applied to a cross-flow wind turbine with a diameter of 150 mm, and the relations between the performance of the cross-flow wind turbine and the inclination angles of the casing against the wind flow direction were investigated. They showed that the symmetrical casing with an inclination angle of 15° could improve the performance of the wind turbine by 70% in the power coefficient. In addition, the effect of adding the side boards, which were standing perpendicular to the wind direction, on the nozzle and the diffuser was also examined. It was shown that the performance of the wind turbine became higher by about 1.9 times compared with the bare wind turbine when the side board was placed on the most upstream side of the nozzle (another side board was placed on the most downstream side of the diffuser) and the inclination angle of the casing was $30°$.

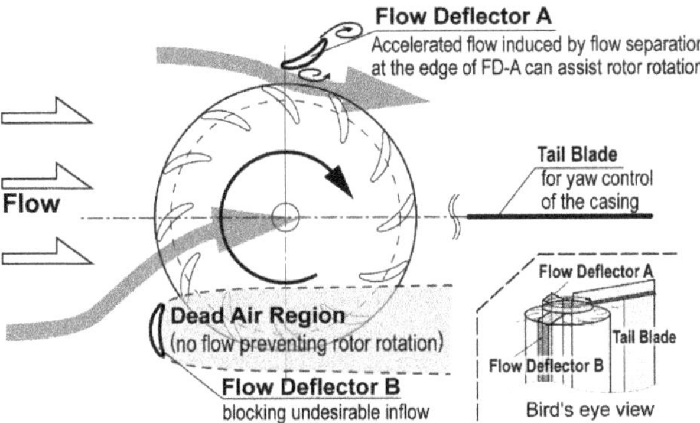

Figure 1. Schematic illustration of the flow of cross-flow wind turbine with wind-collecting casing with two flow deflectors.

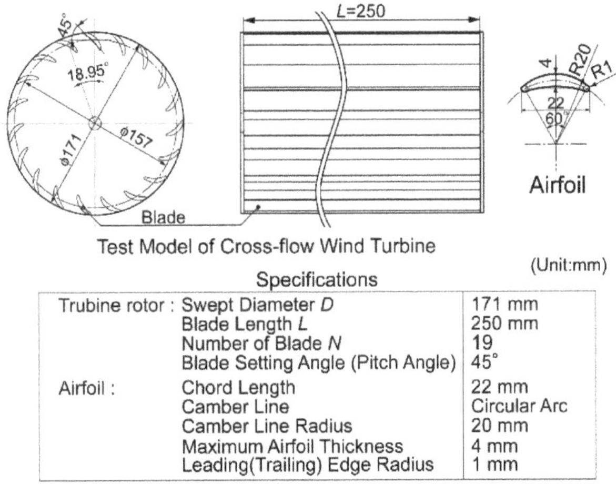

Specifications		
Trubine rotor :	Swept Diameter D	171 mm
	Blade Length L	250 mm
	Number of Blade N	19
	Blade Setting Angle (Pitch Angle)	45°
Airfoil :	Chord Length	22 mm
	Camber Line	Circular Arc
	Camber Line Radius	20 mm
	Maximum Airfoil Thickness	4 mm
	Leading(Trailing) Edge Radius	1 mm

Figure 2. Cross-flow wind turbine test model and airfoil.

In the studies [9–11,13,14] mentioned above, a device to improve the inflow of the cross-flow wind turbine has been added to the turbine rotor to increase the power output of the wind turbine. Similar to these studies, our study aimed to achieve higher power output by improving the flow into the cross-flow turbine rotor by the addition of surrounding structures. The wind-collecting casing studied in this study is a device that covers the rotor of a cross-flow wind turbine and here is called a wind-collecting casing (simply called a casing). Figure 1 shows the schematic of a cross-flow wind turbine with a wind-collecting casing that was studied [8] and the flow of the cross-flow wind turbine with the casing. The casing consists of several flow deflector plates (hereinafter a "flow deflector") and a tail blade [15]. In Figure 1, the casing has two flow deflectors with the shape of a circular-arc airfoil. One of the flow deflectors is located very close to the top of the rotor with a negative angle of attack with respect to the direction of the incoming wind (called FD-A), and the other is located upstream of the lower half of the rotor in the figure (called FD-B). These flow deflectors of FD-A and FD-B have different effects on flow improvement, and the combination of these flow deflectors is a feature of the proposed wind-collecting casing. In addition, compared to the ring diffuser introduced above by Shimizu et al. [9], the additional structure composed of two flow deflectors is smaller than that of the ring diffuser. This contributes to reducing the upstream flow velocity reduction and can make the structure of the casing simple.

According to the study of a wind turbine with a shroud, which has a diffuser shape, by Ohya et al. [12], when a diffuser was set in a uniform flow, the flow speed at the inlet of the diffuser was increased compared with that of the ambient wind. Moreover, by adding a brim (flange) to the diffuser at the rear edge of the diffuser, the flow-accelerating effect was enhanced extremely. In their research, using a propeller-type wind turbine with a diameter $D = 720$ mm, the optimal form of the flanged diffuser (called a wind lens) with an axial length of $1.25D$ was examined. From the results, a flanged diffuser with an axial length of $1.25D$, a diffuser opening angle of $12°$, and a flange height of $0.5D$ were applied to the wind turbine, and the output performance of the wind turbine increased by about 4 to 5 times compared to that of the bare wind turbine. Furthermore, Ohya et al. also examined the compact type of the flanged diffuser [16]. Even with a compacted flanged diffuser with an axial length of $0.22D$ and a flange height of $0.1D$, the output performance of the wind turbine could be increased by about 2.5 times, and the wind turbine with the compacted flanged diffuser was proposed as a more practical one. In addition, in the continuous study of the wind turbine with the compact flanged diffuser shortened in the axial direction by Oka et al. [17], it was shown that the accelerated flow caused by the flanged diffuser was particularly strong near the inner surface of the flanged diffuser from the results of the numerical study. Based on the results of these studies, the position of the flow deflector of FD-A was determined to obtain an accelerated flow to the rotor of a cross-flow wind turbine. This accelerated flow assists the rotor to rotate in the vicinity of the flow deflector of FD-A.

In addition, considering the flow of a cross-flow wind turbine without a wind-collecting casing based on Figure 1, the blades on the upper side of the rotor in the figure move in the direction of the wind flow into the rotor. On the other hand, the blades on the lower side of the rotor move in the opposite direction of the wind direction. Therefore, the flow into the lower half of the rotor is undesirable for the wind turbine. In our previous study of a cross-flow wind turbine using the separation flow near the edge of a structure such as a building [18], the power output of a cross-flow wind turbine was obviously improved by arranging the position of the rotor to be near the edge of a structure where the flow separation generated, and only the blades moving in the upwind direction in the half swept area of the rotor were in the separation zone. Based on these results, we considered the arrangement of the flow deflectors of FD-B so that the lower half of the rotor in Figure 1 was in the dead air region. The flow deflectors of FD-B are for blocking the inflow of wind into the lower half of the rotor and reducing the aerodynamic resistance to the blade moving upstream.

As shown in the lower right of Figure 1, the actual wind-collecting casing added to the cross-flow rotor proposed by us is a type of casing consisting of multiple flow deflectors and a tail blade. This casing was fixed to the rotating shaft of the wind turbine by bearings above and below the rotor. As mentioned above, the structure was designed to improve the flow suitable for a cross-flow rotor by flow deflectors to increase the power output of the wind turbine and to maintain the improved flow condition for the wind turbine in any wind direction by a tail blade to change the orientation of the casing only in response to changes in wind direction. In the previous study of high power output by utilizing the separation flow at the edge of a building or other structure, the wind direction was limited to one direction, but the introduction of the tail blade solved the problem of wind direction limitation. With the aim of further improving the casing, in this study, we examined the possibility of increasing the power output of a cross-flow wind turbine by using a wind-collecting casing with three flow deflectors. This casing is an improvement over the previous casing with two flow deflectors as shown in Figure 1. As mentioned above, the flow deflectors rotate with the tail blade so that the casing faces the wind in the correct direction even when the wind direction changes. However, when the wind direction changes rapidly, there is likely to be a misalignment between the wind direction and the orientation of the casing, and the misalignment causes a reduction in the effect of output performance improvement. The reduction in output performance increases as the width of the flow deflector shortens [19,20]. Therefore, to maintain a high output performance improvement effect even when the wind direction changes, the number of flow deflectors was increased in this casing.

In this study, a test model of cross-flow wind turbine with a casing composed of three flow deflectors was created by using a 3D printer and using a wind tunnel, the output performance tests were conducted by changing the casing orientation with respect to the wind direction in several ways. In addition, numerical fluid analyses using OpenFOAM were conducted under the same conditions as the output performance test. The obtained results are discussed in terms of the effectiveness of the proposed casing in improving output performance and also its superiority over the casing with two flow deflectors in our previous study against changes in wind direction.

2. Test Model of Cross-Flow Wind Turbine with Three Flow Deflectors

Figure 2 shows the geometry and dimensions of the tested cross-flow wind turbine model and airfoils. The typical diameter of the test cross-flow turbine was 171 mm, and the number of blades was 19. The airfoil was a circular-arc airfoil with a 22 mm chord length, a blade length of 250 mm, and an angle of attachment of 45°.

Figure 3 shows the shape and dimensions of the cross-flow wind turbine test model with three flow deflectors. In the experimental investigation of the performance improvement effect of the casing, only the three flow deflectors were added to the cross-flow rotor, and for simplicity, no tail blade was added. All the shapes of the three flow deflectors added to the rotor were the same, and a similar figure to the circular-arc airfoil used for the wind turbine blades, with dimensions 1.5 times larger. The arrangement of the flow deflectors in the figure is in the case when the wind flows from the left side of the figure same as in Figure 1, and this arrangement was used as the reference condition.

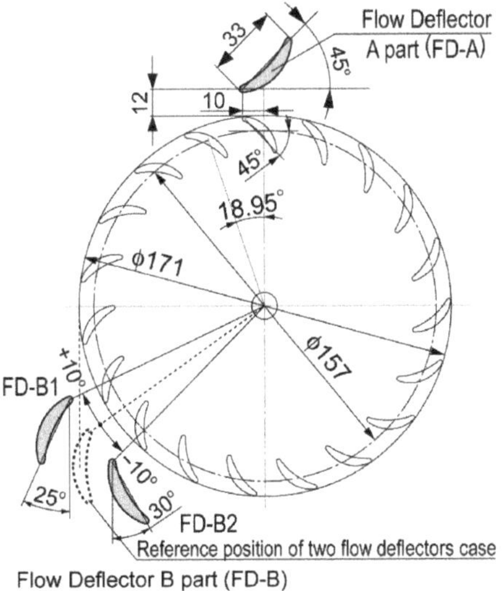

Figure 3. Cross-flow wind turbine with a casing having three flow deflectors.

As shown in Figure 3, the position of the flow deflector of FD-A is the same as in Figure 1. Each of the two flow deflectors of FD-B was positioned back and forth with respect to the position of the single flow deflector of FD-B in the casing shown in Figure 1.

In Figure 3, the position of the flow deflector of FD-B in Figure 1 is also indicated by a dashed line. The two flow deflectors of FD-B were one flow deflector B1 on the upstream side and the other one B2 on the downstream side. As shown in Figure 3, the position of the flow deflector B1 was shifted upstream around the center of the wind turbine axis to reduce the deceleration of incoming wind based on the position of the flow deflector indicated by the dashed line. Another flow deflector B2 was shifted downstream around the center of the rotor axis. This was to block the inflow of wind that would prevent the rotor from rotating, considering the case that the wind direction changes to from the lower left direction in the figure.

3. Output Performance Test and Numerical Flow Analysis

3.1. Geometric and Inlet Flow Conditions

In this study, we evaluated the effect of the misalignment between the wind direction and the direction of the casing (the inclination angle of the casing as a set of three flow deflectors relative to the wind direction) on the performance improvement effect of the casing composed of three flow deflectors by means of output performance test and numerical fluid analysis. Figure 4 shows a schematic of the method used to change the inclination angles of the three flow deflectors (hereinafter an "inclination angle") in the output performance test. The three flow deflectors in gray indicate the reference positions (inclination angle 0°) shown in Figure 3.

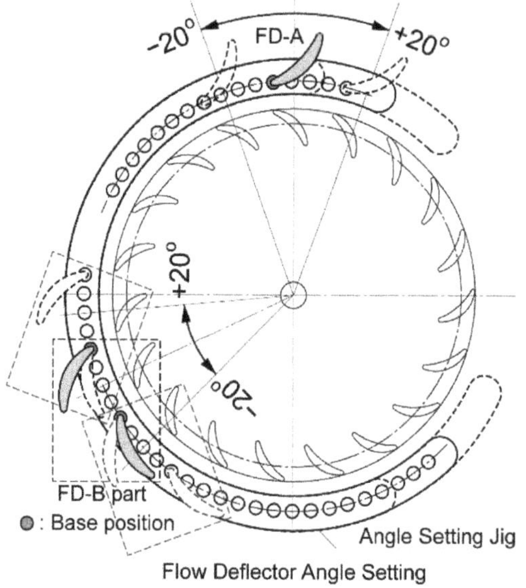

Figure 4. Inclination angle setting of three flow deflectors by jig and range of inclination angles for output performance test experiments.

A total of nine inclination angle conditions around the axis of the wind turbine rotor were examined from −20° to +20° at 5-degree intervals. Since the three flow deflectors must be rotated together around the wind turbine axis to change the inclination angle, a jig as shown in Figure 4 was used to fix the three flow deflectors in the experiment easily, and the inclination angle was set by rotating the jig. Table 1 shows all the conditions for the output performance test and numerical fluid analysis.

Even for the numerical fluid analysis, the conditions of inclination angle and inlet flow velocity were the same as in the output performance test described earlier, as shown in Table 1.

Table 1. Conditions of inlet wind speed and inclination angles of wind-collecting casing for output performance test experiment and numerical flow simulation.

Items	Values
Inlet wind speed, U_∞	6 m/s
Inclination angles of three flow deflectors, θ	$-20°\sim+20°$ (with 5-degree intervals)

3.2. Experimental Method for Output Performance Test

Figure 5 shows the blower-type wind tunnel and test section used for the output performance test. Figure 6 shows a test model of a cross-flow wind turbine with three flow deflectors fixed to the test section. The cross-sectional dimensions of the outlet section of the wind tunnel were 680 mm × 680 mm, and in the rectifier section, two wire meshes with an aperture ratio of 60% and a wire diameter of 0.14 mm, a honeycomb grid with a cell size of 6.35 mm, and a thin household non-woven filter were installed to rectify the flow at the exit of the wind tunnel. The wind turbine test model was installed at a position where the center of the wind turbine rotor was 500 mm downstream of the wind tunnel exit. Torque and rotation detectors (SS-050 and MP-981, respectively, Ono Sokki Co., Ltd., Yokohama, Japan) and a rotation control motor (P50B0502DXS00, Sanyo Denki Co., Ltd., Tokyo, Japan) were connected to the rotor shaft. The rotor speed was controlled, and torque and rotation speed measurements were collected by PC for measurement. Torque and rotation speed were measured with a sampling time of 3 s and a sampling frequency of 100 Hz.

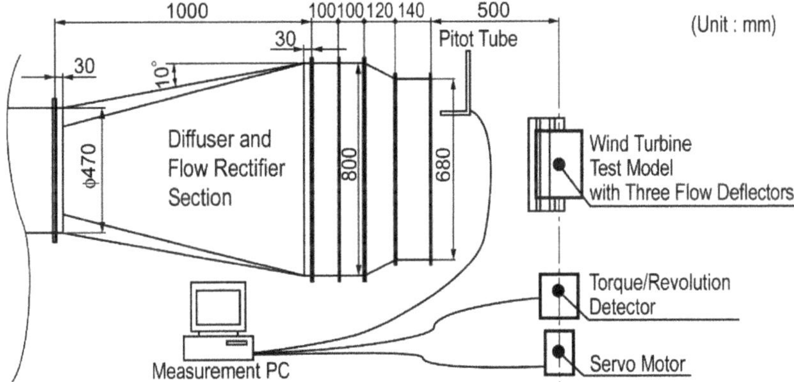

Figure 5. Schematic of output performance test equipment and the test section.

The torque value measured in the output performance test included the friction torque such as at the bearings, so the torque value T was corrected by the friction torque of the shaft friction measurement conducted as a preliminary experiment under the same wind speed condition as in the output performance test. The tip speed ratio λ and power coefficient Cp shown in Equations (1) and (2), respectively, were used to evaluate the output performance of the wind turbine test model.

$$\text{Tip speed ratio } \lambda : \ \lambda = \frac{r\omega}{U_\infty}, \tag{1}$$

$$\text{Power coefficient } Cp : \ Cp = \frac{T\omega}{\frac{1}{2}\rho U_\infty^3 A}, \tag{2}$$

where r is the rotor radius, ω is the rotation angular velocity, U_∞ is the inlet wind speed, T is the rotor torque, ρ is the air density, and A $(=D \times L)$ is the rotor swept area.

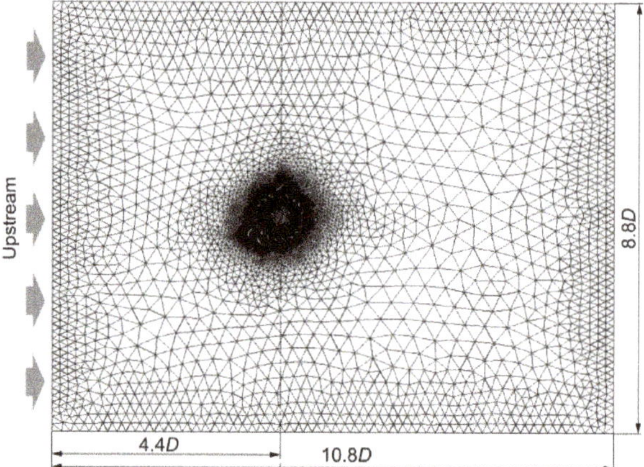

(**a**) 2D computational region and whole mesh.

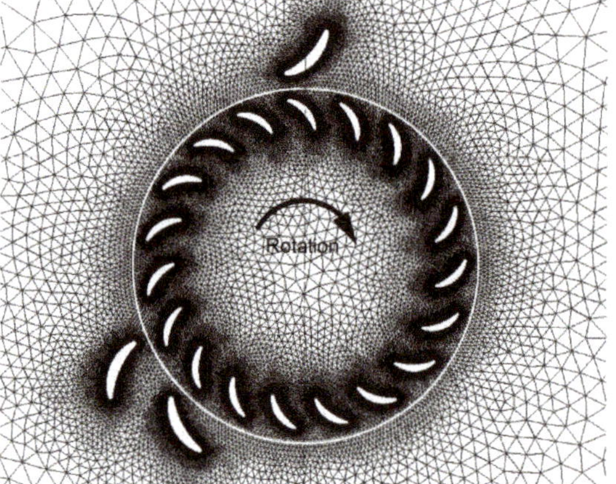

(**b**) Zoomed-in view of mesh near cross-flow wind turbine.

Figure 8. 2D computational region, whole mesh, and zoomed-in view of the mesh near the cross-flow wind turbine with three flow deflectors (in case of reference setting positions of flow deflectors).

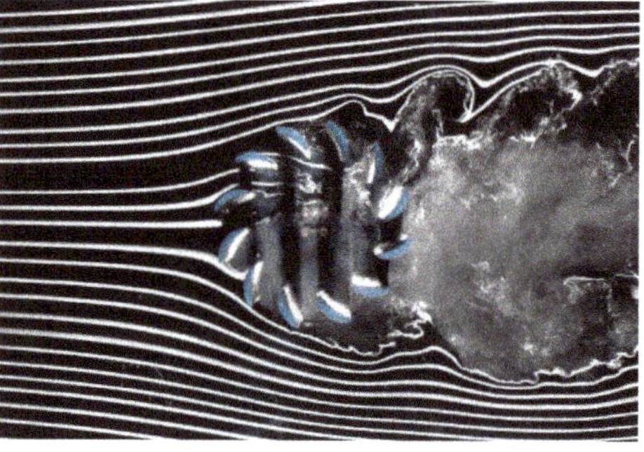

(**a**) Shutter speed 1/4000 s (momentary streamlines).

Figure 9. *Cont.*

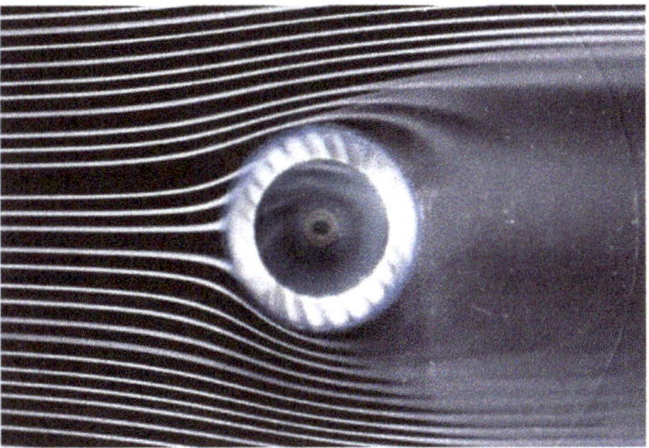

(**b**) Shutter speed 1 s (averaged streamlines).

Figure 9. Smoke streamlines of a cross-flow wind turbine (D = 114 mm, λ = 0.4, and U_∞ = 6 m/s).

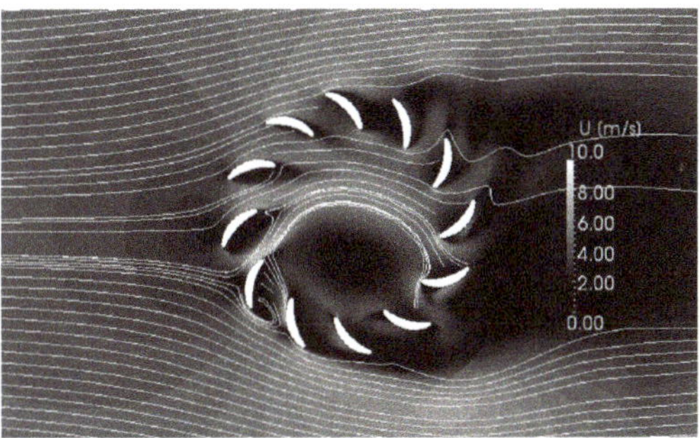

Figure 10. Velocity distribution and streamlines of a cross-flow wind turbine obtained by OpenFOAM numerical calculation (D = 114 mm, λ = 0.436, and U_∞ = 6 m/s).

4. Results and Consideration

4.1. Results of Output Performance Test

Figure 11a,b show the performance curves of the test model of a cross-flow wind turbine with three flow deflectors (hereinafter "TFD") obtained from the output performance test conducted under the conditions shown in Table 1. Figure 11a compares the performance curves of the wind turbine with TFD at inclination angles from 0° to +20°, and Figure 11b compares the performance curves at inclination angles from −20° to 0°. In each figure, the performance curve of the bare cross-flow wind turbine without TFD is also shown for comparison. In each figure, the vertical axis is the power coefficient, and the horizontal axis is the tip speed ratio.

Figure 11a,b show that the output performance of the cross-flow wind turbine with TFD was clearly better than that of the bare cross-flow wind turbine for all inclination angle conditions. That is, the maximum value of the power coefficient Cp_{max} was higher, and the range of tip speed ratios λ over which the power coefficient Cp is positive became wider. When the inclination angle was positive in Figure 11a, the maximum power coefficient Cp_{max} was the highest at an inclination angle of +5° followed by an inclination angle of 0°, and for other angles, Cp_{max} decreased as the inclination angle increased. At an inclination angle of +5°, Cp_{max} = 0.194, which is about 64% higher than Cp_{max} = 0.118 for the bare cross-flow wind turbine. The no-load tip speed ratio also exceeded 1.0 at all inclination angle cases. On the other hand, when the inclination angle was negative in Figure 11b, Cp_{max} was the highest at 0° and decreased as the inclination angle decreased. However, even in these cases, the maximum power coefficient was more than 30% higher than that of the bare cross-flow wind turbine.

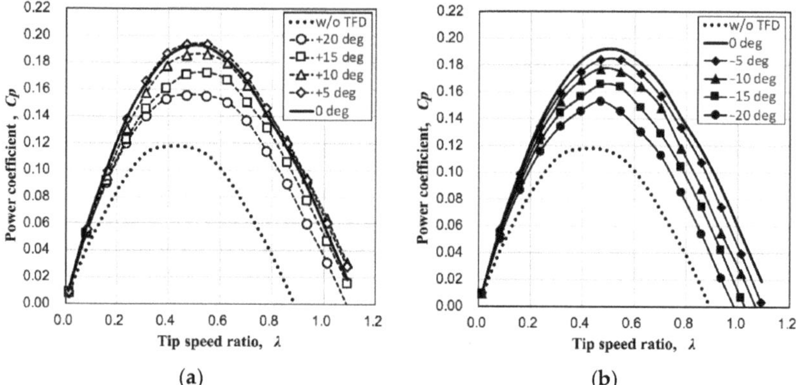

Figure 11. Performance curves of cross-flow wind turbine with three flow deflectors compared with the case with no flow deflector (w/o TFD), (**a**) inclination angles $0°$ to $+20°$; (**b**) inclination angles $-20°$ to $0°$.

Figure 12 shows the relationship between the maximum power coefficient Cp_{max} and the inclination angle obtained from the output performance curves of the cross-flow wind turbine with TFD shown in Figure 11a,b. As a comparison, the maximum power coefficient of the bare cross-flow wind turbine without TFD is shown as a dashed line, and the results of the cross-flow wind turbine with two (dual) flow deflectors (hereinafter "DFD") [8] shown in Figure 1 are also shown as a dotted line. Note that the results for DFD case were obtained through new experiments for this study. The experiments in the case with DFD were conducted together with the experiments with TFD using the same measurement system as in the case with TFD.

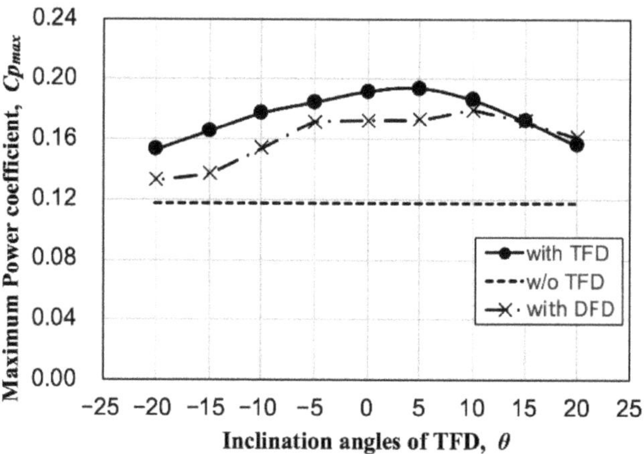

Figure 12. Relation between power coefficient Cp_{max} and inclination angles of three flow deflectors θ by output performance test experiment.

Comparing the TFD case and the DFD case, it was clear that the TFD was more effective in improving the output performance of the cross-flow wind turbine than the DFD. Especially when the inclination angle was negative, the TFD case showed superiority in output performance improvement, so the additional flow deflector is expected to extend the range of inclination angle where the output performance of the cross-flow wind turbine can be improved. As described in Figure 11, the maximum power coefficient Cp_{max} was highest at an inclination angle of $+5°$, and the highest Cp_{max} was between inclination angles of $0°$

and +5°. The degree of decrease in the maximum power coefficient was different between the positive and negative inclination angles, and the decrease in the maximum power coefficient for the negative inclination angle was more gradual than that for the positive angle. In addition, the maximum power coefficient at an inclination angle of $-10°$ to $+10°$ was 90% or more of the maximum power coefficient at an inclination angle of $+5°$. Within this inclination angle range, the maximum power coefficient was able to be kept 50% to 60% higher than that of the bare cross-flow wind turbine with no flow deflector.

4.2. Results of Numerical Flow Analysis

Figure 13 shows the power coefficients *Cp* for each inclination angle at a tip speed ratio $\lambda = 0.497$ obtained from the numerical fluid analysis, with the inclination angle on the horizontal axis as in Figure 12. As shown in Figure 11, the maximum power coefficient of the cross-flow wind turbine with the three flow deflectors (TFDs) was around $\lambda = 0.497$ for all inclination angles. Therefore, the numerical results shown in Figure 13 were considered to capture the trend in the variation of the maximum power coefficient for each inclination angle shown in Figure 12 obtained by experiment at the same tip speed ratio $\lambda = 0.497$, and in Figure 13, the output performance test results shown in Figure 12 are also compared.

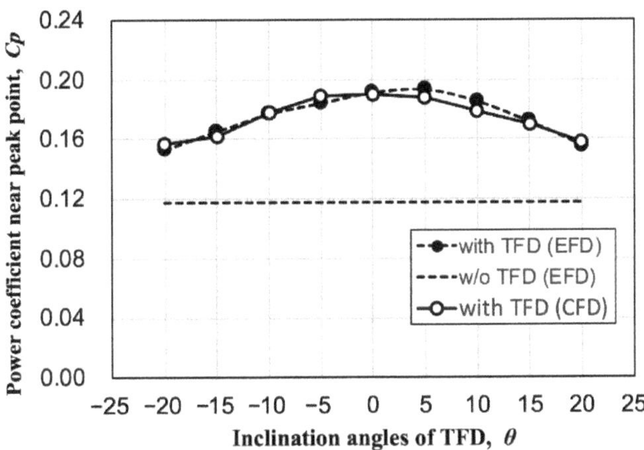

Figure 13. Relation between power coefficient *Cp* near peak point and inclination angles of three flow deflectors θ by CFD comparing with EFD results (maximum power coefficient, Cp_{max}).

Comparing the numerical analysis results (CFD) and the output performance test results (EFD) in Figure 13, the distribution of power coefficients by the CFD results was very close to the EFD results, so it was considered that the numerical analysis could simulate the flow of the cross-flow wind turbine with three flow deflectors adequately. However, the power coefficient of CFD was slightly higher than that of EFD for an inclination angle of $-5°$ (the deviation was about 1.4%), and that of CFD was slightly lower than that of EFD for an inclination angle of $+5°$ and $+10°$ (the deviation was about 1.3% to 1.8%). Otherwise, the results of CFD and EFD were almost the same. Next, the flow fields, that is, velocity and pressure distributions, are shown in Figures 14–17. Figure 14 shows the flow fields of the bare cross-flow wind turbine with no flow deflector, and Figures 15–17 show the flow fields of the cross-flow wind turbine with three flow deflectors (TFD) for the inclination angles of 0° and ±15°. For these inclination angles, the effect of the flow deflectors clearly differs depending on the inclination angle as shown in Figure 13. Figure 15 shows the case of inclination angle 0°, Figure 16 of inclination angle +15°, and Figure 17 of inclination angle −15°. In addition, in Figure 14, each blade is numbered.

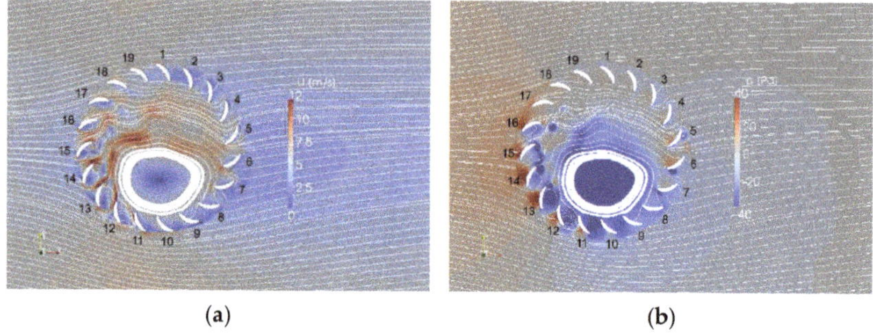

(a) (b)

Figure 14. Velocity and pressure distribution and streamlines in case of the bare cross-flow wind turbine with no flow deflector, (**a**) velocity distribution, U; (**b**) pressure distribution, p.

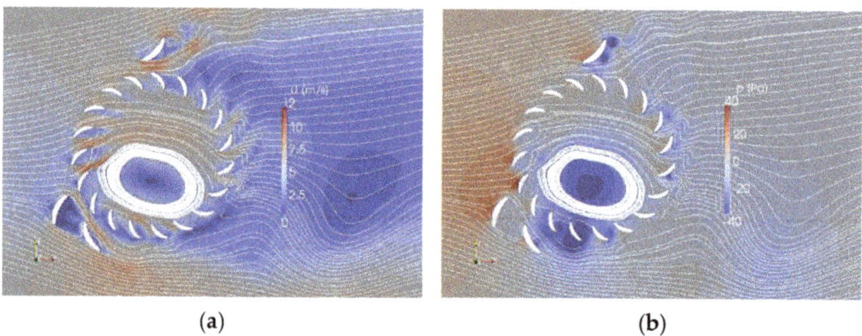

(a) (b)

Figure 15. Velocity and pressure distribution and streamlines of the cross-flow wind turbine with TFD of inclination angle $\theta = 0$ deg. (**a**) velocity distribution, U; (**b**) pressure distribution, p.

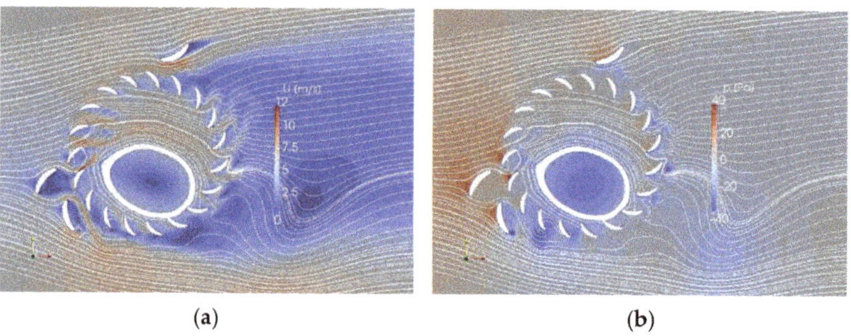

(a) (b)

Figure 16. Velocity and pressure distribution and streamlines of the cross-flow wind turbine with TFD of inclination angle $\theta = +15$ deg. (**a**) velocity distribution, U; (**b**) pressure distribution, p.

Comparing the flow velocity distributions in Figure 14 for the case with no flow deflector and Figures 15–17 for the cases with TFD, the effect of three flow deflectors is seen as a difference in the velocity of the flow downstream from the rotor. In Figures 15–17, relatively high velocity flow out of the rotor can be seen between blades 3 and 7, while such high velocity flow can hardly be seen in Figure 14. Based on this, a comparison of the pressure distribution shows that in Figures 15–17 for the cases with TFD, the pressure difference between the inside and outside of the rotor on each blade surface along the boundary of the zigzag lines is clear between blades 3 and 7 in the downstream section of the rotor, with higher pressure on the inside and lower pressure on the outside. Next, focusing on the upstream side of the rotor, in the pressure distribution, there is a very high pressure distribution on the upper (rotor outside) surface side of blades 11 to 16 or 17 in Figure 14, except for some blades (blades 18 and 19) on the upper side of the rotor in the upstream section. On the other hand, the same high pressure distribution in Figures 15–17 is observed for about 3 blades only between blades 14 and 16 in the upstream section in Figures 15 and 16, and for about 4 blades only (blades 13 to 16) in the upstream section in Figure 17. Accordingly, we can see that the flow entering the rotor in Figures 15–17 flows downstream without significant meandering compared to Figure 14.

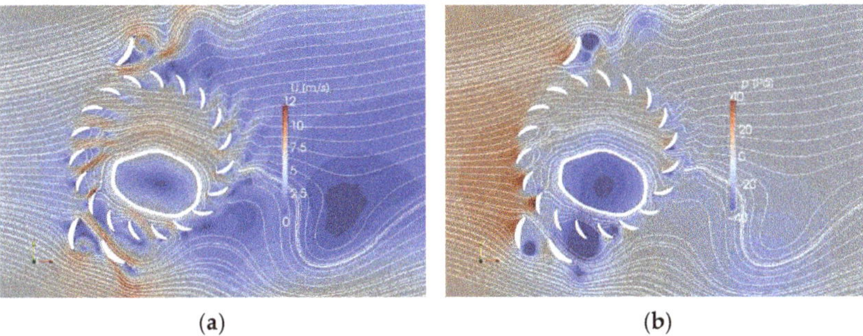

(a) (b)

Figure 17. Velocity and pressure distribution and streamlines of the cross-flow wind turbine with TFD of inclination angle $\theta = -15$ deg. (**a**) velocity distribution, U; (**b**) pressure distribution, p.

Next, focusing on the flow around the flow deflectors of FD-A and FD-B each, an accelerated flow on the rotor side of the flow deflector FD-A clearly appears in Figures 15–17. In particular, in Figures 15 and 17, where the inclination angles are $0°$ and $-15°$, respectively, the pressure distribution on the blade surface near the FD-A is clearly higher on the inside of the rotor and lower on the outside (blade 1 in Figure 15 and blade 19 in Figure 17); these pressure distributions contribute to the rotor rotation. In addition, the direction of the streamline toward the leading edge of the FD-A changes from a flow away from the rotor in Figure 14, which is the case without TFD, to a flow slightly toward the rotor in Figures 15 and 17. It can be inferred that this flow contributes to the fast flow toward the rear side of the rotor without meandering the flow in the rotor. In Figure 16, where the FD-A has shifted toward the downstream side, the pressure difference on both sides of the blade (blades 1 and 2) near the flow deflector is smaller than in Figures 15 and 17, and the contribution of this pressure difference to the rotor rotation is also lower.

For the flow deflectors of FD-B, focusing on the flow velocity distribution in front of the rotor in each Figure, the vertical position in the Figure, where the velocity is slowed down the most, differs depending on whether the rotor has the FD-B or not. In Figure 14 for the bare cross-flow wind turbine, the position is almost at the center of the rotor (in front of blade 15), but in Figures 15–17 for the cases with the FD-B, the position is shifted to the lower side of the rotor. Similarly, the position of the streamlines flowing almost horizontally into the rotor on the upstream side is near the center of the rotor (in front of blade 15) in Figure 14, while in Figures 15–17, it is moved to the lower side (near between blades 14 and 15) from the center of the rotor and near the center of the rotor the streamlines are flowing slightly upward. It is considered that this helps the flow of wind entering into the blade rows on the upstream side of the rotor in flowing into the blade rows on the downstream side without meandering in the rotor. On the contrary, in Figure 14, the entering flow into the blade rows on the upstream side is clearly turbulent and the flow in the rotor is meandering.

In particular, in Figures 15 and 16 with inclination angles of $0°$ and $+15°$, respectively, there are two flow deflectors (FD-B) upstream of the blades (blades 11 to 13 in Figure 15, blades 12 to 14 in Figure 16) moving in the upwind direction in the lower half of the rotor, these flow deflectors are blocking the entering flow that usually prevents the rotor from rotating. In the flow velocity distribution in Figure 14, local high velocity flow is observed on the outer edges of the blades (blades 11 to 14) moving upstream, which corresponds to the position just downstream of the FD-B in Figures 15 and 16, and such local high velocity flow on the outer edges of the blades is not seen in Figures 15 and 16 for the cases with the FD-B and the blades are moving without flow resistance in the lowest flow velocity region. While, in the pressure distributions in Figures 15 and 16, those blades (blades 11, 12, and 13) are in a large low pressure region and there is almost no pressure difference between both sides of each blade. However, in Figure 17 with an inclination angle of −15 degrees, the FD-B moves to the lower side compared to Figures 15 and 16 so that one or two blades on the upstream side and the lower half of the rotor are exposed to undesirable incoming wind that prevents the rotor from rotating. As the pressure distribution in Figure 17 shows, the pressure on the upper surface of blade 13 where the wind strikes directly is clearly high. In Figure 13 mentioned before, as the negative inclination angle increases, the power coefficient becomes lower than for the positive inclination angle, and this is considered to be due to the reduction of the FD-B effect.

From the above discussion of the flow field, it can be said that to improve the power output of a cross-flow wind turbine, the inflow into the upper half of the rotor should be improved by properly blocking the inflow in the lower half of the rotor, and the flow should be sent directly to the blade rows on the downstream side of the rotor. In addition, the results shown in Figure 13 indicate that the flow of the cross-flow wind turbine becomes optimum at an inclination angle of $0°$ to $+5°$ for the casing consisting of a flow deflector of FD-A and two flow deflectors of FD-B, resulting in a high output performance improvement effect. Furthermore, the effect could be kept at a wide range of casing inclination angles by increasing the number of the flow deflector from two to three.

5. Conclusions

In this study, we examined the possibility of increasing the output power of a cross-flow wind turbine by adding a casing consisting of three flow deflectors. Output performance experiments and numerical flow analyses were conducted on a wind turbine with three flow deflectors, and the relationships among the output performance, the flow field, and the inclination angle of the casing against the wind direction within a range of ± 20 degrees were discussed. As a result, the following conclusions were obtained.

(1) The casing composed of three flow deflectors could improve the output performance of the cross-flow wind turbine by about 60% at the maximum output point.

(2) It was shown that a higher output performance improvement effect was obtained by increasing the number of flow deflectors from two to three compared to our previous study and that a higher output performance improvement effect was maintained even when the misalignment between the wind direction and the direction of the casing, i.e., the inclination angle, was increased. The inclination angle at which a high output performance improvement effect (approximately 50% or more) could be obtained with a casing with three flow deflector plates was in the range of −10 to +10 degrees.

(3) From a comparison of the flow fields obtained by numerical flow analyses for the inclination angles, it can be said that the direction of flow improvement for obtaining higher output performance from a cross-flow wind turbine is to change the flow direction slightly upward so that the inflow flows smoothly into the blade row on the upstream side of the rotor, and in the rotor to send the flow to the downstream blade row without meandering to increase the work of the downstream blade row.

Author Contributions: Conceptualization, T.T.; methodology, T.T., R.Y., and T.M.; validation, T.T., R.Y., and T.M.; resources, T.T.; data curation, T.T. and R.Y.; writing—original draft preparation, T.T.; writing—review and editing, T.T. and T.M.; visualization, T.T.; supervision, T.T.; project administration, T.T.; funding acquisition, T.T. All authors have read and agreed to the published version of the manuscript.

Funding: This study was possible partially thanks to the support of JSPS the Grants-in-Aid for Scientific Research (C) number JP21K03887.

Institutional Review Board Statement: Not applicable.

Informed Consent Statement: Not applicable.

Data Availability Statement: Not applicable.

Acknowledgments: The authors are grateful to Yuta Kawahara for assistance with experiments and numerical analyses at the time of his enrollment at National Institute of Technology, Kurume College.

Conflicts of Interest: The authors declare no conflict of interest.

Nomenclature

A	Rotor swept area
Cp	Power coefficient
Cp_{max}	Maximum power coefficient
D	Rotor diameter
DFD	Two (dual) flow deflectors
FD-A	Flow deflector A part
FD-B	Flow deflector B part
FD-B1	Upstream flow deflector of FD-B
FD-B2	Downstream flow deflector of FD-B
L	Blade length
N	Number of blade
p	Pressure
r	Rotor radius
T	Rotor shaft torque
TFD	Three flow deflectors
U_∞	Inlet wind speed
U	Flow velocity
β	Blade pitch angle
θ	Inclination angles of a set of three flow deflectors
λ	Tip speed ratio
ρ	Air density
σ	Solidity (blade chord length/blade pitch)
ω	Rotation angular velocity of rotor

References

1. Eldridge, F.R. *Wind Machines*; Van Nostrand Reinhold Co.: New York, NY, USA, 1980.
2. Johnson, G.L. *Wind Energy Systems*; Prentice-Hall, Inc.: Englewood Cliffs, NJ, USA, 1985.
3. Yoshimoto, H.; Awashima, Y.; Kitakoji, Y.; Suzuki, H. Development of floating offshore substation and wind turbine for Fukushima FORWARD. In Proceedings of the International Symposium on Marine and Offshore Renewable Energy, Tokyo, Japan, 28–30 October 2013; p. 28.
4. Report of Japan Wind Power Association (In Japanese), Agency for Natural Resources and Energy, March 2021. Available online: https://www.meti.go.jp/shingikai/enecho/denryoku_gas/saisei_kano/pdf/028_05_00.pdf (accessed on 10 August 2022).
5. Izumi, U.; Naotsugu, I.; Guo-zhong, C. Design Configuration and Performance Evaluation of Cross-flow Wind Rotors. *J. Jpn. Sol. Energy Soc.* **1998**, *20*, 36–41. (In Japanese)
6. Shoichi, T.; Yukimaru, S.; Izumi, U. An Experimental Study of Cross Flow Wind Turbine. *Proc. 75th Regul. Meet. Jpn. Soc. Mech. Eng.* **1998**, *98*, 291–292. (In Japanese)
7. Tadakazu, T.; Shinichiro, N. Influence of Number of Blade and Blade Setting Angle on the Performance of a Cross-flow Wind Turbine. *Trans. Jpn. Soc. Mech. Eng.* **2007**, *73*, 225–230. (In Japanese) [CrossRef]
8. Tadakazu, T.; Hiroo, M.; Masayuki, F. Study on Upscaling of Cross-flow Wind Turbine with Two Flow Deflectors as Wind Collector. In Proceedings of the Japan Council for Renewable Energy, Yokohama, Japan, 1–4 June 2018. [CrossRef]
9. Yukimaru, S.; Minoru, T.; Jinsaku, S. Development of a High-Performance Cross-Flow Wind Turbine (On the Effects of Ring-Diffusers and Multiple-Guide Vanes on the Power Augmentation for a Cross-Flow Wind Turbine). *Trans. Jpn. Soc. Mech. Eng.* **1998**, *64*, 202–207. (In Japanese) [CrossRef]
10. Takahiro, K.; Hiroaki, N.; Tomohiro, K.; Hiroko, F.; Akito, N.; Nobuyosho, K. Study of Performance of a Cross-Flow Wind Turbine Located Above a Windbreak Fence and the Associated Flow Field. *Proc. 21st Natl. Symp. Wind. Eng.* **2010**, *21*, 221–226. [CrossRef]
11. Mohamed, H.; Takaaki, K.; Takahiro, K. Investigating the effects of wind concentrator on power performance improvement of crossflow wind turbine. *Energy Convers. Manag.* **2022**, *255*, 115326. [CrossRef]

12. Yuji, O.; Takashi, K.; Akira, S.; Kenichi, A.; Masahiro, I. Development of a shrouded wind turbine with a flanged diffuser. *J. Wind. Eng. Ind. Aerodyn.* **2008**, *96*, 524–539. [CrossRef]
13. Toru, S.; Junichiro, F.; Yuichi, T. Study on Performance Improvement of Cross-Flow Wind Turbine with Symmetrical Casing. *J. Environ. Eng.* **2009**, *4*, 490–501. [CrossRef]
14. Toru, S.; Junichiro, F.; Masaaki, T. Performance and Flow Condition of Cross-Flow Wind Turbine with a Symmetrical Casing Having Side Boards. *Int. J. Fluid Mach. Syst.* **2016**, *9*, 169–174. [CrossRef]
15. Tadakazu, T.; Kohki, S. Design of the Tail Blade composing Wind Collector Casing without Mechanical Control to improve the Performance of Cross-flow Wind Turbine. *Proc. Jpn. Wind. Energy Symp.* **2018**, *40*, 405–408. (In Japanese) [CrossRef]
16. Yuji, O.; Takashi, K. A Shrouded Wind Turbine Generating High Output Power with Wind-lens Technology. *Energies* **2010**, *3*, 634–649. [CrossRef]
17. Nobuhiro, O.; Masato, F.; Kenta, K.; Kazutoyo, Y. Optimum aerodynamic design for wind-lens turbine. *J. Fluid Sci. Technol.* **2016**, *11*, JFST0011. [CrossRef]
18. Tadakazu, T.; Shinichiro, N.; Genki, U. Improving ambient wind environments of a cross-flow wind turbine near a structure by using an Inlet Guide Structure and a Flow Deflector. *J. Therm. Sci.* **2005**, *14*, 242–248.
19. Tadakazu, T.; Takeshi, M. Dual Flow Deflector Casing for Performance Improvement of Cross-flow Wind Turbine. In Proceedings of the Japan Council for Renewable Energy, Yokohama, Japan, 1–4 June 2014.
20. Tadakazu, T.; Takeshi, M. Study of a Casing with Two Flow Deflector Plates for Performance Improvement of a Cross-flow Wind Turbine by CFD Analyses. In Proceedings of the 15th World Wind Energy Conference (WWEC2016), Small wind and hybrid system, Tokyo, Japan, 1–4 November 2016.
21. Tadakazu, T.; Hiroo, M.; Masayuki, F. Comparative Study on Upscaling of Cross-flow Wind Turbine with Two Flow Deflectors of Wind Collector Casing. *Wind Energy* **2018**, *42*, 25–31. (In Japanese) [CrossRef]
22. Tadakazu, T.; Takeshi, M.; Shinichiro, N. Flow Patterns of a Cross-flow Wind Turbine by Flow Visualization Measurements on the Performance Curves. *Proc. Jpn. Soc. Mech. Eng. Annu. Meet.* **2007**, *2*, 369–370. (In Japanese)

Article

Investigations of Vertical-Axis Wind-Turbine Group Synergy Using an Actuator Line Model

Ji Hao Zhang, Fue-Sang Lien * and Eugene Yee

Mechanical and Mechatronics Engineering, University of Waterloo, 200 University Avenue West, Waterloo, ON N2L 3G1, Canada
* Correspondence: fue-sang.lien@uwaterloo.ca

Abstract: The presence of power augmentation effects, or synergy, in vertical-axis wind turbines (VAWTs) offers unique opportunities for enhancing wind-farm performance. This paper uses an open-source actuator-line-method (ALM) code library for OpenFOAM (turbinesFoam) to conduct an investigation into the synergy patterns within two- and three-turbine VAWT arrays. The application of ALM greatly reduces the computational cost of simulating VAWTs by modelling turbines as momentum source terms in the Navier–Stokes equations. In conjunction with an unsteady Reynolds-Averaged Navier–Stokes (URANS) approach using the k-ω shear stress transport (SST) turbulence model, the ALM has proven capable of predicting VAWT synergy. The synergy of multi-turbine cases is characterized using the power ratio which is defined as the power coefficient of the turbine cluster normalized by that for turbines in isolated operation. The variation of the power ratio is characterized with respect to the array layout parameters, and connections are drawn with previous investigations, showing good agreement. The results from 108 two-turbine and 40 three-turbine configurations obtained using ALM are visualized and analyzed to augment the understanding of the VAWT synergy landscape, demonstrating the effectiveness of various layouts. A novel synergy superposition scheme is proposed for approximating three-turbine synergy using pairwise interactions, and it is shown to be remarkably accurate.

Keywords: actuator line method; OpenFOAM; synergy; vertical-axis wind turbine; wind energy

Citation: Zhang, J.H.; Lien, F.-S.; Yee, E. Investigations of Vertical-Axis Wind-Turbine Group Synergy Using an Actuator Line Model. *Energies* **2022**, *15*, 6211. https://doi.org/10.3390/en15176211

Academic Editors: Yoshifumi Jodai and Yutaka Hara

Received: 15 August 2022
Accepted: 24 August 2022
Published: 26 August 2022

Publisher's Note: MDPI stays neutral with regard to jurisdictional claims in published maps and institutional affiliations.

1. Introduction

Wind energy is a vital element in creating a sustainable energy future, and wind turbines is a thriving area for contemporary research. Horizontal-axis wind turbines (HAWTs) and vertical-axis wind turbines (VAWTs) are two mature technologies available today for wind-energy extraction, though the former is adopted almost universally and occupies the vast majority of the market share [1]. In this area, Liu et al. [2] performed a fluid-structure interaction study on a novel hybrid Darrieus-Modified-Savonius VAWT design, Jain & Abhishek [3] comprehensively investigated the impact of turbine parameters on the performance of a VAWT with dynamic blade pitching, Johari et al. [4] conducted an experimental comparison of the performances of a small-scale HAWT and VAWT, Du et al. [5] provided an extensive review of H-Darrieus VAWT literature, and Ghasemian et al. [6] reviewed computational fluid dynamic (CFD) simulation techniques for Darrieus VAWTs. From these works, the comparative disadvantages of the VAWTs are identified to be lower aerodynamic efficiencies [2–4], self-starting issues encountered at lower wind speeds [2,3,5,6], and structural difficulties in large-scale designs [1]. These shortcomings notwithstanding, the VAWT archetype remains interesting to study and refine because of its unique, intrinsic properties: namely, operational feasibility at smaller scales [3,7] and omni-directional power generation [8]. Such traits confer on VAWTs higher suitability for operation in distributed energy settings such as urban areas, where low-noise requirements and robustness to volatile wind directions restrict the utilization of HAWTs. As a consequence, it is this applicability that continues to drive research on this subject, with efforts to augment the power performance of VAWTs directed towards unique turbine/blade designs [2], guide vanes [6,9], blade-pitch control [10–12], and intra-cycle rotational speed control [13].

Recent discoveries within the last decade have demonstrated the additional benefits of the VAWT. Experimental investigations conducted by Kinzel et al. [14] found that the stream-wise velocity could recover to 95% of the freestream at only $6D$ (where D is the rotor diameter) downstream of a VAWT, compared to some $14D$ required by a HAWT. This essentially enables tighter array packing within wind farms, while still mitigating the detrimental effects of turbine wakes on power generation. In 2010, Dabiri [7] found that H-bladed Darrieus VAWTs operating in close proximity to one another exhibited larger power coefficients (C_P) than their isolated equivalents, revealing the existence of a mutually beneficial, power-enhancing phenomenon driven by inter-turbine interaction. We refer to this favorable effect as "synergy" in this work for conciseness, as this term has been adopted in prior literature (e.g., by Hezaveh et al. [15]).

Subsequent works, both experimental and numerical, have confirmed the presence of synergy. Ahmadi-Baloutaki et al. [16] tested two- and three-turbine arrays in wind tunnel experiments, finding synergy both among adjacent pairs (turbines placed side-by-side perpendicular to the incident wind direction) and among staggered configurations. Zanforlin & Nishino [17] studied synergy in a pair of VAWTs using the unsteady Reynolds-Averaged Navier–Stokes (URANS) approach with a k-ω shear stress transport (SST) turbulence model, establishing a precedent for using numerical methods and turbulence models to model synergy. Lam & Peng [18] confirmed the presence of synergy by measuring the wake characteristics of a pair of VAWTs and proposed the idea that small synergistic clusters (composed of 2 and/or 3 turbines) can be used as building blocks when optimizing placement in a larger wind farm. Peng [19] used a dynamic torque-driven approach (in contrast to the more commonly prescribed rotational velocity approach) to model the operation of VAWTs and determined that an array of five turbines could still manifest synergy. Shaaban et al. [20] and Barnes & Hughes [21] used URANS coupled with a turbulence model to study synergy among arrays consisting of up to 6 and 16 turbines, respectively. Brownstein et al. [22] performed a detailed experimental characterization of the 3-D flow field surrounding pairs of VAWTs, establishing synergy patterns such as the degradation of a downwind turbine's performance as it enters an upwind turbine's wake. In 2021, Hansen et al. [23] performed a comprehensive study of synergy in two- and three-turbine arrays by parameterizing configurations using the relative angle and spacing, thereby contributing valuable insight on the functional relationship between turbine positioning and synergy.

The implication of this abundantly supported synergy phenomenon is the unprecedented opportunity to arrange VAWTs closer together in order to produce greater power, a concept unthinkable for HAWTs. As a consequence, it is of crucial importance to model synergistic effects in large VAWT farms to exploit this feature as part of micro-siting and optimization efforts. For these applications, the conventional full-order or blade-resolved computational fluid dynamics (CFD) of VAWTs using URANS [24–28] or large-eddy simulation (LES) [10,29,30], in which blade-level dynamics are explicitly resolved in the mesh, can be prohibitively expensive at the wind-farm scale. Therefore, the development of *reduced-order models* involving various forms of approximations or simplifications can become invaluable for wind-farm-level modelling. Sanderse et al. [31] surveyed the pertinent literature to find that the common approaches are momentum-based streamtube models, free/fixed vortex models, and actuator models. Delafin et al. [32] compared the accuracy of a Double Multiple Streamtube (DMST) model, a free-vortex model, and a full-order CFD model, finding that the reduced-order models led to significant errors. The actuator models are more robust, since they are used in conjunction with CFD and can employ a variety of representations for the rotor/blades, such as the actuator disk model [31], the actuator surface or cylinder model [30,33], and the actuator line model [34]. Among these, the Actuator Line Model (ALM) enables the highest fidelity approximation, owing to the fact that the blade motion being dynamically incorporated in the solver [31,34]. This accuracy advantage is saliently demonstrated in a comparison of the Actuator Swept-Surface Model (ASSM) and the ALM performed by Shamsoddin & Porté-Agel [30]. As a result, we chose to explore the capabilities of the ALM in the context of the prediction and characterization of VAWT synergy.

Originally, the ALM was developed by Sørensen & Shen [35] in 2002 to model HAWT performance. In this formulation, blade-element momentum (BEM) theory describes the forces acting on blade elements as a function of the local relative velocity (u_{rel}), the angle of attack (α), and the lift and drag coefficients (C_l and C_d, respectively) obtained from airfoil data tables. These blade forces are then reversed and applied to the flow field as momentum source terms to the Navier–Stokes equations after being smoothed by a Gaussian regularization kernel (η) to avoid singularity effects [35]. In recent years, researchers have applied ALM to the modelling of VAWTs by adapting the formulation while maintaining the underlying theory. Shamsoddin & Porté-Agel [30], Hezaveh et al. [36], Creech et al. [37], and Abkar [38] are important investigations that use ALM and LES to study VAWT-wake characteristics and performance, typically with an emphasis on the former. While the results of these investigations crucially demonstrate the applicability of ALM to VAWTs, the LES studies broadly feature coarse grids (often with a cell size far greater than the blade chord) and time-averaging and thus do not prioritize the simulations of the blade-level dynamics. The seminal works of Bachant et al. [39] and Zhao et al. [34] demonstrated that a reduced-order approach consisting of ALM and URANS (with a turbulence model) is both valid and promising for future VAWT investigations. Some recent studies focus on enhancing the fidelity of VAWT ALM (with LES and URANS) by using cubic spline smoothing on the angle of attack combined with a novel inflow velocity sampling procedure [40], evaluating advanced sub-models for tip effects (Glauert correction versus Dag & Sørensen) [41], and developing a cohesive and upgraded framework to account for higher-order aerodynamic effects [42].

To our best knowledge, there are only two published investigations that apply VAWT ALM to the study of wind turbine array synergy: namely, Hezaveh et al. [15] in 2018 and Raj V et al. [43] in 2021. While it is a groundbreaking approach that leverages the computational efficiency of ALM to investigate synergy, the ALM-LES study by Hezaveh et al. [15] utilizes a grid size relative to the blade chord length (c) of $2.22c$ (implying significant spatial averaging) and does not validate the aerodynamic forces to prove synergy. Raj V et al. [43] used ALM and URANS to confirm that the coefficients of power for VAWTs in a two-turbine configuration exceed that of isolated rotors. The use of ALM-LES in the study of VAWTs is standard practice, since LES can directly resolve most of the turbulent energy spectrum and compute small-scale dynamics using a subgrid-scale (SGS) model. According to Bachant et al. [39], this enables the capture of the important mean swirling motion near the turbine, as well as the turbulence effects produced by the blade–tip vortex shedding and dynamic stall. However, while, in theory, ALM-LES is the superior approach given a sufficiently fine grid resolution, such a case could be prohibitively expensive. For example, Bachant et al.'s [39] ALM-LES simulation with the UNH-RVAT turbine utilizes a reasonably refined grid consisting of 64 cells per turbine diameter but requires 100 times the computational effort of an ALM-URANS simulation that is similarly refined. As a consequence, there is potential for ALM-URANS to offer a more cost-effective solution to the modelling of VAWT synergy while achieving reasonable accuracy, even though turbulence phenomena would be less accurately resolved relative to ALM-LES. The 3-D full-order or blade-resolved CFD results are significantly more expensive, with a 3-D full-order URANS simulation being four orders of magnitude more expensive computationally than ALM-URANS [39], and LES simulations would be even more computationally costly. Given this context, we seek to contribute value to the current research landscape by applying ALM-URANS to the study of a wide variety of two- and three-turbine configurations in order to characterize the fundamental patterns that arise from the emergence and magnitude of VAWT synergy.

Following this Introduction, in Section 2, we present the ALM formulation (Section 2.1), the turbine geometry and CFD parameters (Section 2.2), the two-turbine group-configuration design (Section 2.3), the three-turbine group configuration design (Section 2.4), and our novel synergy superposition scheme (Section 2.5). In Section 3, we start with a validation of the presently used ALM approach using results from various methodologies/sources (Section 3.1) and then present the two-turbine group results (Section 3.2), three-turbine group results (Section 3.3), the effectiveness of the synergy superposition scheme (Section 3.4), and the key conclusions of our investigations (Section 4).

2. Methodology

2.1. Actuator Line Model Formulation

We use the open-source ALM library developed by Bachant et al. [39] for OpenFOAM (referred to as turbinesFoam [44]), to perform the ALM-URANS calculations. This library grants ease-of-access to researchers aiming to implement vertical-axis (wind) turbine ALM in a computationally performative manner, and it is easily extendable to multi-turbine simulations. In the original work of Bachant et al. [39], this framework is validated using experimental results for the UNH-RVAT and RM2 turbines, showing particularly good agreement in terms of C_P versus the tip-speed ratio (or TSR, which is defined as the rotor-tip speed divided by the freestream-incident wind speed) in the former case. Mendoza et al. [45] utilized this library in an ALM-LES study that established high degrees of accuracy in the cyclic variations of the angle of attack and rotor normal force compared to experiments. Mendoza & Goude [46] compared the ALM-LES approach to two different vortex models and some experimental measurements, also demonstrating good agreement with the measured normal forces. Therefore, the validity of this formulation and the correctness of the code implementation have been consistently demonstrated through these works.

In the current section, we will provide a brief summary of the mathematical formulation of Bachant et al.'s [39] ALM implementation. Each blade in the turbine is discretized in the span-wise direction into actuator elements (blade elements), and each element is assigned a control point (the actuator point) located at the mid-span and the quarter-chord location within the element. Each actuator point governs the forces for the entire associated actuator element. Figure 1 shows a sample blade element rotating with angular velocity ω in the counterclockwise (CCW) direction due to an incident wind in the positive x-direction. The blade is fixed along a circular trajectory at a distance R (corresponding to the rotor radius) from the center of the turbine, with the mounting point being the quarter-chord location or actuator point. The azimuthal angle θ of the blade is defined as the CCW starting from the positive x-axis and varies due to the constant rotational velocity ω. The blade's rotation induces a relative velocity of ωR in the trajectory's tangential direction, and the blade also experiences the local flow vector $\vec{u}$ (which is predominantly comprised of the stream-wise component). Thus, the relative velocity $\vec{u}_{rel}$ at the actuator point is the vector difference between ωR and $\vec{u}$. Once obtained, $\vec{u}_{rel}$ determines the angle of attack, α, along with the lift and drag forces (F_l and F_d, respectively) experienced by this actuator element. At an appropriate, chord-based Reynolds number, Re_c, α is used to look up the sectional airfoil aerodynamic coefficients (C_l and C_d) sourced from the Sheldahl & Klimas [47] report. At this stage, the forces can be computed as follows [39]:

$$F_l = \frac{1}{2}\rho A_{elem} C_l |\vec{u}_{rel}|^2 , \tag{1}$$

$$F_d = \frac{1}{2}\rho A_{elem} C_d |\vec{u}_{rel}|^2 , \tag{2}$$

where ρ is density of the air, A_{elem} is the blade element planform area (span times chord), and $\vec{u}_{rel}$ is the local relative velocity.

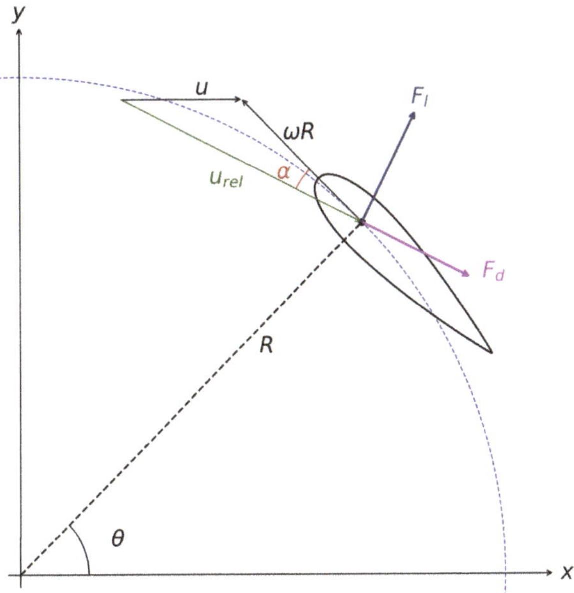

Figure 1. Diagram of a VAWT blade in operation illustrating the ALM formulation, where flow velocities and forces are located at the quarter-chord location ($c/4$).

After these forces are obtained and converted into the desired components (viz. the force components along *x-y* Cartesian coordinate directions), they are convolved with an isotropic, three-dimensional (3-D) Gaussian regularization kernel at each cell neighboring the actuator point in order to distribute and smooth the forces. The Gaussian kernel used assumes the following form:

$$\eta = \frac{1}{\epsilon^3 \pi^{3/2}} \exp\left[-\left(\frac{|\vec{r}|}{\epsilon}\right)^2\right], \tag{3}$$

where η is the kernel value (calculated individually for every relevant cell), ϵ is the Gaussian width (controlling the central concentration of the force distribution), and $|\vec{r}|$ is the distance from the actuator point to the cell centroid. The ϵ parameter can be tuned to obtain a kernel that most accurately captures the actual blade dynamics. To this purpose, Bachant et al. [39] uses the maximum of three values, which are based on chord length, mesh size, and the momentum thickness due to drag, to determine ϵ. In practice, the property that determines the value of ϵ is the mesh size, namely, $\epsilon = 2C_{mesh}\Delta x$, where $C_{mesh} = 2$ is a calibration factor, and Δx is the approximate cell size (cube root of the cell volume). At each cell, within some threshold radius of the actuator point, the cell-specific η is multiplied with the forces to perform the projection. This threshold radius is set to be $c + \epsilon(\ln(1.0/0.001))^{1/2}$ [44], which encapsulates a sufficiently high percentage of the total (probability) mass of the kernel in the present cases. This means the corresponding percentage of the forces at the actuator point can be distributed to cells within this region, while neglecting the insignificant contributions from all other cells for the sake of computational efficiency. The $\eta \cdot F$ value (for each component) is then injected into the solver as a momentum source term for the Navier–Stokes equations.

To investigate synergy, we first extract the power coefficient computed at each time step of the ALM simulation and calculate a mean based on the last two periodically converged revolutions. The power coefficient C_P is defined as follows:

$$C_P = \frac{P}{\frac{1}{2}\rho DLU^3},$$

(4)

where P is the instantaneous turbine power, ρ is the air density, D is the wind turbine diameter, L is the blade span, and U is a reference velocity (typically taken to be the freestream incident velocity) [7]. The power coefficient C_P value will first be calculated for a baseline case consisting of an isolated turbine. Afterwards, two- and three-turbine groupings (or array configurations) can be simulated, and the mean C_P of each turbine in the cluster can be determined. To measure synergy quantitatively, we follow the conventions of Hansen et al. [23] and define a *power ratio* for each turbine, as well as for the entire turbine cluster, as follows:

$$power\ ratio = \frac{C_P}{C_{P,iso}},$$

(5)

where C_P is the turbine cluster power coefficient in the multi-turbine case, and $C_{P,iso}$ is the C_P of turbine in isolated operation (e.g., for a three-turbine configuration, this would be three times the power coefficient C_P for a single (isolated) turbine). As a consequence, if the power ratio is greater than the unity, then we can confirm the presence of synergy.

2.2. Turbine Geometry and CFD Setup

The turbine that will be studied in our work is a 2-bladed H-type VAWT, which was measured in the wind tunnel experiments conducted by LeBlanc & Ferreira [48] and used in the ALM investigations reported by Zhao et al. [34]. This turbine was specifically chosen owing to the availability of both experimental data and ALM results, so it lends itself well to a validation of Bachant et al.'s [39] implementation. Table 1 lists the important turbine properties and the operating point for a TSR of $\lambda = 3.7$ (for an inflow wind speed of 4.01 m s^{-1}). No additional structure (e.g., shaft and/or struts) outside of the blades is included for consistency with the simulations conducted by Zhao et al. [34]. We chose a single, fixed TSR for our simulations in order to focus our investigative efforts on the relationship between the turbine array configuration and the multi-turbine synergy. Furthermore, the value of $\lambda = 3.7$ is specifically chosen because it avoids the occurrence of dynamic stall [34], a phenomenon that hinders aerodynamic efficiency [26]. Therefore, dynamic stall correction and other add-on models provided by Bachant et al. [39] can be disabled to allow for a direct comparison with Zhao et al.'s [34] results (which did not employ such models), while still ensuring appropriate simulation accuracy. The airfoil data table used for all the ALM calculations in this study is the $Re_c = 10^6$ dataset for the NACA 0021 airfoil retrieved from Sheldahl & Klimas [47], which is consistent with that used by Zhao et al. [34].

Table 2 displays the CFD parameters for all ALM simulation cases conducted using the turbinesFoam library. These parameters are identical to Zhao et al.'s [34] setup with the exception of an enlarged domain, which, in terms of the rotor diameter, is about $91.2D \times 60.8D \times 1.9D$. This is in accordance with the guidelines set out by Balduzzi et al. [49] and followed by Bachant et al. [39]: namely, to use a domain size of at least $W = 60D \times L = 90D$ in order to ensure that the boundaries of the computational domain are consistent with an open field and mitigate blockage effects (which are found to inflate ALM forces).

Figure 2 illustrates the domain and mesh for the current ALM simulations. Note that all CFD cases are 3-D URANS with the k-ω SST turbulence model, which was chosen to ensure consistency with Zhao et al.'s [34] setup and direct comparability of ALM results. We retain the boundary conditions used by Bachant et al. [39,44]. In each case, the isolated turbine or groups of turbines will be centered at a location $20D$ downstream of the inlet plane, which efficiently utilizes the domain space and consistently minimizes the impact of the lateral boundaries on the ALM results. A step-wise mesh refinement process is used to set the cell size in the vicinity of the virtual VAWTs. More specifically, in snappyHexMesh, two refinement regions followed by an additional region per turbine are used to refine the mesh procedurally from level 2 to level 4. The background mesh is defined such that the level 4 mesh (innermost cells of the mesh) has a size of about 0.05 m in each direction (consistent with Zhao et al. [34]), which covers a region of W = 4 m × L = 6 m. Note that the boundary-layer mesh near the blade surfaces is not required, since the presence of the blades is modelled virtually in ALM. The next two levels are defined over conjoined refinement regions encompassing all turbines, with lengths and widths based on the maximum dimensions of the inner region and the inclusion of two meters of padding on each side. The coarse global mesh has a cell size of about 0.8 m. It is challenging and expensive to systematically verify the mesh size in an ALM simulation, since the physics being emulated are sensitive to the combination of the cell size and the Gaussian kernel width ϵ. These two parameters must be investigated in conjunction, since different Gaussian widths require different spatial and temporal CFD resolutions [34]. Therefore, a separate grid-independence study was not performed in this case, and, instead, we follow Zhao et al.'s [34] mesh settings for the 2-bladed VAWT being investigated herein.

Table 1. List of ALM and wind turbine properties and parameters.

Property	Value
Turbine diameter (D)	1.48 m
Number of blades	2
Airfoil type	NACA 0021
Blade chord (c)	0.075 m
Blade span (L)	1.5 m
Blade pitch (β)	0°
Tip-speed ratio (λ)	3.7
Rotational speed (ω)	20.05 rad s^{-1}

Table 2. List of CFD case properties.

Parameter	Value
Inlet (freestream) velocity	4.01 m s^{-1}
Turbulence model	k-ω SST
Inlet turbulence kinetic energy	0.24 m^2 s^{-2}
Inlet specific dissipation rate	1.78 s^{-1}
Density (ρ)	1.207 kg m^{-3}
Time step size	0.003 s
Domain size: $L_x \times L_y \times L_z$	135 m × 90 m × 2.85 m

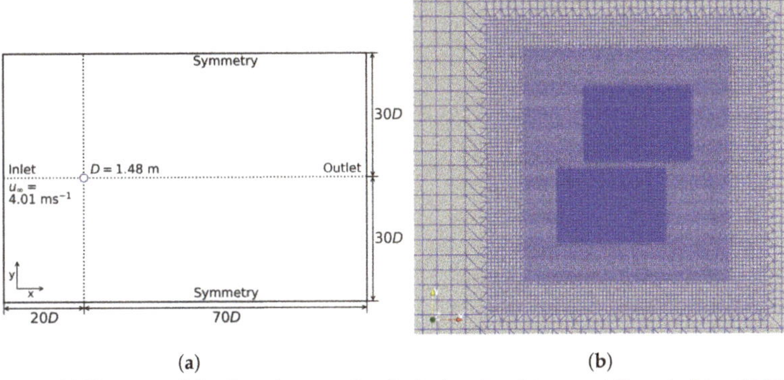

Figure 2. (**a**) Diagram of the domain setup for the isolated and two- and three-turbine ALM cases. The lateral boundaries are set to be symmetrical, as shown, and the top and bottom boundaries (parallel to the *x*-*y* plane) are also set to be symmetrical. (**b**) Sample mesh near the virtual VAWTs in a two-turbine ALM case. The innermost cells are approximately 0.05 m in size.

2.3. Two-Turbine Group (Array) Configurations

The design of the configurations of the two-turbine groups that will be simulated using ALM is vital. Previous literature loosely explored the configuration space, with Hansen et al. [23] being the only study to systematically test two-turbine cases. Hansen et al. labels the two turbines as rotor 1 (R1) and rotor 2 (R2) and parameterizes R2's relative position to R1 using an array angle β and a turbine spacing *dist*. For some fixed $dist = 1.375D$, $2D$, and $3D$, the normalized power-generation performance of R1 and R2 can be plotted as a function of β in the range $[-90°, 90°]$, where a magnitude of $90°$ corresponds to the adjacent arrangement for which the turbines are side-by-side facing the incident wind direction. Hansen et al. [23] claimed that "this is the first attempt at numerically investigating the efficiency augmentations of VAWTs for more than 20 different layouts" [23]. This is a landmark achievement for optimizing VAWT layouts through the consideration of the unique synergistic phenomenon, and we seek to expand (extend) considerably these results using a reduced-order model based on ALM simulations, rather than a full-order model based on blade-resolved CFD simulations.

Figure 3 illustrates the suite of wind-turbine-array layouts that will be simulated using ALM. For ease of reference, we label the turbines T1 and T2, where T1 is the leading (or, upwind) turbine in non-adjacent (side-by-side) arrangements. Due to the fact that only the relative positions matter in the layout, we fix T1 at $(0,0)$ (origin of the Cartesian coordinate system) and vary T2's position in a staggered grid around T1 at a downwind x-spacing of $0.5D$ and a crosswind y-spacing of $1D$. Figure 3 resembles half of a donut shape, because the left half of the configuration space (where T2 would be the leading or upwind turbine) is eliminated owing to symmetry. The "hole" surrounding T1 is present in order to avoid simulating pairwise interactions where the spatial proximity of the two turbines is so close that there is no reliable way to ascertain the validity of ALM. For instance, ALM is unable to capture the collision of blades when the turbines overlap. An upper limit is also set to allow for a feasible yet useful configuration space, enabling the meaningful evaluation of synergy up until the separation distance sufficiently weakens such interaction effects. We will simulate all 78 of the shown two-turbine layouts with co-rotating turbines (viz. with the blades of both turbines rotating CCW), as well as a reduced grid size of 30 counter-rotating cases (viz. with T1 rotating CCW and T2 rotating clockwise or CW), forming a total of 108 two-turbine cases that will be simulated. To our knowledge, this is the largest number of cases simulated to date, using either full-order modeling based on a blade-resolved CFD or reduced-order modeling based on ALM.

2.4. Three-Turbine Group (Array) Configurations

It is more difficult to parameterize the three-turbine configurations in a comprehensive manner due to the additional degree of freedom. Therefore, based upon the synergy patterns derived from the two-turbine results, three promising shapes are used within the three-turbine groupings—it is expected that these designs will manifest a net cluster-level synergy. In Figure 4, with an incident wind direction from left to right, the V, Reverse V, and Line shapes (configurations) are defined. All turbines will be co-rotating with respect to each other in the CCW direction. The definition of the spacing parameters x_{sep} (in the downwind direction) and y_{sep} (in the crosswind direction) is consistent across the three array configurations, with $x_{sep} = (0.34D, 0.50D, 0.68D)$ for the V and for the Reverse V configurations and $x_{sep} = (0, 0.34D, 0.50D, 0.68D)$ for the Line configuration. Finally, $y_{sep} = (3D, 4D, 5D, 6D)$ for all these three-turbine array configurations. These constitute a total of 40 ALM cases for the three-turbine groups.

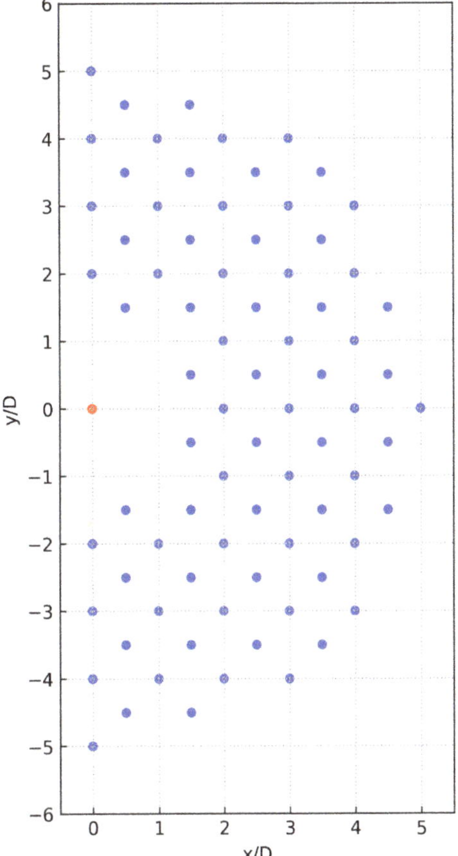

Figure 3. Turbine locations for two-turbine arrangements, where turbine T1 is fixed at $(0,0)$ (orange point), and turbine T2 is situated relative to T1 in a staggered grid (blue points).

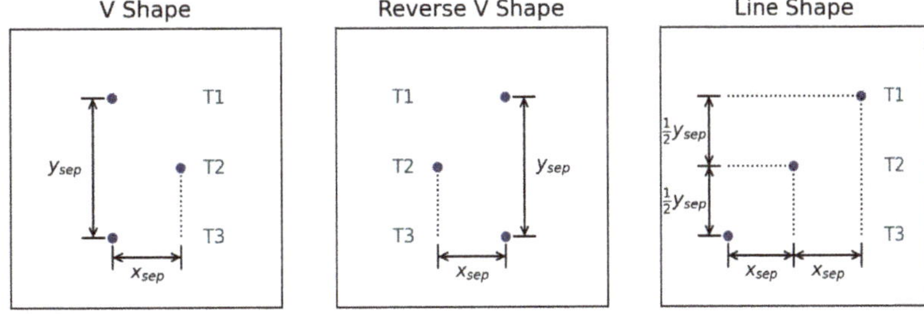

Figure 4. Diagrams of the V, Reverse V, and Line shapes (configurations) for three-turbine cluster arrangements, illustrating conventions for defining x_{sep}, y_{sep}, and turbine labels in each case.

2.5. Synergy Superposition Scheme

A novel proposition in this study is the idea of a "synergy superposition scheme". Inspired by the synergy patterns which will be discussed in the next section, we formulated the hypothesis that three-turbine cluster synergy may be approximated as a linear superposition of all the possible pairwise VAWT interactions contained therein. The specific procedure to test this hypothesis involves identifying every relative pairwise sub-arrangement in each of the 40 three-turbine configurations and simulating those ALM cases. Following on from this analysis, the superposed estimate of some turbine's power ratio is formulated as follows.

$$\left(\frac{C_P}{C_{P,iso}} \right)_s = 1 + \sum_{i=1}^{n} \left[\left(\frac{C_P}{C_{P,iso}} \right)_i - 1 \right], \tag{6}$$

where the subscript s denotes the superposed (approximate) power ratio of some turbine in the three-turbine group, and the summation is computed for $n = \binom{n_t}{2}$ pairs indexed as i (here, n_t is the number of turbines in the array so, for example, $n = \binom{3}{2} = 3$ for a three-turbine cluster). The power coefficient C_P of the i-th turbine depends on the turbines in each pair that correspond to the same relative positioning in the greater cluster. Essentially, this assumes that the deviation of the power ratio from 1 is a quantity that can be linearly superposed via summation. This scheme is a simple one, and the formulation is arbitrarily chosen to serve as a starting point towards demonstrating a proof-of-concept.

3. Results and Discussion

3.1. Validation

We now present a validation of the currently used turbinesFoam ALM library for the isolated VAWT being studied. The first simulation result in Figure 5 is obtained using a separate ALM code we developed for Fluent using the User-Defined Functions (UDF) framework. This implementation is entirely based on an interpretation of Zhao et al.'s [34] approach, which used a 2-D Gaussian kernel for force projection and smoothing. For comparison, Zhao et al.'s [34] ALM results, the experimental measurements obtained by Leblanc & Ferreira [48], and the ALM simulation results obtained using Bachant's [39] OpenFOAM code (turbinesFoam) are displayed. Finally, the last result used in the comparison is a 2-D full-order blade-resolved CFD simulation (URANS with the k-ω SST turbulence model) of the same 2-bladed turbine. This case was generated and conducted in ANSYS Fluent 2020 R2 with a steady-state initialization run (using a Moving Reference Frame) followed by a transient run (using a Sliding Mesh procedure), closely following a similar CFD case undertaken by Abdalrahman [11]. Each of the two blades is meshed with 600 cell divisions around the profile and 50 layers of inflation cells, beginning at a first layer with a thickness of 1.08×10^{-5} m, guaranteeing that the non-dimensional wall-normal distance y^+ is less than five throughout the simulation. All other setup parameters are consistent with those shown in Table 2, consequently, conform with what has been documented by Zhao et al. [34].

A comparison of the results in Figure 5 shows that the general trends of the turbine force components F_x and F_y in the x- and y-directions, respectively, to concur generally across all the simulation and experimental measurements, although some discrepancies in magnitude exist (especially near the peaks). The values for F_x obtained from the ALM simulations conducted by Zhao et al. [34] and from the experimental measurements of Leblanc & Ferreira [48] are most notably underpredicted by the current turbinesFoam ALM simulations, the Fluent UDF ALM simulations, and the full-order URANS CFD simulations, although these three simulations results conform very well with one another. It is worth noting that, between the two ALM simulations, the cases are standardized in terms of mesh, Gaussian width (ϵ), and airfoil data ($Re_c = 10^6$ from Sheldahl & Klimas [47] in accordance with the setup of Zhao et al. [34]). However, for the values of F_y, the results from these three simulations are in excellent conformance with experimental measurements, whereas those of Zhao et al. [34] over-predict the transverse (F_y) forces.

There may be several reasons for the mis-prediction of the current ALM results. Firstly, the airfoil data is synthesized and not experimentally obtained [39], and the static airfoil data would likely be unable to fully represent the loading on dynamically pitching airfoils, as is the operational reality for the VAWT blades. Mendoza & Goude [46] also demonstrated a tendency for these airfoil data tables to under-predict the turbine blade forces when compared to the lift and drag coefficients generated by the XFOIL program. Scheurich

& Brown's [50] study found that using only static airfoil data led to significant under-predictions of the sectional tangential force coefficient near the extremes in the range of the angle of attack (viz., $|\alpha| \leq 15°$), whereas a dynamic stall model was able to produce large aerodynamic force coefficients. This points to some uncertainty as to whether a value of TSR of $\lambda = 3.7$ is truly sufficient to avoid dynamic stall. Abhishek [3] also found that a lack of correction for virtual camber, which arises in symmetrical airfoils operating in a curvilinear flow (blade rotation), can account for an under-prediction of C_l by 0.4–0.65. Finally, a similar magnitude of under-prediction can be found in Bachant et al.'s [39] validation for the RM2 turbine. In view of this, we hereby proceed with the current approach, since the two ALM simulation results are in good conformance with the full-order blade-resolved CFD simulations results, and the mis-predictions are within reasonable limits. Finally, it is noted that synergy is a relative measure of performance normalized by $C_{P,iso}$, so the accurate prediction of the absolute value of C_P may not be as critical.

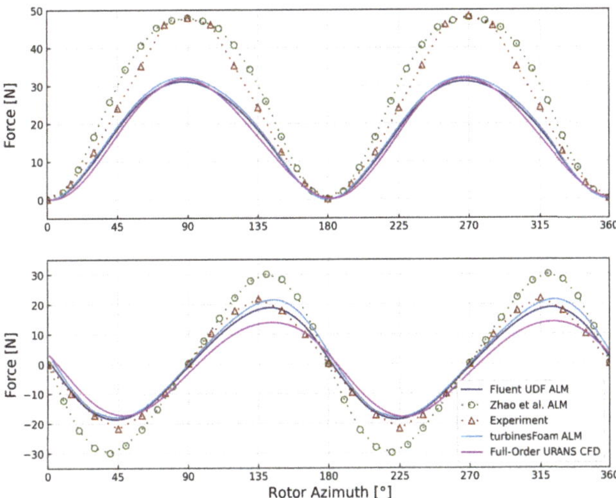

Figure 5. Turbine forces F_x (**top**) and F_y (**bottom**) in the x- and y-directions, respectively, as functions of rotor azimuth for an ALM implementation in Fluent User-Defined Functions (UDF), Zhao et al.'s [34] implementation, experimental measurements obtained by Leblanc & Ferreira [48], ALM simulation using Bachant et al.'s [39] turbinesFoam library, and a full-order (blade-resolved) CFD simulation for a TSR value of $\lambda = 3.7$. Results are taken at a periodically converged revolution for the ALM runs.

3.2. Two-Turbine Group (Array) Results

Using the current ALM methodology, the isolated turbine-power coefficient C_P is determined to be 0.5458. We find that the stream-wise velocities are enhanced or sped-up in a region adjacent to the rotor and surrounding the wake, implying that $u/U_0 > 1$ (where u is the stream-wise velocity component, and U_0 is the freestream (incident) velocity). A sample of the flow field illustrating this velocity speed-up effect is presented in Figure 6. This is consistent with the previous findings of Zanforlin & Nishino [17], Lam & Peng [18], Brownstein et al. [22], and Hansen et al. [23]. Using this baseline $C_{P,iso}$, the heatmaps in Figure 7 (showing T1 and T2 power ratios) and Figure 8 (showing the cluster mean power ratio) can be obtained, which include the 78 array layouts for the pairwise co-rotating turbines.

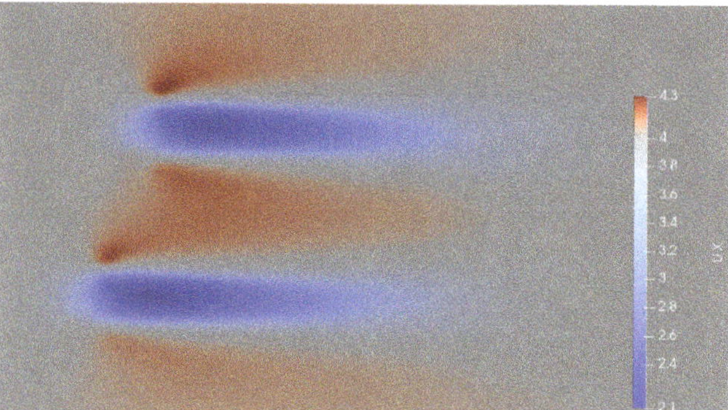

Figure 6. stream-wise velocity (u) magnitude contour of a two-turbine ALM case at 3.15 s of flow time, exhibiting high-velocity regions (colored in red) characterized by $u > U_0$ (freestream velocity) around the turbines and wakes.

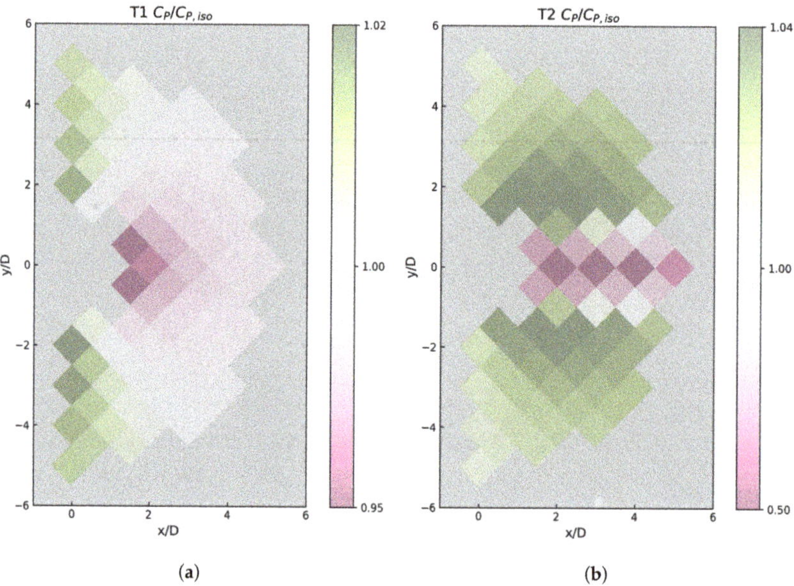

(a)   (b)

Figure 7. Heatmap of power ratios ($C_P/C_{P,iso}$) of (**a**) T1 and (**b**) T2 in a co-rotating pair of VAWTs, where the value of each cell corresponds to the configuration which has T2 located at that cell relative to T1 at the origin.

There are three sets of data to present for the two-turbine cases, namely, T1, T2, and the cluster mean power ratios (sum of C_P of both turbines divided by $2C_{P,iso}$). Each set is visualized in its own heatmap, where T1 is fixed at the coordinates $(0,0)$ and a coordinate of $(x/D, y/D)$ indicates a layout with T2 located there (similar to Figure 3). Each diamond-shaped cell's color-coded power ratio value is defined by the layout, which contains T2 located at that cell's center. This provides a convenient overview of the synergy landscape for a pair of turbines when the T2 position is parameterized in a staggered grid fashion. Between the heatmaps presented in Figures 7 and 8, it is clear that a wide range of layouts is indeed capable of generating synergy (viz., cells are colored green), as predicted by the ALM simulations. This serves to demonstrate that there appear to be no fundamental barriers which may inhibit the ALM simulations from expressing VAWT synergy.

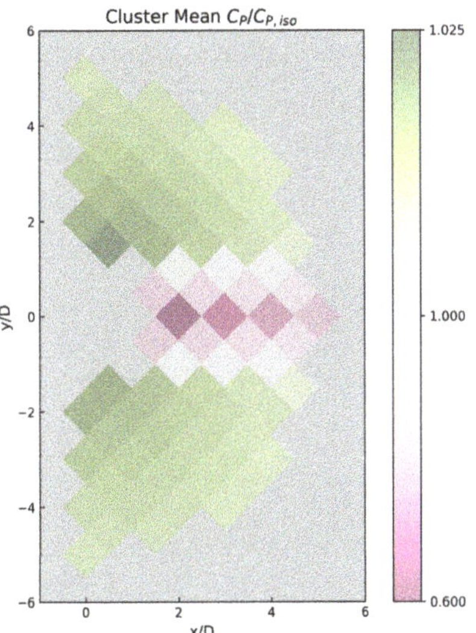

Figure 8. Heatmap of cluster mean power ratios ($C_P/C_{P,iso}$) for a co-rotating pair of VAWTs.

Additionally, we may also extract some useful synergy patterns from a careful perusal of these plots. In Figure 7a, T1 (the leading or upwind turbine in non-adjacent layouts) benefits the most from synergy when T2 is adjacent to T1 (i.e., they are in the same column) or slightly downstream from those configurations. Notice the broad patterns are mirrored across the x-axis, but there are subtle asymmetries, which are likely due to the inherent wake asymmetry unique to VAWTs. The power performance of T1 also diminishes not only when T2 lies directly within T1's wake (y/D in the range $[-0.5, 0.5]$), but also in a sector of approximately $90°$ or more. This provides evidence of a blockage effect, which occurs when a downwind turbine (T2) blocks the flow through the upwind turbine, decreasing its power-generation performance. This blockage effect is well-documented in the works of Zanforlin & Nishino [17], Shaaban et al. [20], Brownstein et al. [22], and Hansen et al. [23]. Using the heatmap visualization of T1, it is possible to observe the gradual transition from a power detriment (due to blockage) to a synergistic interaction across the configuration space.

Simultaneously, the effects of these layout changes on T2 (the downwind turbine) can be ascertained from a careful examination of Figure 7b. Interestingly, T2 experiences synergy almost universally, with the only (predictable) exception being the cases where T2 is situated in the velocity-deficit region of T1's wake. Another useful observation is that T2 generally experiences greater synergy—about 2% more than T1 in terms of the most favorable synergistic arrangements. The most favorable synergistic cases for T2 also deviate from T1's preference for adjacent configurations and are instead located along a band just outside of the wake region (at $|y|/D = 1.5$) and at some distance downstream of T1 ($x/D \geq 0.5$). In fact, this correlates with the high-velocity flow acceleration regions found adjacent to the turbine wake, which is a phenomenon also observed by Hansen et al. [23]. Moreover, further evidence for an accelerated fluid flow [51] around two rotating VAWTS arranged side-by-side (yielding increased power generation or synergy) has been provided using flow visualization [52] and CFD simulations to study the dynamic fluid-body interaction [53]. While we do not claim to contribute to the fundamental understanding of the physics that drive synergy, it is conceivable that higher ALM forces can be produced when the stream-wise velocity is augmented, since the force is a function of the magnitude of $\bar{u}_{rel}$.

Finally, the cluster mean power ratio heatmap in Figure 8 is useful for assessing the synergistic potential of the entire two-turbine group. Due to the generally larger synergy magnitudes of T2, cluster mean patterns are expectedly similar—almost all the layouts are advantageous over the isolated, baseline case. However, the competition effect remains crucial in determining the optimal configuration, which is a compromise between the adjacent arrangements (favored by T1) and downstream arrangements (favored by T2). The optimal synergy is found at a T2 oriented at 71.6° and spaced at 1.58D relative to T1, which produces a cluster-level power ratio of 1.0247. This means the cluster would produce 2.47% more power compared to two individually operating (isolated) turbines. The synergy patterns presented herein agree well with Hansen et al.'s results, namely, a maximum in T1 synergy in the adjacent configurations, an asymmetric bias of synergy in favor of the bottom-half of the layouts (T2 below T1), and T1's power-generation performance decrease due to the proximity of T2 to T1's wake [23]. This overall consistency lends credence to our methodology, and these results offer a unique perspective on the relationship between turbine layout and synergy, which complements and elaborates upon previous investigations.

The 30 counter-rotating cases are presented in Figures 9 and 10. These exhibit essentially the same macro-patterns with small differences in the synergy magnitudes for some layouts, possibly attributable to the opposite blade-rotation scenarios. This is also consistent with Hansen et al.'s [23] results. We find the optimal configuration in the counter-rotating case to be identical to the co-rotating case (71.6° angle and 1.58D spacing between T1 and T2), but the power ratio is 1.0266, which is slightly larger than the co-rotating case. Overall, across all configurations, neither the co-rotating nor the counter-rotating cases dominate in cluster-level synergy—in fact, the favorable cases are evenly split with a less than 1% difference. This indicates there is no practical preference for co-rotating or counter-rotating arrays after considering the likely variations of the incoming wind direction.

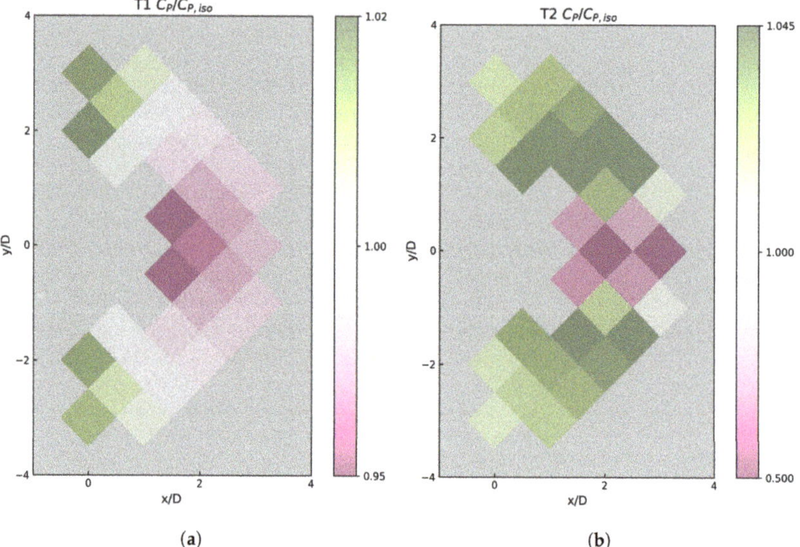

Figure 9. Heatmap of the power ratios ($C_P/C_{P,iso}$) of (**a**) T1 and (**b**) T2 in a counter-rotating pair of VAWTs, where the value of each cell corresponds to the configuration which has T2 located at that cell relative to T1 at the origin.

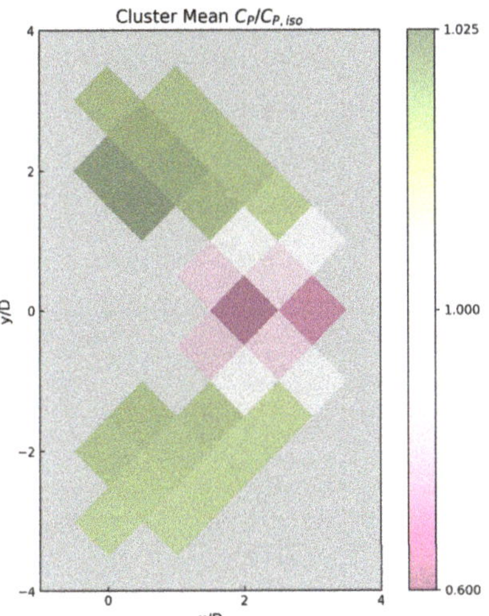

Figure 10. Heatmap of cluster mean power ratios ($C_P/C_{P,iso}$) for a counter-rotating pair of VAWTs.

3.3. Three-Turbine Group (Array) Results

For the three-turbine group results, the power ratios as a function of the spacing parameters (x_{sep} and y_{sep}) are plotted in Figure 11 (V), Figure 12 (Reverse V), and Figure 13 (Line). The power ratios of T1, T2, T3, and the cluster are divided into the respective subplots in each figure. We are able to identify general synergy patterns that hold across all cases despite the differences in the array layout shapes. Specifically, the leading (upwind) and trailing (downwind) turbines share common functional relationships with the spacing parameters. Referring to Figure 4, T1 and T3 are the leading turbines in the V shape, T2 is leading turbine in the Reverse V shape, and T3 is leading turbine in the Line shape. Similarly, T2 is the trailing turbine in the V shape, T1 and T3 are trailing turbines in the Reverse V shape, and T1 is trailing turbine in the Line shape. It is important to note that T2 is an "intermediate" turbine, which is positioned between the most upwind and downwind turbines, and the implications of this will be discussed shortly.

In fact, the two-turbine group synergy patterns are useful even in the context of the interpretations of the characteristic synergy patterns apparent in three-turbine groups. For example, with regard to the trailing turbines, the functional relationships between the subplots are qualitatively similar. There is a consistent decrease of the power ratio with an increase in the crosswind separation y_{sep}, corresponding to an intensification of synergy in the $|y|/D = 1.5$ bands of Figures 7b and 9b. Again, a plausible reason for this is that the downwind turbine is exploiting the accelerated flow ($u/U_0 > 1$) that is most pronounced in this region. As the downwind separation x_{sep} increases, the power ratio also increases. While it may seem counter-intuitive, this is essentially reflecting the pattern that the optimal downwind turbine placement is not at the adjacent location ($x_{sep} = 0$), but rather increases with x_{sep} until some maximum before decreasing. The currently investigated range has an upper bound of $x_{sep} = 0.68D$, which is not sufficiently far downstream to reveal this convex variation with increasing downstream separation of the turbines.

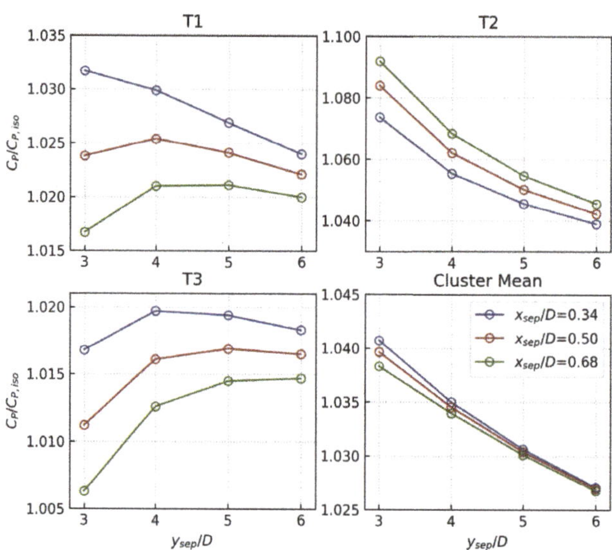

Figure 11. Power ratio ($C_P/C_{P,iso}$) as a function of inter-turbine x and y separations for the V shape cases. Subplots present the T1, T2, T3, and the cluster mean power ratios.

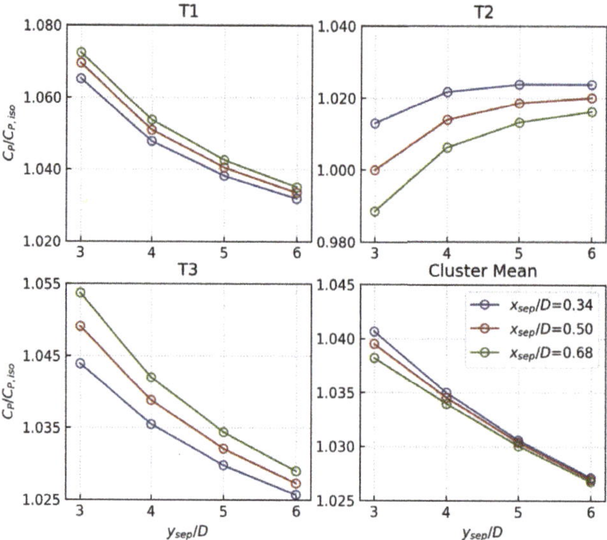

Figure 12. Power ratio ($C_P/C_{P,iso}$) as a function of inter-turbine x and y separations for the Reverse V shape cases. Subplots present the T1, T2, T3, and the cluster mean power ratios.

The situation for the leading turbines can be explained in a similar manner. To begin, the power ratio decreases with x_{sep} consistently, matching the pattern seen in Figures 7a and 9a. When T2 begins to move downwind of T1, the synergistic benefit for T1 weakens, and the detrimental dynamic of the blockage effect strengthens, both resulting in the deterioration of the power-generation performance of T1. In terms of y_{sep}, the patterns also vary with x_{sep} and can change from a straightforward, strictly decreasing pattern (e.g., $x_{sep}/D = 0.34$ in the T1 subplot of Figure 11) to a convex (parabolic) variation that exhibits a maximum value (e.g., for $x_{sep}/D \geq 0.50$ in the same subplot). There appears to be a strong connection between this functional relationship and the presence of a contest between the synergistic effect and the negative blockage effect, both due to the interaction with a downwind T2. For the intermediate turbine (T2 in Figure 13), the aforementioned dynamics of the leading and trailing turbines are both in play, creating competition. The competing dynamics are evidenced in the compactness of the variation of the data points, both in terms of x_{sep} and y_{sep}. We see that, as a consequence, the most favorable x_{sep} distance for power production does not follow a strictly monotonic ascending or descending pattern (viz., $x_{sep} = 0.34D$ has the largest power ratio, followed by $0.5D$, $0.68D$, and then, disruptively, 0, for $y_{sep} = 3D$). In fact, this contention parallels the trend in the cluster mean of all three array configurations, wherein the synergy patterns are "averages" of the individual leading and trailing turbine dynamics that constitute the cluster.

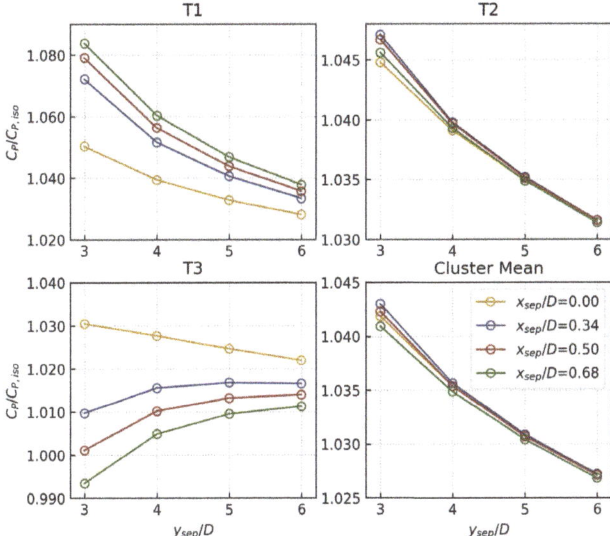

Figure 13. Power ratio ($C_p/C_{p,iso}$) as a function of inter-turbine x and y separations for the Line shape cases. Subplots present the T1, T2, T3, and the cluster mean power ratios.

Overall, the three-turbine group cases simulated herein demonstrate their general feasibility as successful synergy-producing configurations. There are limitations to studying the power ratio as a function of the x and y spatial configuration parameters, since the underlying synergistic principles do not necessarily conform to this framework. For instance, T1's synergy patterns are better contextualized with angle and spacing parameters (as performed by Hansen et al. [23]), but the correlation of T2 synergy to a high-velocity band surrounding the wake can be more saliently recognized when the layouts are organized on a Cartesian grid. Additionally, heatmaps do not extend well to three-turbine cases, where the larger number of degrees of freedom obscures efforts for a holistic visualization. Despite this, valuable insight has been generated, and these can provide useful design heuristics if not rules in VAWT wind-farm micro-siting. For example, the cluster mean power ratios in the V and Reverse V turbine groupings are nearly identical (with no practical difference in value). While this is not entirely surprising, because these shapes are mirrored across the y-axis with respect to each other, this result seems to suggest that there may be a "conservation" of synergy. In other words, inserting additional downwind turbines to exploit the flow-field of an upwind turbine would proportionally diminish the overall synergy due to blockage. This is an interesting idea that warrants further investigation.

3.4. Synergy-Superposition Scheme Results

Here we present the results of the synergy-superposition scheme, which is an entirely novel idea (to the best of our knowledge). Using the scheme and procedure detailed above, the approximate power ratios are calculated and compared to the complete three-turbine ALM simulation results. The percent differences are displayed in Figures 14–16 for the V, Reverse V, and Line turbine groupings, respectively. Overall, the error magnitudes are bounded within 0.38%, 0.32%, and 0.35% for the V, Reverse V, and Line shapes, respectively. This is a remarkably close agreement, demonstrating the validity of the superposition scheme within the scope of this work. It is important to caution the reader that the error plots exhibited here may not contain meaningful information or patterns beyond the noise associated with the numerical (round-off) errors expected of the simulations. Furthermore, while this scheme appears promising, its generalizability remains to be determined using larger turbine groups and especially using full-order blade-resolved CFD simulations for validation. Perhaps the first-order, linear effects superposed here are adequate for predicting three-turbine group synergistic interactions (for the current turbine and operating conditions), but larger errors may arise in extending this to an N-turbine group ($N \geq 4$) due to the emergence of elusive higher-order dynamical interactions.

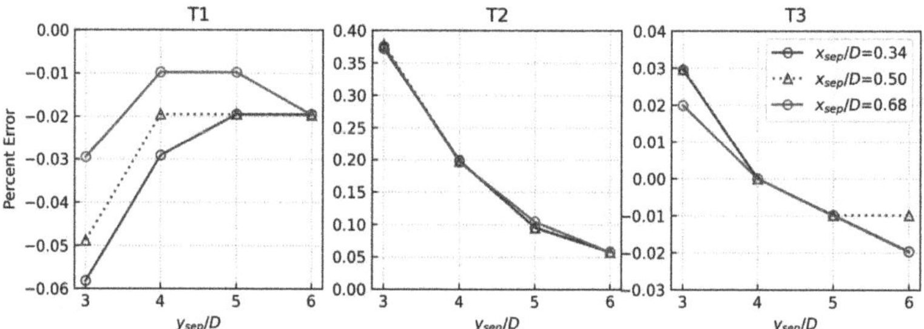

Figure 14. Plot of percent errors when using pairwise power ratio superposition to approximate three-turbine interactions for the V shape configurations. A positive error indicates the superposition over-predicted the power ratio, and a negative error indicates an under-prediction.

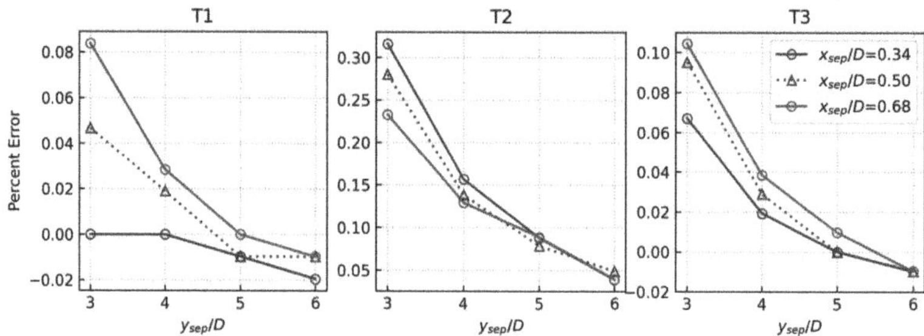

Figure 15. Plot of percent errors when using pairwise power ratio superposition to approximate three-turbine interactions for the Reverse V shape configurations. A positive error indicates the superposition over-predicted the power ratio and a negative error indicates an under-prediction.

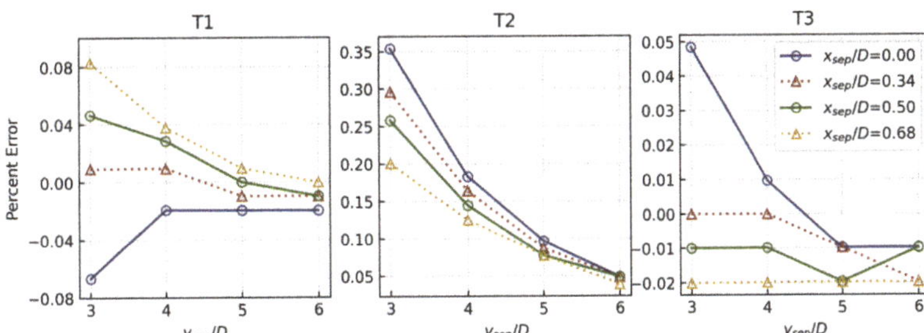

Figure 16. Plot of percent errors when using pairwise power ratio superposition to approximate three-turbine interactions for the Line shape configurations. A positive error indicates the superposition over-predicted the power ratio and a negative error indicates an under-prediction.

4. Conclusions

In this work, we have presented a methodology for using an ALM code (turbinesFoam) to evaluate synergy in VAWT groups consisting of two and three turbines. The code has been validated and used in previous investigations, and a comparison has been performed with 2 other ALM simulations codes, with a 2-D full-order blade-resolved URANS CFD simulation, and with available experimental measurements. A key result obtained herein was the successful demonstration that ALM simulations are capable of predicting VAWT synergy, yielding generally very good conformance to the pairwise turbine synergy patterns documented in the previous literature. Furthermore, we used a total of 108 two-turbine and 40 three-turbine ALM simulations to map the variation of turbine power-generation performance (characterized using a power ratio) arising from the systematic variation in the turbine array layout parameters. We find that a downwind turbine benefits almost universally from operating in the proximity of an upwind turbine (except in the wake regions), while the upwind turbine's synergy is equalized or diminished due to the blockage from this downwind turbine. It is also shown that co- and counter-rotating pairs are both capable of generating synergy, and neither offers an appreciable (significant) advantage over the other. Within the currently used methodology, it is possible to identify an optimal two-turbine layout that maximizes the cluster-level synergy. This pattern generalizes well to three-turbine cases and contributes value by explaining three-turbine synergistic interactions in a more granular way. We push the envelope further by proving the effectiveness and accuracy of a novel synergy-superposition scheme, confirming that a three-turbine group's synergy can be adequately modeled as a linear superposition of all the pairwise interactions in the three-turbine array.

These results could be enhanced by future work in the following ways. First, the isolated turbine performance could be more rigorously validated. This may be achieved by obtaining a high-fidelity airfoil coefficient dataset (possibly driven by full-order URANS CFD simulations), exploring various additional correction models (e.g., for dynamic stall), and testing different Gaussian kernel formulations. It is recommended that multi-turbine full-order blade-resolved CFD simulations be performed to thoroughly and systematically evaluate the accuracy of ALM simulations in predicting VAWT power-generation performance. Furthermore, it would be valuable to extend the present investigation to larger turbine groups (four or more turbines in the array), to more diverse turbine geometries or heterogeneous wind farms, and to different TSRs. The current model is also suitable for extension towards the evaluation of synergy in complex terrain situations, for which a different lower-boundary condition can be imposed to represent the complex terrain. The virtual turbines can also be freely arranged both in the horizontal plane and the vertical direction, using the ALM technique, in order to capture the synergy in sophisticated cases such as an urban siting. The use of a low-cost, effective reduced-order approach such as an ALM simulation is expected to be invaluable to the optimal design of large-scale VAWT farms—involving, as such, the leveraging of the unique opportunities provided by turbine synergistic interactions.

Author Contributions: Conceptualization, J.H.Z., F.-S.L. and E.Y.; methodology, J.H.Z., F.-S.L. and E.Y.; software, J.H.Z.; validation, J.H.Z.; formal analysis, J.H.Z.; investigation, J.H.Z., F.-S.L. and E.Y.; resources, F.-S.L.; data curation, J.H.Z.; writing—original draft preparation, J.H.Z.; writing—review and editing, E.Y. and F.-S.L.; visualization, J.H.Z.; supervision, F.-S.L. and E.Y.; project administration, F.-S.L. and E.Y.; funding acquisition, F.-S.L. All authors have read and agreed to the published version of the manuscript.

Funding: This research was funded by the Natural Sciences and Engineering Research Council of Canada (NSERC) Discovery Grants program Grant No. 50503-10234.

Institutional Review Board Statement: Not applicable.

Informed Consent Statement: Not applicable.

Conflicts of Interest: The authors declare no conflict of interest.

Abbreviations

The following abbreviations are used in this paper:

ALM	actuator line model
BEM	blade element momentum
CFD	computational fluid dynamics
HAWT	horizontal-axis wind turbine
TSR	tip-speed ratio
URANS	unsteady Reynolds-Averaged Navier–Stokes
VAWT	vertical-axis wind turbine

References

1. Möllerström, E.; Gipe, P.; Beurskens, J.; Ottermo, F. A historical review of vertical axis wind turbines rated 100 kW and above. *Renew. Sustain. Energy Rev.* **2019**, *105*, 1–13. [CrossRef]
2. Liu, K.; Yu, M.; Zhu, W. Enhancing wind energy harvesting performance of vertical axis wind turbines with a new hybrid design: A fluid-structure interaction study. *Renew. Energy* **2019**, *140*, 912–927. [CrossRef]
3. Abhishek, P.J.A. Performance prediction and fundamental understanding of small scale vertical axis wind turbine with variable amplitude blade pitching. *Renew. Energy* **2016**, *97*, 97–113.
4. Johari, M.K.; Jalil, M.A.A.; Shariff, M.F.M. Comparison of horizontal axis wind turbine (HAWT) and vertical axis wind turbine (VAWT). *Int. J. Eng. Technol.* **2018**, *7*, 74–80. [CrossRef]
5. Du, L.; Ingram, G.; Dominy, R.G. A review of H-Darrieus wind turbine aerodynamic research. *J. Mech. Eng. Sci.* **2019**, *233*, 23–24. [CrossRef]
6. Ghasemian, M.; Ashrafi, Z.N.; Sedaghat, A. A review on computational fluid dynamic simulation techniques for Darrieus vertical axis wind turbines. *Energy Convers. Manag.* **2017**, *149*, 87–100. [CrossRef]
7. Dabiri, J.O. Potential order-of-magnitude enhancement of wind farm power density via counter-rotating vertical-axis wind turbine arrays. *J. Renew. Sustain. Energy* **2011**, *3*, 043104. [CrossRef]
8. Paraschivoiu, I. *Wind Turbine Design: With Emphasis on Darreius Concept*; Presses Internationales Polytechniques: Montreal, QC, Canada, 2009.
9. Nobile, R.; Vahdati, M.; Barlow, J.F.; Mewburn-Crook, A. Unsteady flow simulation of a vertical axis augmented wind turbine: A two-dimensional study. *J. Wind. Eng. Ind. Aerodyn.* **2014**, *125*, 168–179. [CrossRef]
10. Elkhoury, M.; Kiwata, T.; Aoun, E. Experimental and numerical investigation of a three-dimensional vertical-axis wind turbine with variable-pitch. *J. Wind. Eng. Ind. Aerodyn.* **2015**, *139*, 111–123. [CrossRef]
11. Abdalrahman, G. Pitch Angle Control for a Small-Scale Darrieus Vertical Axis Wind Turbine with Straight Blades (H-Type VAWT). Ph.D. Thesis, University of Waterloo, Waterloo, ON, Canada, 2019.
12. Tavernier, D.D.; Ferreira, C.; van Bussel, G. Airfoil optimisation for vertical-axis wind turbines with variable pitch. *Wind Energy* **2019**, *22*, 547–562. [CrossRef]
13. Strom, B.; Brunton, S.L.; Polagye, B. Intracycle angular velocity control of cross-flow turbines. *Nat. Energy* **2017**, *2*, 17103. [CrossRef]
14. Kinzel, M.; Mulligan, Q.; Dabiri, J.O. Energy exchange in an array of vertical-axis wind turbines. *J. Turbul.* **2012**, *13*, N38. [CrossRef]

15. Hezaveh, S.H.; Bou-Zeid, E.; Dabiri, J.O.; Kinzel, M.; Cortina, G.; Martinelli, L. Increasing the power production of vertical-axis wind-turbine farms using synergistic clustering. *Bound.-Layer Meteorol.* **2018**, *169*, 275–296. [CrossRef]
16. Ahmadi-Baloutaki, M.; Carriveau, R.; Ting, D.S. A wind tunnel study on the aerodynamic interaction of vertical axis wind turbines in array configurations. *Renew. Energy* **2016**, *96*, 904–913. [CrossRef]
17. Zanforlin, S.; Nishino, T. Fluid dynamic mechanisms of enhanced power generation by closely spaced vertical axis wind turbines. *Renew. Energy* **2016**, *99*, 1213–1226. [CrossRef]
18. Lam, H.F.; Peng, H.Y. Measurements of the wake characteristics of co- and counter-rotating twin H-rotor vertical axis wind turbines. *Energy* **2017**, *131*, 13–26. [CrossRef]
19. Peng, J. Effects of aerodynamic interactions of closely-placed vertical axis wind turbine pairs. *Energies* **2018**, *11*, 2842. [CrossRef]
20. Shaaban, S.; Albatal, A.; Mohamed, M.H. Optimization of H-Rotor Darrieus turbines' mutual interaction in staggered arrangements. *Renew. Energy* **2018**, *125*, 87–99. [CrossRef]
21. Barnes, A.; Hughes, B. Determining the impact of VAWT farm configurations on power output. *Renew. Energy* **2019**, *143*, 1111–1120. [CrossRef]
22. Brownstein, I.D.; Wei, N.J.; Dabiri, J.O. Aerodynamically interacting vertical axis wind turbines: Performance enhancement and three-dimensional flow. *Energies* **2019**, *12*, 2724. [CrossRef]
23. Hansen, J.T.; Mahak, M.; Tzanakis, I. Numerical modelling and optimization of vertical axis wind turbine pairs: A scale up approach. *Renew. Energy* **2021**, *171*, 1371–1381. [CrossRef]
24. Hamada, K.; Smith, T.; Durrani, N.; Qin, N.; Howell, R. Unsteady flow simulation and dynamic stall around vertical axis wind turbine blades. In Proceedings of the 46th AIAA Aerospace Sciences Meeting and Exhibit, Reno, Nevada, 7–10 January 2008.
25. Howell, R.; Qin, N.; Edwards, J.; Durrani, N. Wind tunnel and numerical study of a small vertical axis wind turbine. *Renew. Energy* **2010**, *35*, 412–422. [CrossRef]
26. Buchner, A.J.; Lohry, M.W.; Martinelli, L.; Soria, J.; Smits, A.J. Dynamic stall in vertical axis wind turbines: Comparing experiments and computations. *J. Wind. Eng. Ind. Aerodyn.* **2015**, *146*, 163–171. [CrossRef]
27. Orlandi, A.; Collu, M.; Zanforlin, S.; Shires, A. 3D URANS analysis of a vertical axis wind turbine in skewed flows. *J. Wind. Eng. Ind. Aerodyn.* **2015**, *147*, 77–84. [CrossRef]
28. Zuo, W.; Wang, X.; Kang, S. Numerical simulations on the wake effect of H-type vertical axis wind turbines. *Energy* **2016**, *106*, 691–700. [CrossRef]
29. Li, C.; Zhu, S.; Xu, Y.; Xiao, Y. 2.5D large eddy simulation of vertical axis wind turbine in consideration of high angle of attack flow. *Renew. Energy* **2013**, *51*, 317–330. [CrossRef]
30. Shamsoddin, S.; Porté-Agel, F. Large Eddy Simulation of vertical axis wind turbine wakes. *Energies* **2014**, *7*, 890–912. [CrossRef]
31. Sanderse, B.; van der Pijl, S.P.; Koren, B. Review of computational fluid dynamics for wind turbine wake aerodynamics. *Wind Energy* **2011**, *14*, 799–819. [CrossRef]
32. Delafin, P.L.; Nishino, T.; Kolios, A.; Wang, L. Comparison of low-order aerodynamic models and RANS CFD for full scale 3D vertical axis wind turbines. *Renew. Energy* **2017**, *109*, 564–575. [CrossRef]
33. Ning, A. Actuator cylinder theory for multiple vertical axis wind turbines. *Wind. Energy Sci.* **2016**, *1*, 327–340. [CrossRef]
34. Zhao, R.; Creech, A.C.W.; Borthwick, A.G.L.; Venugoal, V.; Nishino, T. Aerodynamic analysis of a two-bladed vertical-axis wind turbine using a coupled unsteady RANS and Actuator Line Model. *Energies* **2020**, *13*, 776. [CrossRef]
35. Sørensen, J.N.; Shen, W.Z. Numerical modeling of wind turbine wakes. *J. Fluids Eng.* **2002**, *124*, 393–399. [CrossRef]
36. Hezaveh, S.H.; Bou-Zeid, E.; Lohry, M.W.; Martinelli, L. Simulation and wake analysis of a single vertical axis wind turbine. *Wind Energy* **2017**, *20*, 713–730. [CrossRef]
37. Creech, A.C.W.; Borthwick, A.G.L.; Ingram, D. Effects of support structures in an LES actuator line model of a tidal turbine with contra-rotating rotors. *Energies* **2017**, *10*, 726. [CrossRef]
38. Abkar, M. Impact of subgrid-scale modeling in actuator-line based large-eddy simulation of vertical-axis wind turbine wakes. *Atmosphere* **2018**, *9*, 257. [CrossRef]
39. Bachant, P.; Goude, A.; Wosnik, M. Actuator line modeling of vertical-axis turbines. *arXiv* **2018**, arXiv:1605.01449v4.
40. Melani, P.F.; Balduzzi, F.; Bianchini, A. A robust procedure to implement dynamic stall models into actuator line methods for the simulation of vertical-axis wind turbines. *J. Eng. Gas Turbines Power* **2021**, *143*, 111008 . [CrossRef]
41. Melani, P.F.; Balduzzi, F.; Bianchini, A. Simulating tip effects in vertical-axis wind turbines with the actuator line method. *J. Phys. Conf. Ser.* **2022**, *2265*, 032028. [CrossRef]
42. Melani, P.F.; Balduzzi, F.; Ferrara, G.; Bianchini, A. Tailoring the actuator line theory to the simulation of Vertical-Axis Wind Turbines. *Energy Convers. Manag.* **2021**, *243*, 114422. [CrossRef]
43. Raj V, S.; Solanki, R.S.; Chalamalla, V.K.; Sinha, S.S. Numerical simulations and analysis of flow past vertical-axis wind turbines employing the actuator line method. In Proceedings of the 26th National and 4th International ISHMT-ASTFE Heat and Mass Transfer Conference, Tamil Nadu, India, 17–20 December 2021.
44. Bachant, P.; Goude, A.; Wosnik, M. turbinesFoam: v0.0.8. Zenodo. 2018. Available online: https://zenodo.org/record/1210366 (accessed on 3 August 2021). [CrossRef]
45. Mendoza, V.; Bachant, P.; Ferreira, C.; Goude, A. Near-wake flow simulation of a vertical axis turbine using an actuator line model. *Wind Energy* **2019**, *22*, 171–188. [CrossRef]

46. Mendoza, V.; Goude, A. Validation of actuator line and vortex models using normal forces measurements of a straight-bladed vertical axis wind turbine. *Energies* **2020**, *13*, 511. [CrossRef]
47. Sheldahl, R.E.; Klimas, P.C. *Aerodynamic Characteristics of Seven Symmetrical Airfoil Sections through 180-Degree Angle of Attack for Use in Aerodynamic Analysis of Vertical Axis Wind Turbines*; Technical Report SAND8O-2114; Sandia National Laboratories: Albuquerque, NM, USA, 1981.
48. LeBlanc, B.P.; Ferreira, C.S. Experimental determination of thrust loading of a 2-bladed vertical axis wind turbine. *J. Phys. Conf. Ser.* **2018**, *1037*, 022043. [CrossRef]
49. Balduzzi, F.; Bianchini, A.; Maleci, R.; Ferrara, G.; Ferrari, L. Critical issues in the CFD simulation of Darrieus wind turbines. *Renew. Energy* **2016**, *85*, 419–435. [CrossRef]
50. Scheurich, F.; Brown, R.E. Effect of dynamic stall on the aerodynamics of vertical-axis wind turbines. *AIAA J.* **2011**, *49*, 2511–2521. [CrossRef]
51. Zheng, H.D.; Zheng, X.Y.; Zhao, S.X. Arrangement of clustered straight-bladed wind turbines. *Energy* **2020**, *200*, 117563. [CrossRef]
52. Jodai, Y.; Hara, Y. Wind tunnel experiments on interaction between two closely spaced vertical-axis wind turbines in side-by-side arrangement. *Energies* **2021**, *14*, 7874. [CrossRef]
53. Hara, Y.; Jodai, Y.; Okinaga, T.; Furukawa, M. Numerical analysis of the dynamic interaction between two closely spaced vertical-axis wind turbines. *Energies* **2021**, *14*, 2286. [CrossRef]

Article

Investigation into the Aerodynamic Performance of a Vertical Axis Wind Turbine with Endplate Design

Shern-Khai Ung [1], Wen-Tong Chong [2], Shabudin Mat [3], Jo-Han Ng [1], Yin-Hui Kok [1] and Kok-Hoe Wong [1,*]

[1] Department of Mechanical Engineering, Faculty of Engineering and Physical Sciences, University of Southampton Malaysia, Iskandar Puteri 79200, Malaysia
[2] Department of Mechanical Engineering, Faculty of Engineering, Universiti Malaya, Kuala Lumpur 50603, Malaysia
[3] Institute for Vehicle Systems and Engineering (IVeSE), Universiti Teknologi Malaysia, Skudai 81310, Malaysia
* Correspondence: k.h.wong@soton.ac.uk; Tel.: +60-165363421

Abstract: For the past decade, research on vertical axis wind turbines (VAWTs) has garnered immense interest due to their omnidirectional characteristic, especially the lift-type VAWT. The H-rotor Darrieus VAWT operates based on the lift generated by aerofoil blades and typically possesses higher efficiency than the drag-type Savonius VAWT. However, the open-ended blades generate tip loss effects that reduce the power output. Wingtip devices such as winglets and endplates are commonly used in aerofoil design to increase performance by reducing tip losses. In this study, a CFD simulation is conducted using the sliding mesh method and the k-ω *SST* turbulence model on a two-bladed NACA0018 VAWT. The aerodynamic performance of a VAWT with offset, symmetric V, asymmetric and triangular endplates are presented and compared against the baseline turbine. The simulation was first validated with the wind tunnel experimental data published in the literature. The simulation showed that the endplates reduced the swirling vortex and improved the pressure distribution along the blade span, especially at the blade tip. The relationship between TSR regimes and the tip loss effect is also reported in the paper. Increasing VAWT performance by using endplates to minimise tip loss is a simple yet effective solution. However, the improvement of the power coefficient is not remarkable as the power degradation only involves a small section of the blades.

Keywords: endplate; wingtip device; blade tip losses; Darrieus VAWT; CFD

Citation: Ung, S.-K.; Chong, W.-T.; Mat, S.; Ng, J.-H.; Kok, Y.-H.; Wong, K.-H. Investigation into the Aerodynamic Performance of a Vertical Axis Wind Turbine with Endplate Design. *Energies* **2022**, *15*, 6925. https://doi.org/10.3390/en15196925

Academic Editors: Yoshifumi Jodai and Yutaka Hara

Received: 24 August 2022
Accepted: 16 September 2022
Published: 21 September 2022

Publisher's Note: MDPI stays neutral with regard to jurisdictional claims in published maps and institutional affiliations.

1. Introduction

As global warming and environmental issues continue to be aggravated, people have begun to shift their focus from non-renewable resources to sustainable and renewable resources for power generation. Wind energy has gained traction and become a prominent source of renewable energy. According to the Global Wind Energy Council (GWEC) [1], 93.6 GW of wind capacity was installed in 2021 globally, leading to a total wind capacity of 837 GW, which is a 12.4% increase from 2020. Wind turbines (WTs) of scales up to megawatts are widely deployed in onshore and offshore regions to capture wind energy. There are two common types of wind turbines—horizontal axis wind turbines (HAWTs) and vertical axis wind turbines (VAWTs), in which their rotor rotates on a different axis, as the name suggests. Currently, HAWTs make up most of the WTs in the market due to their higher efficiency compared to VAWTs [2–5]. However, a large megawatt-scale HAWT has its downsides and limitations, such as an increase in cost and weight with blade length [3,6]. Meanwhile, the omnidirectional characteristics and simplified machine design of VAWTs have attracted researchers' interest, leading to rigorous research and development in the past few years [3,5].

VAWTs are further categorised into the lift type (Darrieus) and the drag type (Savonius), corresponding to their operating principle. The former has higher efficiency and is driven by the pressure difference across the aerofoil blade, while the latter has a better

---

self-start ability and generates torque from the drag force acting on the concave blades [7]. It is crucial to understand the complex flow characteristics of VAWTs, which are very different from HAWTs. The VAWT suffers from tip losses at both ends of the blade, blade–wake interaction in the downwind region, and dynamic stalls caused by the continuous variation in the angle of attack throughout a revolution [8–10]. According to the study of a three-bladed VAWT model performed by Howell et al. [11], the interaction with wake can affect the pressure distribution on the blade. Vortex strength is found to be correlated with the lift generated by the blades, where the tip vortex generated is the strongest right after the blade has attained maximum lift. The tip vortex becomes weaker in the downstream region, and the overall power output of the VAWT is reduced compared to an ideal 2D case.

To weaken these blade tip losses, wingtip devices such as endplates, elliptical terminations, and winglets are commonly utilised in the aerospace and automotive sector to improve aerodynamic performance. These wingtip devices reduce spanwise flow due to the pressure difference between both sides of an aerofoil and improve the aerodynamic characteristic of the wing or blade [9,12–14]. For instance, Jung et al. [13] numerically investigated the effect of an endplate on a wing-in-ground (WIG) craft's wingtip over the free surface. With a NACA4406 profile blade in an aspect ratio of 2, their simulation revealed that the endplate improves the pressure distribution on both sides of the wing, especially on the pressure side. The blockage effect introduced by the endplate enhances the air cushion effect on the wing, leading to a 14% and 124% improvement in the power coefficient (C_P) near the leading edge and the trailing edge, respectively. An increase of up to 46% in the lift-to-drag ratio is also observed when the wing is close to the free surface; however, these improvements become negligible as the wing is further away from the ground due to a weaker WIG effect. Although a larger vortex is generated at the endplate tip due to flow separation, the lateral motion of the tip vortex helps to reduce the induced drag.

Wingtip devices are also implemented in HAWTs to improve aerodynamic efficiency and power output. Johansen and Sørensen [15] studied the effect of winglet parameters on a HAWT's performance. All configurations with winglets lead to power enhancements ranging from 0.98% to 2.77%. Twist angles have negligible effects on power and thrust, while a smaller curvature radius leads to larger power increments. With a 30° sweep angle imposed on the winglet, the performance improvement is reduced compared to no sweep angle.

However, the beneficial effect of wingtip devices is highly dependent on the wing parameters and use case scenario. For example, the effect of an endplate on a VAWT blade that rotates around its vertical axis is different from its application in an airplane or a car, where the oncoming wind direction is more consistent. Thus, the existing research data from other fields cannot represent a VAWT's characteristics. Extensive research is required to find the optimised wingtip device design for VAWTs, as a wrongly designed wingtip device may adversely affect the VAWT's performance [16].

Laín et al. [4] studied the effects of winglets on a straight-bladed Darrieus water turbine and found that both symmetric and asymmetric winglets can improve turbine performance. The symmetric winglet particularly brings 20% power enhancement, in which the improvement mainly occurs at the maxima that coincides with the location where maximum lift is obtained. However, the power augmentation leads to a larger thrust acting on the blades and, thus, requires extra attention to the structural strength when winglets are employed. Besides that, the long trailing vortex generated in the bare blade is weakened with the asymmetric winglet, while the symmetric blade eliminates the trailing vortex. The detachment of vorticity still occurs with winglets but is delayed compared to the bare blade.

Syawitri et al. [14] reviewed a range of passive flow control devices (PFCDs) to enhance the lift-type VAWT's performance via minimising flow separation and reducing dynamic stalls without external power input. Winglets weaken the tip vortex by preventing flow mixing between the pressure and suction sides at the blade tip. This can lead to a 10–19% improvement in the low tip speed ratio (TSR) regime and a 6.7–10.5% improvement in the medium regime. However, the C_P improvement diminished as the regime of TSRs increased. Other PFCDs, such as inward dimple and gurney flap, can also enhance the C_P of VAWTs, but every device performs differently across different TSR regimes. For instance, a 2D simulation conducted by Mousavi et al. [17] on a Darrieus VAWT with NACA0021 aerofoil found that the gurney flap installed on either the suction or pressure side of the blade results in power enhancement in the low TSR range (i.e., 0.6 to 1.6). The reverse was observed at TSRs of 1.8 to 3. The study also discovered that a gurney flap installed at an angle towards the blade surface, in general, offers better performance over the standard configuration.

The effect of the curvature radius of a winglet attached to a three-bladed VAWT was studied by Malla et al. [2]. They found that the aerodynamic performance increases as the curvature radius decreases, and this result coincides with the findings in [15] on HAWTs. The lift coefficient showed an increase in the upstream region, while the opposite results were observed in the downstream region. The largest pressure difference on the blade happened at a 90° azimuthal angle, and the addition of winglets led to lower minimum pressure on the suction side and a larger area of maximum pressure on the pressure side.

Besides winglets, endplates, which are sometimes referred to as bulkheads in the literature [16,18], are also implemented in both lift- and drag-type VAWTs. Premkumar et al. [19] set up an experiment to test a helical Savonius drag-type VAWT with and without a circular endplate. The presence of the endplate lowered the torque coefficient and raised the power coefficient of the VAWT across wind speeds of 3–6 m/s due to a higher TSR achieved. With a conventional Savonius VAWT, the 3D CFD conducted by Kassab et al. [20] showed that a larger pressure difference is achieved when a circular endplate is installed, as it acts as a physical barrier to block spanwise flow. The endplate also smoothens flow through the turbine but generates more vortices at the blade tip. At the optimum TSR of 0.8, the average lift and drag coefficients were elevated by 400% and 180%, respectively, while the C_P saw a 42.5% improvement over the bare VAWT.

Amato et al. [16] analysed and compared the effect of the endplate, elliptic tip, and winglets on a lift-type VAWT. All configurations managed to improve C_P over the baseline rotor; however, the winglet with a 90° cant angle performed the best over TSRs 2.65 to 3.7, with an 11.85% increment at 620 rpm. The 50% longer version of the same winglet showed less improvement due to the decrease in the pressure torque on the profile.

Miao et al. [12] studied the effect of 20 blade tip device designs, including winglets and endplates, on the aerodynamic performance of an H-rotor Darrieus VAWT. With the same aspect ratio and sweep area, they found that the flat endplate with an offset of $0.18c$ from the blade profile improves C_P, but the improvement degrades at higher TSR due to larger drag. Reduction of drag can be achieved by implementing a streamlined shape on the top surface, and the study showed that only the streamline-shaped endplate and a novel Winglet-H could improve the C_P at TSRs 1.85, 2.29, and 2.52. This study revealed that the blade tip devices predominantly affect the blade tip region without pronounced effects on the pattern of wake propagation of the VAWT.

Daróczy et al. [21] investigated the effect of different parameters of a winglet and compared it with the baseline and endplate cases. At TSR 2.6, all configurations barely improved C_P at the region 0.1 m away from the blade tip. Overall, the endplate slightly improved the C_P by 0.65%, while all six winglet configurations tested in this study not only failed to improve the turbine performance but were subpar to the baseline VAWT. This further emphasises the importance of implementing a well-optimised winglet or wingtip device for a VAWT.

Jiang et al. [18] performed a numerical simulation on a single-bladed rotor with an aspect ratio of 15. They observed the tip loss effect until $3c$ from the blade tip, where the torque coefficient remained negative throughout the cycle at $0.05c$. The implementation of the endplate smoothens the flow at the blade tip and improves the work done by the blades. A larger endplate shows a greater enhancement in blade work. However, it induces more drag, which offsets a large portion of the gains achieved. Thus, the optimal size of the endplate is found to be 0.35c. However, this configuration deteriorates the performance at low TSRs. The same observation is reported in [14], where flow control devices performed differently across all TSR regimes.

Gosselin et al. [9] performed simulations on a single-bladed NACA0015 turbine with aspect ratios (ARs) 7 and 15, where the larger AR blade attained a better C_P of 62.5%. This indicates that a short blade suffers greater tip loss effects due to a large portion of the blade being influenced by the tip effects. Therefore, low AR blades may benefit more from wingtip devices. The addition of endplates reduces the blade tip effect and produces uniform pressure distribution across the length of the blade. However, a larger endplate induces more drag and ultimately exceeds the power enhancement on the VAWT.

Nathan and Thanigaiarasu [22] investigated the effect of aerofoil-shaped, rectangular and full circular endplates on the aerodynamic performance of a three-bladed VAWT. The aerofoil-shaped and rectangular endplate managed to elevate the C_P by up to 3%. However, the extra mass introduced by the full circular endplate, which covers the entire rotor, failed to bring positive results. The full circular endplate also induced the largest load on the blades at 80% to 90% span, where 90% span is closer to the blade tip.

The results presented by Mishra et al. [23] through experiments and simulations showed that endplates levitated the C_P of a three-bladed VAWT with a NACA0018 blade profile. Endplates also increased the rotational speed of the turbine at the same wind velocity compared to the baseline turbine. Opposite results were observed with winglets due to the larger overall drag induced. To provide a clearer view of the literature review, the wingtip devices discussed above are summarised in Table 1 below.

Table 1. A summary of wingtip devices from the literature review.

Author	Wingtip Device	Parameter	Application	Main Finding
Jung et al. [13]	Endplate	Distance from the free surface	WIG craft	14% and 124% improvement in C_P at the leading edge and trailing edge, respectively. Improvement degrades as the distance from the free surface increases.
Johansen and Sørensen [15]	Winglet	Winglet height, sweep angle, curvature radius, twist angle	HAWT	Power improved by 0.98% to 2.77%. A smaller curvature radius generates larger power improvements.
Laín et al. [4]	Winglet	Symmetric and asymmetric winglet	Darrieus water turbine	Symmetric winglet brought a larger improvement at 20%.
Syawitri et al. [14]	Winglet and others	-	Lift-type VAWT	10–19% power improvement in the low TSR regime and 6.7–10.5% at the medium regime.
Mousavi et al. [17]	Gurney Flap	Placement of flap and angle	Two-bladed lift-type VAWT	Gurney flap enhances C_P at low TSRs. Angled gurney flap provides superior performance over standard gurney flap configuration.
Malla et al. [2]	Winglet	Curvature radius	Lift-type VAWT	Aerodynamic performance increases as the curvature radius decreases.
Premkumar et al. [19]	Endplate	Presence of endplate	Savonius drag-type helical VAWT	Endplate lowers the torque coefficient and raises the power coefficient.
Kassab et al. [20]	Endplate	Presence of endplate	Savonius drag-type VAWT	The average lift and drag coefficients were elevated by 400% and 180%, respectively, while the C_P saw a 42.5% improvement over the bare VAWT at TSR 0.8.
Amato et al. [16]	Winglet, elliptical termination and endplate	Winglet—cant angle and winglet length	Single-bladed lift-type VAWT	Winglet with 90° cant angle achieved 11.85% increment at 620 rpm.
Miao et al. [12]	Winglet and endplate	Wingtip device designs	H-rotor Darrieus VAWT	Improvements from the endplate degrade as TSR increases. Only the streamlined-shaped endplate and a novel Winglet-H can improve C_P over a larger range of TSRs.
Daróczy et al. [21]	Winglet and endplate	Cant angle and sweep angle	Three-bladed lift-type VAWT	All 6 winglet configurations reduced the power output compared to the baseline turbine. Endplate improved power output by 0.65%.
Jiang et al. [18]	Endplate	Endplate offset	Single-bladed lift-type VAWT	The optimal size of endplate is 0.35c.
Gosselin et al. [9]	Endplate	Aspect ratio and endplate shape	Single-bladed lift-type VAWT	Lower AR blades suffer larger tip losses. Endplates reduce the blade tip effect, but overly sized endplates decrease the turbine performance.
Nathan and Thanigaiarasu [22]	Endplate	Endplate Shape	Three-bladed lift-type VAWT	The aerofoil-shaped and rectangular endplates elevated the C_P by up to 3%, while the full circular endplate, which covers the entire rotor, showed negative results.
Mishra et al. [23]	Endplate	Endplate offset	Three-bladed lift-type VAWT	Endplates levitated the C_P and RPM of the turbine.

After reviewing various literature, wingtip devices exhibit the potential to elevate the performance of a VAWT. However, there is limited research on the variation of endplate geometry for the aerodynamic performance of a Darrieus VAWT compared to winglet geometry. Endplates with simple offset from the blade profile have been employed in previous papers, while other possible geometries are rarely proposed. Therefore, this paper aims to explore the effect of different endplate geometries on the aerodynamic performance of a Darrieus VAWT, vortex generation, pressure distribution on the turbine blades and the drag induced. The research findings can provide insight and comparison on the implications of endplate geometry on a VAWT, especially around the blade tip region. In the next section, details of the wind turbine model, simulation setup and validation are presented. Section 3 presents the results and a discussion of the findings obtained from the simulation, and a comprehensive conclusion is deduced in Section 4.

2. Methodology

2.1. Aerodynamics of the H-Rotor Darrieus VAWT

An H-rotor Darrieus VAWT consists of two straight blades with an aerofoil-shaped cross-section. As the name suggests, the rotor rotates about the vertical axis, and thus, the flow characteristics are different from those of HAWTs. Unlike HAWTs, the power output of a VAWT varies with the azimuthal angle. The blade–wake interaction in the downstream region (i.e., $180° \leq \theta \leq 360°$) leads to a much lower power output compared to the upstream region (i.e., $0° \leq \theta \leq 180°$), which receives undisturbed wind. To quantify turbine performance, the dimensionless coefficient of torque (C_T) and coefficient of power (C_P) is derived as shown below:

$$C_T = \frac{T}{\frac{1}{2}\rho A R U_\infty^2} \tag{1}$$

where T is the torque generated, ρ is the air density, A is the swept area, R is the rotor radius and U_∞ is the free stream velocity.

$$C_P = \frac{P}{\frac{1}{2}\rho A U_\infty^3} \tag{2}$$

where P is the power generated. According to the Betz limit, $C_P = 0.593$ is the upper limit tht a wind turbine can achieve.

Another crucial non-dimensional parameter of a wind turbine is the tip speed ratio (TSR). It is the ratio of the velocity at the blade tip to the incoming wind velocity, and it is denoted as λ. The TSR is defined as:

$$\lambda = \frac{R\omega}{U_\infty} \tag{3}$$

where ω is the angular velocity of the rotor. Since power is equivalent to the product of torque and angular velocity, Equation (2) can be simplified into:

$$C_p = \lambda \cdot C_t \tag{4}$$

As aforementioned, VAWTs suffer from tip losses at the blade tip, which can only be captured through 3D simulation [24]. The studies done by Howell et al. [11], Lam et al. [8], and Gosselin et al. [9] show that 2D simulation consistently overestimates C_P while the 3D case predicts C_P values closer to the experimental data. This is due to the 2D model resembling a blade with infinite length; the simulation cannot capture the flow in the span-wise direction and vortical motion at the blade tip. Therefore, a 3D numerical simulation is adopted in this study to properly capture the tip loss effects in VAWTs.

2.2. Rotor Geometry

In this study, a straight-bladed H-rotor Darrieus VAWT is modelled, as illustrated in Figure 1, based on the rotor specifications adopted in the wind tunnel experiment conducted by Watanabe et al. [25]. The VAWT model has a rotor height (H) of 700 mm and a diameter (D) of 700 mm. It comprises two blades with a chord length (c) of 150 mm and no pitch angle. Although omitting the shaft and supporting structure can lead to the overestimation of results [26,27], the current study aims to compare turbine performance based on endplate geometry. Therefore, the two NACA0018 profile blades are modelled in the rotor domain without the centre shaft and spoke to simplify the mesh. As shown in Figure 1, six cut sections with a width of 5 mm each are made on Blade 1 at heights $0.015625H$, $0.03125H$, $0.0625H$, $0.125H$, $0.25H$ and $0.5H$ from the blade tip (i.e., the top end) to extract the C_P values at these locations and observe the blade tip loss effect.

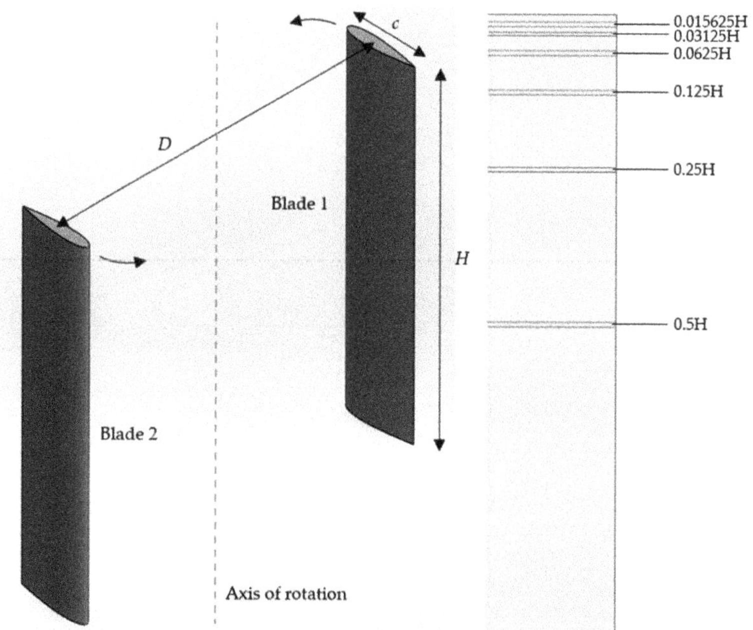

Figure 1. NACA0018 straight-bladed VAWT and cut sections adopted in the current simulation.

2.3. Endplate Geometry

To compare and study the effect of endplates on blade tip losses, the endplates are modelled on both ends of each blade. All endplate configurations have a flat surface with a thickness of $0.03c$, which is also adopted in the literature [12,18]. The aspect ratio of a VAWT blade is the ratio of the blade height to the chord length, and this parameter can affect the power output of a VAWT [9,12,21]. To maintain the aspect ratio for a fair comparison with the baseline rotor, the overall height of the blade, including the endplates, is kept the same as the baseline rotor height. Four different endplates—offset, symmetric V, asymmetric, and triangular endplates—are investigated in this study, as depicted in Figure 2. The offset endplate is modelled with an offset of $0.35c$ from the blade profile, which is claimed to be optimal according to [18], while the symmetric V endplate has a pointed V-shape design. The asymmetric endplate has a larger area allocated, facing the inside of the rotor, in which the flow is more complex [4], whereas the triangular endplate has a similar profile as the symmetric V endplate but with a flat trailing edge. All sharp edges on the endplates are eliminated by applying a 5 mm radius fillet at the trailing edge and a 10 mm radius at the pointed leading edge.

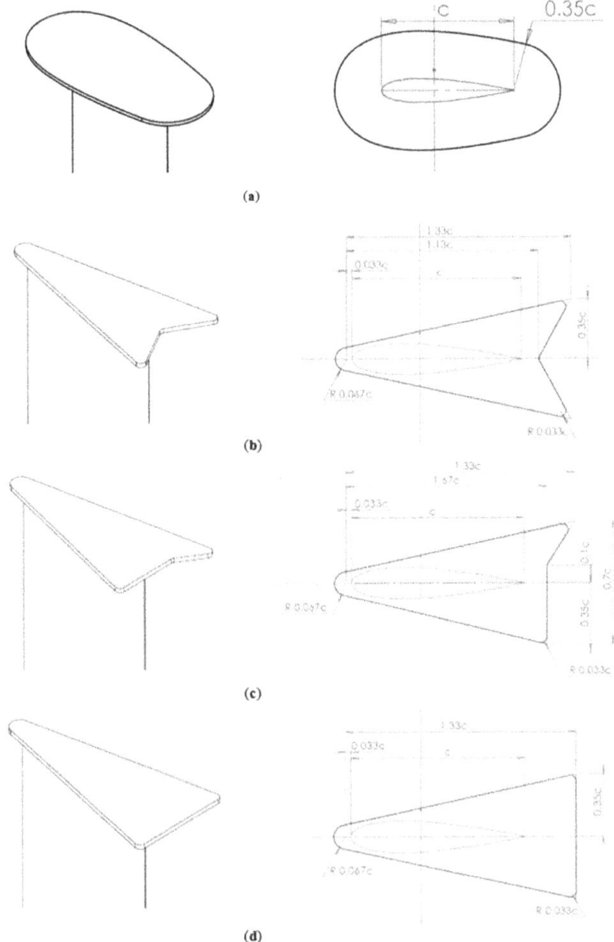

Figure 2. Endplate geometries: (**a**) 0.35*c* offset endplate; (**b**) symmetric V endplate; (**c**) asymmetric endplate; (**d**) triangular endplate.

2.4. Sub-Domain of CFD and Boundary Conditions

The CFD simulation model consists of a cylindrical VAWT rotor and a rectangular tunnel domain. The tunnel domain is modelled with a dimension of 16*D* (length) × 10*D* (width) × 10*D* (height), and the surfaces on the side are set symmetrically. The outlet gauge pressure is set at 0 Pa, and the turbulence intensity and length scale are set to 0.5% and 10.5 mm, respectively. In order to replicate the results obtained from the wind tunnel experiment, the incoming wind is set to be uniform and has a velocity of 6 m/s, as reported by Watanabe et al. [25]. Meanwhile, the rotor domain, located at 6*D* from the inlet, is modelled with a cylindrical enclosure with an offset of 0.1 m on the side and 0.25 m on the top and bottom of the blades/endplates. This is to ensure sufficient space for a gradual transition of mesh between the blade and the boundaries as well as to maintain a higher mesh quality during the meshing stage. The rotor domain is set to rotate at $\omega = 17.14$ rad/s and $\omega = 34.29$ rad/s to achieve TSRs 1 and 2, respectively, for this study. The blockage factor, which is the ratio of the rotor frontal area to the tunnel cross-sectional area, in the simulation is only 1% for the baseline VAWT, which falls within the acceptable range of less than 6–7.5% for numerical simulations [8]. The three boundaries between the rotor and tunnel domain were regarded as interfaces to allow cell zones to connect from one mesh to another using the sliding mesh method. Both sub-domains and their respective boundary conditions applied are shown in Figure 3.

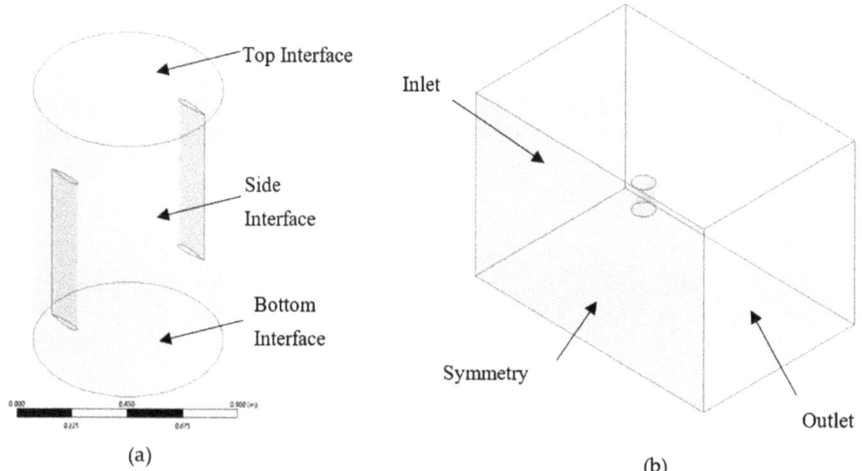

Figure 3. Sub-domains in current CFD simulation. (**a**) Rotor domain; (**b**) tunnel domain.

2.5. Solver Settings

A commercially available simulation software that is based on the finite volume method, Ansys Fluent, is adopted in this study to carry out all the numerical simulations on a VAWT. The pressure-based solver includes relevant variables, such as momentum and pressure, to be taken as the primary variables [28]. The absolute velocity formulation is adopted, while time dependency is set to transient. The current simulation utilises the k-ω SST (shear stress transport) turbulence model [29], which is widely adopted for the simulation of VAWTs [2,9,16,18,23,30,31]. This model combines the standard k-ω model at near-wall regions and the k-ε model at regions beyond the boundary layer [28]. In Gosselin et al.'s [9] investigation, they concluded that the k-ω SST model offers a better dynamic stall representation while simulating a relatively lower cost compared to transition SST. Under the reference value section, the solver is set to compute from the inlet, with the rotor domain being set as the reference zone. This happens while the pressure–velocity coupling runs with the default Semi-Implicit Method for Pressure Linked Equation (SIMPLE) scheme in the solution methods section. The solution gradient was set to least square cell-based with standard pressure. The momentum, turbulent kinetic energy, and the specific dissipation rate were set to a second-order upwind scheme, and the transient formulation was set to a second-order implicit method to obtain results with better accuracy. Default values were maintained for the variables in the under-relaxation factors as well as the fluid settings, where the air density and viscosity are kept at 1.225 kg/m^3 and 1.7894×10^{-5} kg/ms, respectively. The absolute convergence criteria for all residuals are set to 1×10^{-5} to attain species balance [28]. The average C_T achieved by the turbine blades over one revolution is extracted from the simulation and converted into C_P using Equation (4). The simulation continues until the percentage difference of C_P between two consecutive revolutions, calculated using Equation (5), as shown below, is less than 1%.

$$\frac{C_{P,ave(n+1)} - C_{P,ave(n)}}{C_{P,ave(n)}} \times 100\% < 1\% \tag{5}$$

As the rotor domain rotates around its axis during the simulation, a suitable timestep is crucial to ensure the stability of a simulation and achieve convergence in the results. The current simulation adopted a 1° rotation per timestep (equivalent to 1.01811×10^{-3} s per time step and 5.0905×10^{-4} s per time step for TSRs 1 and 2, respectively), which is widely used in VAWT simulations to achieve a balance in result accuracy and simulation time [30]. At each timestep, the simulation will run for a maximum of 20 iterations before proceeding to the next timestep. To analyse the aerodynamic characteristic at different azimuthal angles, the data is saved every 30 timesteps once the results have reached convergence. Before starting the simulation, the solution is initialised with the standard initialisation, which computes from the inlet.

2.6. Model Validation and Mesh Independent Test

All configurations of the rotor are modelled using the SolidWorks software before being imported into the Design Modeller to create the cylindrical rotor domain. For capturing the flow in the rotor domain, the mesh is made to be very dense, with fine elements near the blade surfaces. Face meshing is applied to the blade span surface to obtain a structured mesh. Meanwhile, an inflation layer is applied in the near-wall region to further refine the mesh and capture the boundary layer gradients. The first layer thickness is set to 0.1 mm, with a maximum of 15 layers and a thickness growth rate of 1.2, to achieve a y^+ value of around 1. The y^+ value, as defined below, is a dimensionless parameter that can be used to classify types of boundary layers.

$$y^+ = \frac{u \cdot y}{v} \tag{6}$$

where u is the shear velocity at the nearest wall, y is the absolute cell distance from the nearest wall and v is the fluid local kinematic viscosity. When the y^+ value is less than 5, it is considered a laminar sub-layer [30,32].

In the meantime, the tunnel domain is modelled and meshed separately. The mesh size at the tunnel side interface is set to 10 mm to match the sizing on the rotor interface for a better connection between the two meshes. The generated rotor and tunnel mesh are appended into Ansys Fluent, with the rotor domain being situated within the tunnel domain. There is no overlapping of meshes where both meshes are connected through their respective interfaces.

The refinement of mesh is conducted on the blade surface to ensure accurate results are obtained, as shown in Figure 4, where 85% of the cells are found in the rotor domain in all cases.

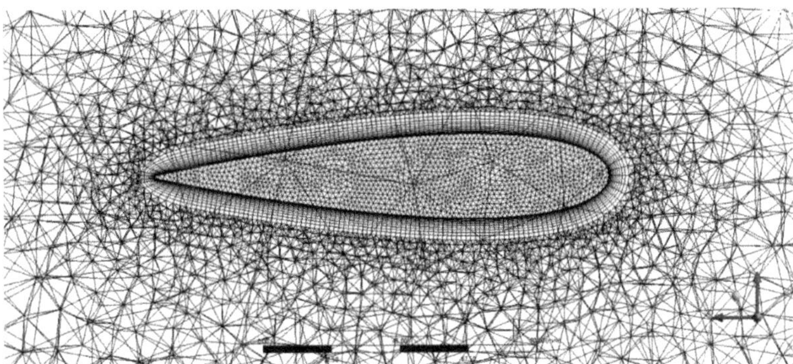

Figure 4. Sectional view of mesh around the blade surface in the rotor domain.

The baseline VAWT model is first validated with the experimental data by Watanabe et al. [25], who studied the same VAWT configuration as the current VAWT model. The VAWT model is simulated over a range of TSRs, from 0.5 to 2.5, with an inlet velocity of 6 m/s, and the result is compared with the experimental data, as illustrated in

Figure 5. The current model replicates the overall trend of the experimental data well, with a slight underestimation of C_P at TSRs 1 to 2. Discrepancies are observed at TSRs 0.75 and 2.5, with a percentage error of around 9–10%. Nevertheless, the VAWT model can produce satisfactory results over the TSR range; hence, the simulation is continued with the mesh independence test.

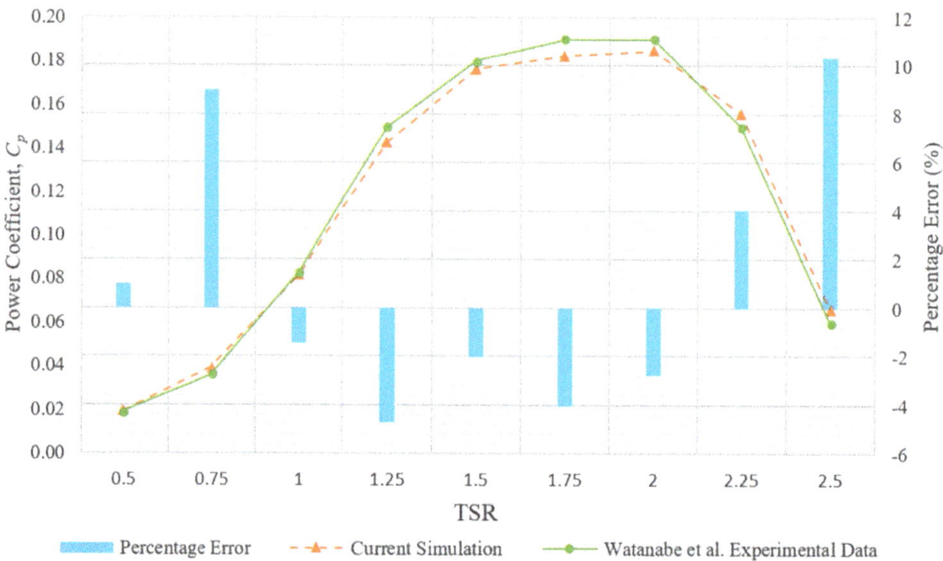

Figure 5. Model validation with experimental data from Watanabe et al. [25].

After establishing an appropriate VAWT model, opting for a fine and dense mesh is desirable to capture more details. However, this demands greater computational power, longer simulation time as well as more storage space. Therefore, a mesh independence test is carried out to obtain a mesh domain that can achieve a balance between accuracy and computational cost. In the mesh independence test, the cell size on the blades is manipulated to observe the effect on the blade torque coefficient while the tunnel mesh remains unchanged. Five meshes with cell sizes of 5, 3, 2, 1.5 and 1.2 mm are generated, corresponding to Meshes A, B, C, D, and E, respectively, as shown in Table 2. Note that in the model validation, the cell size on the blade is set as 2 mm. The detail of each mesh, together with the cycle-averaged power coefficient ($C_{P,ave}$) and the computational time required to complete a revolution, is tabulated in Table 2. The test is carried out at TSR 2, which is the maximum C_P from the experimental data, and a PC equipped with an AMD Ryzen Threadripper 3960X 24-core processor and 64 GB RAM is utilised to run all the simulations.

Figure 6 shows that as the blade element size is refined from 5 to 1.2 mm, the number of cells in the rotor mesh increases from 1.85 to 12.8 million cells. Mesh A shows the largest discrepancy, with a 10% error, while the C_P results converge closer to the experimental data as the number of elements increases. The C_P prediction from the finest mesh, Mesh E, is the closest to the experimental data, with only a 0.35% difference. However, the increase in accuracy is accompanied by a significantly longer computational time. Mesh D managed to return satisfactory results, with a much shorter computational time, with the discrepancy between Meshes D and E being only 0.95%. Therefore, the blade element size of 1.5 mm will be applied to all VAWT configurations investigated in this study.

Table 2. Detail of meshes.

Mesh	Blade Mesh Size (mm)	Rotor Mesh Elements (Million)	Computational Time Per Cycle (hrs)	$C_{P,ave}$	Percentage Error with Consecutive Cycle (%)
A	5.0	1.85	6	0.1622	9.23
B	3.0	3.14	8	0.1787	3.30
C	2.0	5.60	10	0.1848	1.44
D	1.5	8.30	12	0.1875	0.95
E	1.2	12.80	19	0.1893	-

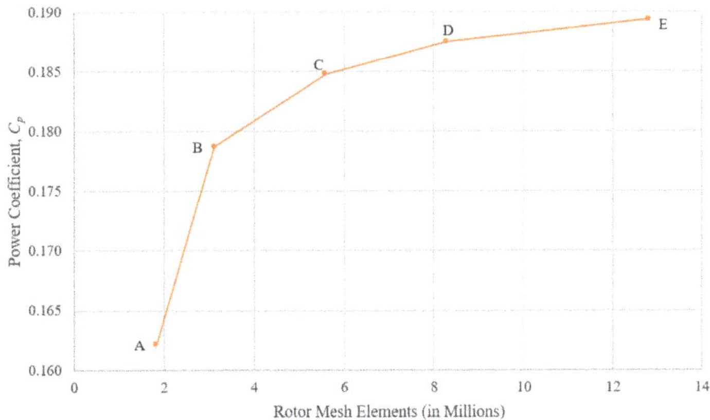

Figure 6. Cycle-averaged C_P at TSR 2 for different blade mesh sizes.

3. Results and Discussions

3.1. Overall C_P Performance

The overall C_P performance is a prime indicator of the efficiency of a wind turbine in extracting wind energy. Therefore, the effectiveness of each endplate geometry can be reflected by comparing its overall C_P performance with the baseline turbine. Table 3 presents the cycle-averaged C_P attained from the CFD simulation for TSRs 1 and 2 for all VAWT configurations. Although the offset endplate had a similar configuration as in [18], it failed to replicate the positive results attained from the literature. Note that supporting struts, which are placed at $4c$ from the top end, are included in [18], while they are absent in the current case. In their study, a 4.25% power enhancement is observed at TSR 3. For TSR 2, the performance difference between their endplate and baseline configurations is negligible, but the supporting strut is relocated to the top end of the blade with the endplate in this case. In the current study, the overall $C_{P,ave}$ showed -22.24% and -0.64% degradation compared to the baseline turbine at TSRs 1 and 2, respectively. Even though direct comparisons cannot be made, the deviation in results may be attributed to the smaller aspect ratio and turbine size adopted in the current simulation. The deterioration of $C_{P,ave}$ is also observed in [12] below TSR 2 for a flat offset endplate with a smaller $0.18c$ offset. Other endplate geometries managed to elevate the overall $C_{P,ave}$ over the baseline turbine in both TSRs, albeit the underwhelming improvement being less than 2%. The marginal performance improvement is due to the endplate mainly improving the flow of a small section near the blade tip, while most of the torque is driven by the remaining blade span. The asymmetric and symmetric V endplates have the best performance in TSRs 1 and 2, respectively.

Table 3. Cycle-averaged C_P performance at TSRs 1 and 2.

VAWT Configuration	TSR 1		TSR 2	
	$C_{P,ave}$	Percentage Improvement (%)	$C_{P,ave}$	Percentage Improvement (%)
Baseline	0.0805	-	0.1875	-
Offset	0.0626	-22.24	0.1863	-0.64
Symmetric V	0.0815	1.31	0.1908	1.73
Asymmetric	0.0820	1.88	0.1902	1.42
Triangular	0.0807	0.27	0.1907	1.68

As shown in Figure 7, the contribution from two turbine blades leads to two C_P maxima and minima in one revolution for the total C_P of the VAWT. All endplate configurations managed to raise the maxima compared to the baseline turbine in both TSRs, and this suggests that the reduction of tip loss effect takes place in this region. The offset endplate showed the highest maxima, while the performance gap between the asymmetric, symmetric V, and triangular endplates is negligible. In TSR 1, the offset endplate showed a sharp decrease in C_P beyond the maxima (i.e., $90° < \theta < 120°$ and $270° < \theta < 330°$), as shown in Figure 7a. Meanwhile, the remaining three endplate configurations closely followed the baseline turbine, with a slightly lower C_P. From Figure 7b, the offset endplate lowers the minima, but the remaining three endplate configurations perform similarly to the baseline turbine. Besides that, all endplate configurations return slightly lower C_P when the blades move toward the upwind direction (i.e., $20° < \theta < 60°$ and $190° < \theta < 240°$) in TSR 2. The C_P reduction in this region is more significant for the offset endplate, while the symmetric V endplate performs more closely to the baseline turbine. It is also observed that the maxima shifts from $70°$ to $100°$ when the TSR increases from 1 to 2.

To investigate the performance differences between each endplate configuration and the baseline turbine, a vorticity diagram is plotted, as illustrated in Figure 8, with the vorticity level set at 0.0005. The azimuthal angle of Blade 1 is set as $\theta = 60°$ and $\theta = 90°$ for TSRs 1 and 2, respectively, as it corresponds to the location right before Blade 1 reaches the peak of C_P maxima, as presented in Figure 7. Overall, the vorticity curl is higher at TSR 2 than at TSR 1 across the board. In TSR 1, vortex shedding is observed in all VAWT configurations. The baseline turbine leaves a long trailing vortex behind both ends of Blade 1; however, the vortex is shredded into smaller structures in all endplate configurations. The vortex structure between the blades is the largest with the offset endplate. Hence, the interaction of the blade with the large vortex structure, as shown in Figure 9, may explain the sharp power degradation at around $\theta = 90°$. In TSR 2, a trailing vortex is present in all VAWT configurations, and the offset endplate once again has the largest vortex structure. A pair of trailing vortex structures is also present at both ends of the turbine, unlike in TSR 1, where it travels closer to the blade centre.

3.2. Blade 1 Performance

In this section, analysis is done on Blade 1 to understand the VAWT behaviour in the presence of different endplate geometries. From Figure 10, the trend is similar to the results found in the overall C_P in both TSRs. The offset endplate attains the highest maxima and is followed by the asymmetric, symmetric V, and triangular endplates, which perform similarly to one another. In the downwind region, the performance of all endplate configurations is slightly inferior compared to the baseline turbine in both TSRs, especially the offset endplate. At TSR 1, although the offset endplate outperforms at maxima ($\theta = 70°$), the sharp drop of C_P at $80° < \theta < 140°$ results in the total performance being worse than the baseline. This is because a greater suction pressure loss happens on the offset endplate; this is probably caused by the wide endplate region on the leading edge. As shown in Figure 11, the baseline and the symmetric V endplates possess similar pressure distributions at the suction side of the blade near the trailing edge. However, the pressure loss is higher along the trailing edge of the offset endplates, which leads to a lower pressure difference between the blade surfaces and causes a sudden drop in the C_P.

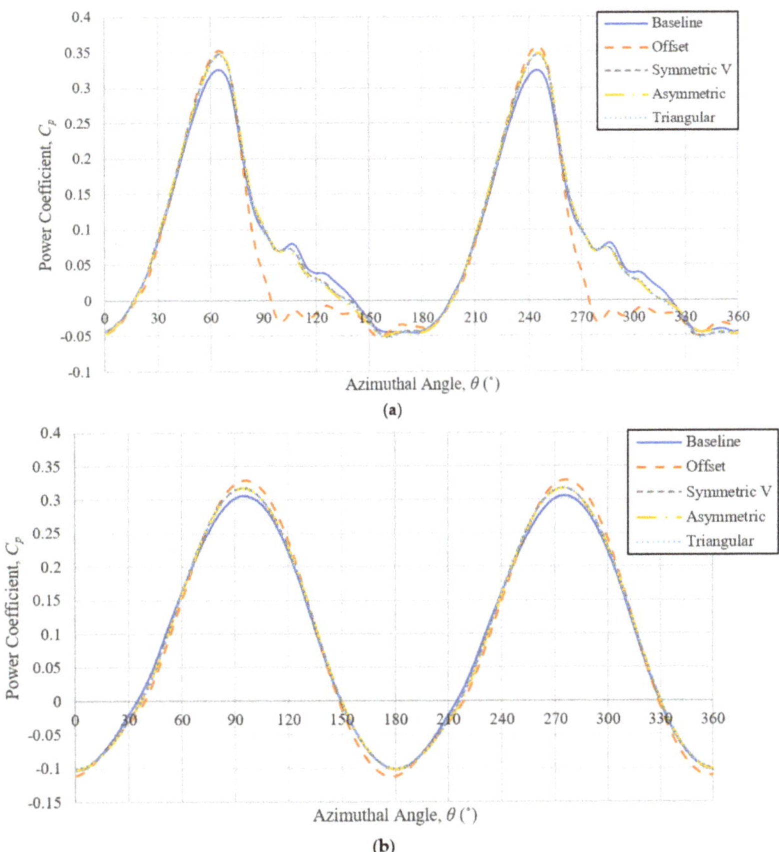

Figure 7. Overall C_P vs. azimuthal angle at (**a**) TSR 1 and (**b**) TSR 2.

As the C_P performance varies with the distance from the blade tip [18], the surface of Blade 1 is sectioned at different blade heights to observe the variation of C_P against blade height. From Table 4, the presence of an endplate greatly improves the power coefficient at the blade tip region in both TSRs 1 and 2. The cycle-averaged C_P increases significantly, beyond 1410%, in TSR 1 compared to the baseline turbine at $0.015625H$. In TSR 2, the asymmetric, triangular, and, especially, offset endplates invert the negative C_P to a positive value. Although the symmetric V endplate shows a slight negative C_P at the blade tip, nevertheless, it is still an improvement compared with the baseline turbine. These results indicate that endplates are very effective at minimising the tip loss effect at the blade tip region.

From Figure 12, the C_P enhancement at the blade tip region is mainly contributed by the higher maxima achieved in the upwind region compared to the baseline turbine. In both TSRs, the offset endplate attains the highest maxima and is followed by the asymmetric, triangular, and symmetric endplates. Although the baseline turbine underperforms in the upwind region, especially in TSR 2, its performance in the downwind region is slightly better than the endplate configurations. Compared to TSR 1, the performance gap between the offset endplate and other endplate configurations in TSR 2 is much more significant. A second smaller local maxima is observed in Figure 12b, where the offset endplate returns a higher C_P than other endplate configurations. It is also observed that the peak of C_P takes place at different azimuthal angles, depending on the endplate geometry in TSR 1. The offset endplate reaches its peak at $\theta = 70°$ and is followed by the other three endplates at $\theta = 100°$ and the baseline at $\theta = 110°$.

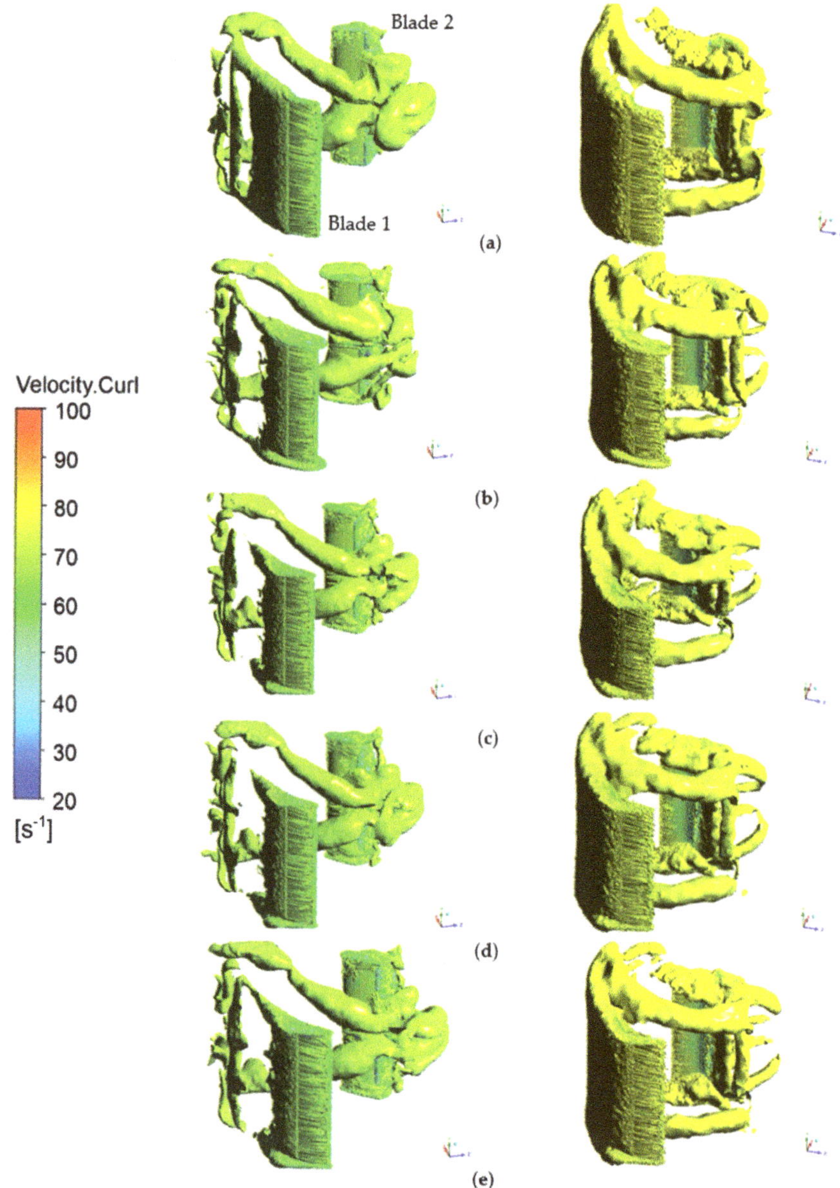

Figure 8. Vorticity of (**a**) baseline, (**b**) offset, (**c**) symmetric V, (**d**) asymmetric, and (**e**) triangular endplates at $\theta = 60°$ in TSR 1 (left column) and $\theta = 90°$ in TSR 2 (right column).

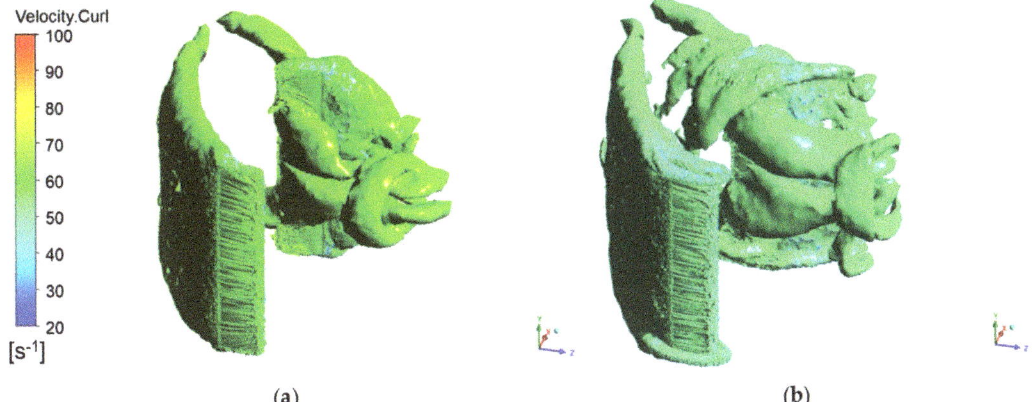

Figure 9. Vorticity of (**a**) baseline; (**b**) offset endplate at $\theta = 90°$ in TSR 1.

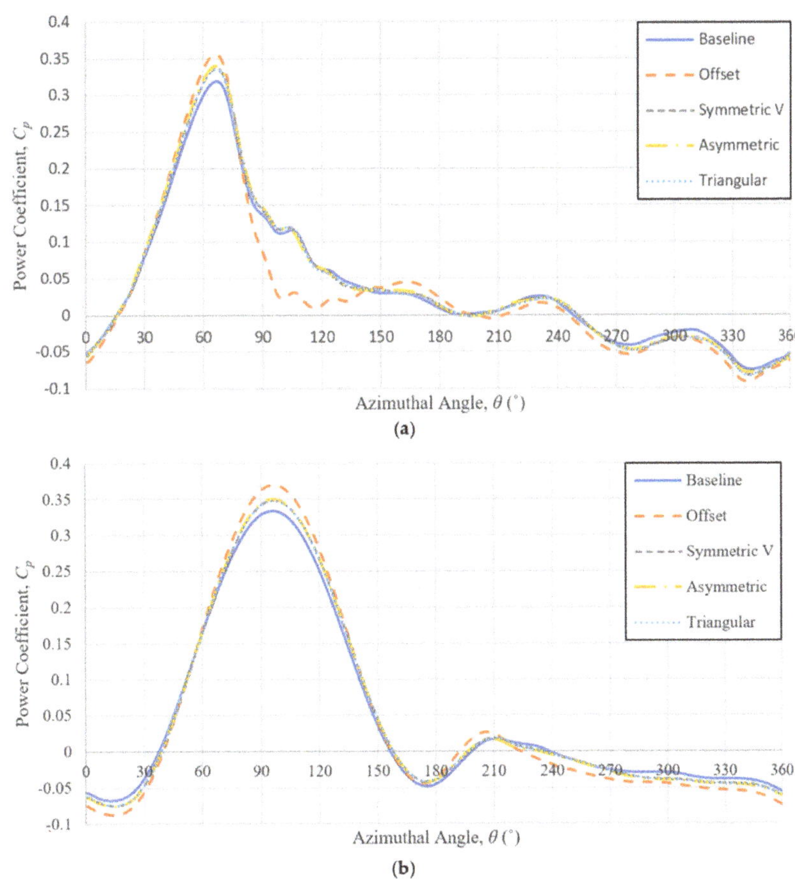

Figure 10. Blade 1 C_P vs. azimuthal angle at (**a**) TSR 1 and (**b**) TSR 2.

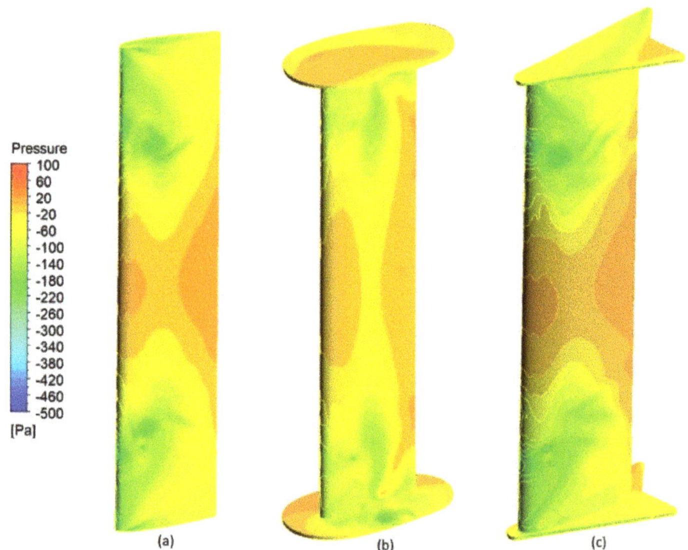

Figure 11. Pressure distribution on the suction surface of (**a**) baseline; (**b**) offset; (**c**) symmetry V endplate at $\theta = 120°$.

Table 4. Cycle-averaged C_P at 0.015625H in TSRs 1 and 2.

VAWT Configuration	TSR 1		TSR 2	
	$C_{P,ave}$	Percentage Improvement (%)	$C_{P,ave}$	Percentage Improvement (%)
Baseline	0.0020	-	−0.1409	-
Offset	0.0304	1410.53	0.0897	163.68
Symmetric V	0.0331	1542.47	−0.0027	98.10
Asymmetric	0.0331	1542.64	0.0036	102.54
Triangular	0.0353	1650.82	0.0077	105.50

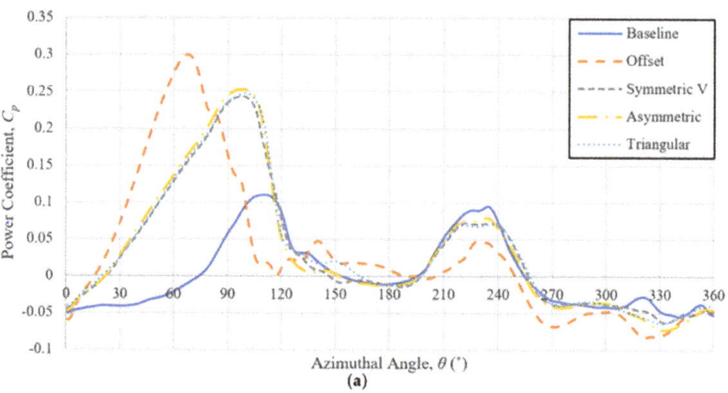

Figure 12. *Cont.*

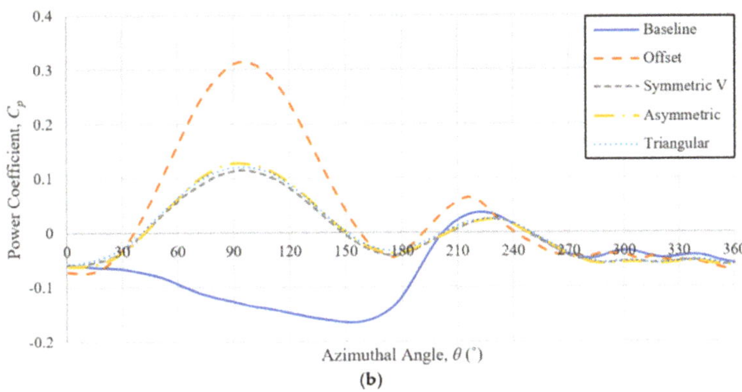

Figure 12. Comparison of C_P at 0.015625H of Blade 1 in (**a**) TSR 1 and (**b**) TSR 2.

Beyond the blade tip, the C_P difference is expected to decrease towards the mid-blade span for all endplate configurations; similar results are presented in [18,33]. This is because the tip loss effect is not significant at the blade centre. However, the results in TSR 1 start to deviate from the previous literature [18,33] beyond 0.125H. As illustrated in Figure 13a, the C_P at 0.25H to 0.5H drops below the level from 0.03125H to 0.125H. Dynamic stall at the low TSR might give rise to a dispute of the results. Besides that, all endplate configurations fail to improve their performance at 0.25H. Although the offset endplate shows inferior performance from 0.0625H to 0.25H, it manages to overcome other configurations at 0.5H.

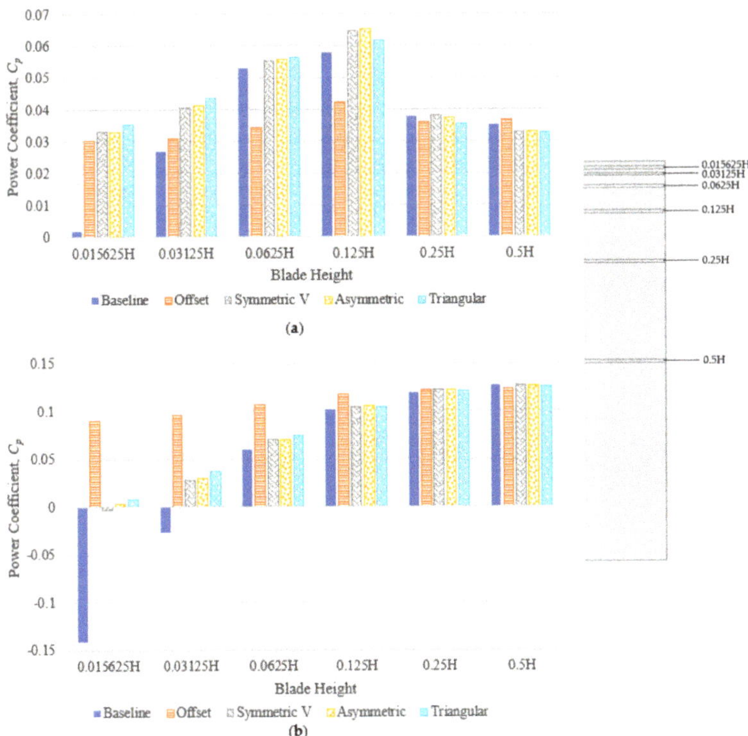

Figure 13. Breakdown of C_P contribution on Blade 1 at different blade heights in (**a**) TSR 1 and (**b**) TSR 2.

The results from TSR 2, presented in Figure 13b, agree well with the previous literature [18,33]. The offset endplate effectively enhances blade performance, especially at the blade tip region. Concurrently, the remaining three endplate configurations show smaller performance improvements before they overcome the offset endplate beyond 0.25H.

To gain further insight into the results, the pressure contour on Blade 1's surface is extracted on $\theta = 60°$ and $\theta = 90°$ for TSRs 1 and 2, respectively, as illustrated in Figures 14 and 15. This corresponds to the location right before Blade 1 reaches the peak of C_P maxima at TSRs 1 and 2. At the blade tip region, the pressure distribution on the baseline turbine is uneven on both sides of Blade 1. The loss of pressure on the pressure side leads to a degraded C_P output near the blade tips. The uniformity of pressure distribution is improved with the presence of the endplate in both TSRs. Notably, the offset endplate offers better uniformisation at the region around the leading edge than all other configurations. This is due to the offset endplate having a larger area allocated around the leading edge in contrast with other endplate configurations that have a pointed design that covers a much smaller area. Thus, the offset endplate is more effective in blocking the spanwise flow, as indicated by the uniform pressure contour around the blade tip region. The pressure distribution around the trailing edge is also enhanced over the baseline turbine, and the results are similar among all endplate configurations.

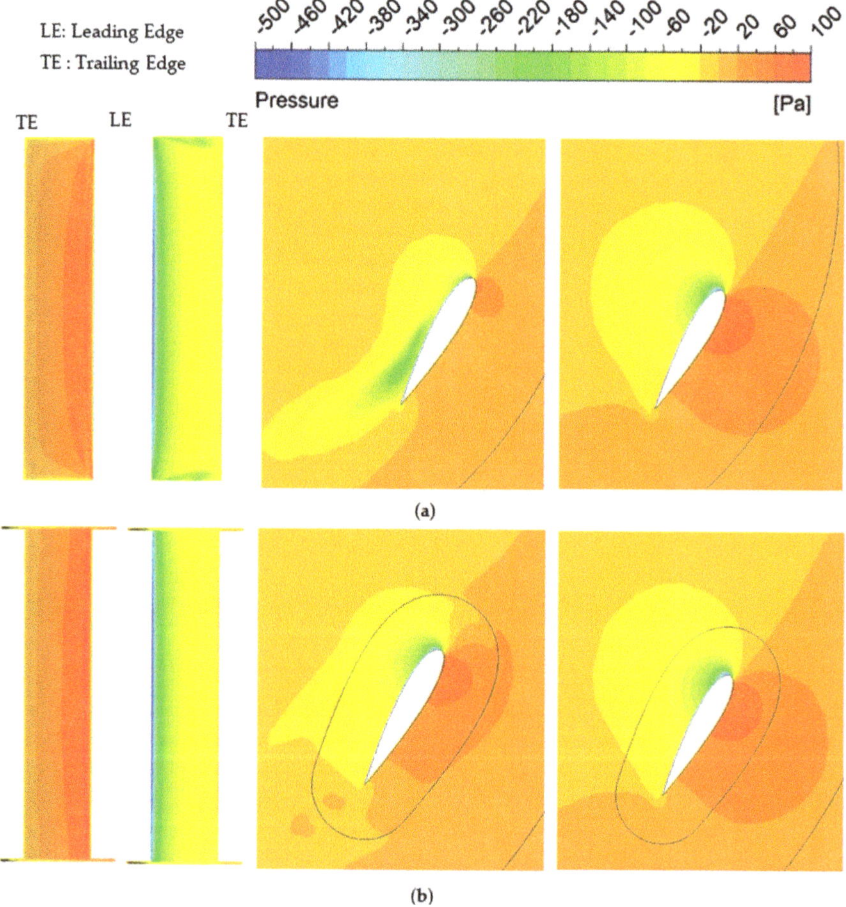

Figure 14. *Cont.*

Figure 14. From left to right is the pressure distribution at the pressure side, suction side, at 0.015625H and 0.5H of Blade 1 at $\theta = 60°$ and TSR 1 for (**a**) baseline, (**b**) offset, (**c**) symmetric V, (**d**) asymmetric and (**e**) triangular endplates.

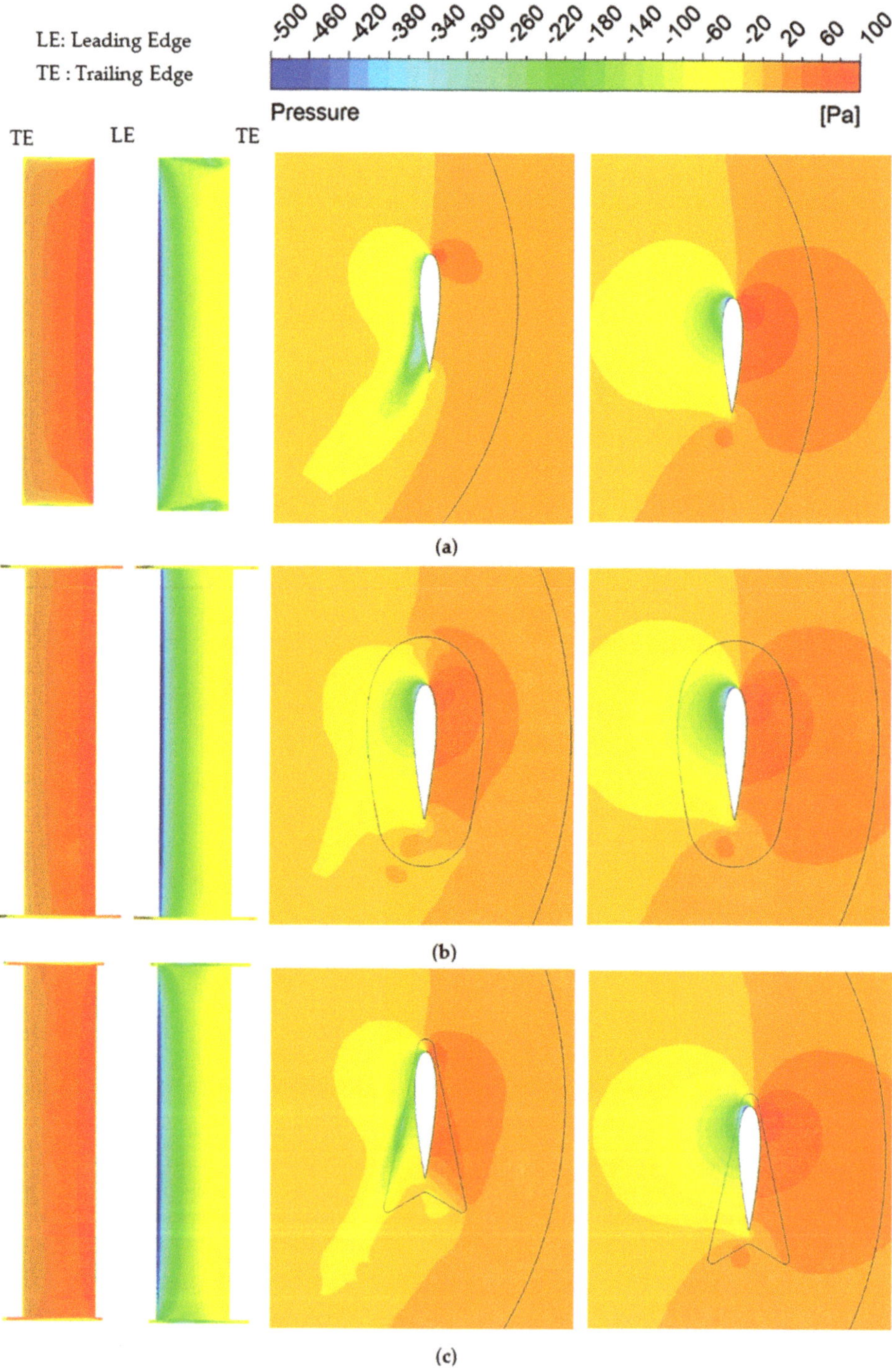

Figure 15. *Cont.*

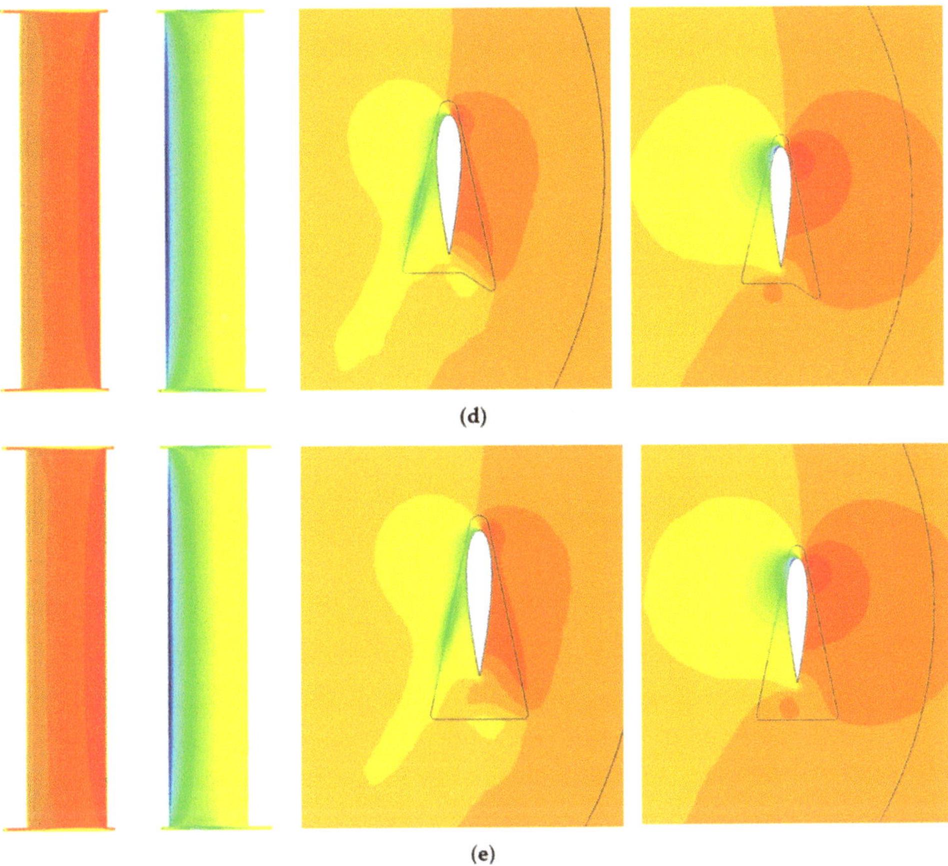

Figure 15. From left to right is the pressure distribution at the pressure side, suction side, at 0.015625*H* and 0.5*H* of Blade 1 at $\theta = 90°$ and TSR 2 for (**a**) baseline, (**b**) offset, (**c**) symmetric V, (**d**) asymmetric and (**e**) triangular endplates.

A cut made on 0.015625*H* revealed a low-pressure region that spans across the chord length of the baseline turbine blade on the suction side in both TSRs. A similar observation is found in symmetric V, asymmetric, and triangular endplate configurations, with the exception of the low-pressure region detaching from the blade surface, unlike the baseline turbine. Such extension of the low-pressure region is absent in the offset endplate. On the pressure side, all endplates extended the high-pressure region towards the trailing edge. This increases the pressure difference between the pressure and suction sides at the blade tip and ultimately improves the C_P in that region. Besides that, different endplates also result in a slight pressure variation at the region behind the trailing edge. At 0.5*H* or the blade centre, the presence of the endplate shows a minimal difference.

3.3. Endplate Performance

A velocity streamline is plotted in Figure 16 for Blade 1 at $\theta = 60°$ and $\theta = 90°$ for TSRs 1 and 2, respectively, to visualise the flow around the blade tip. After the wind passes through the blade, the streamlines become uneven and spread wider, especially at TSR 2. The turbine performance will be adversely affected when the blade encounters unsteady wind in the downwind region. A swirling vortex is generated at the tip of the baseline turbine for both

TSRs, but this phenomenon is suppressed with the implementation of endplates. For TSR 2, the flow is also being deflected slightly upwards after passing Blade 1.

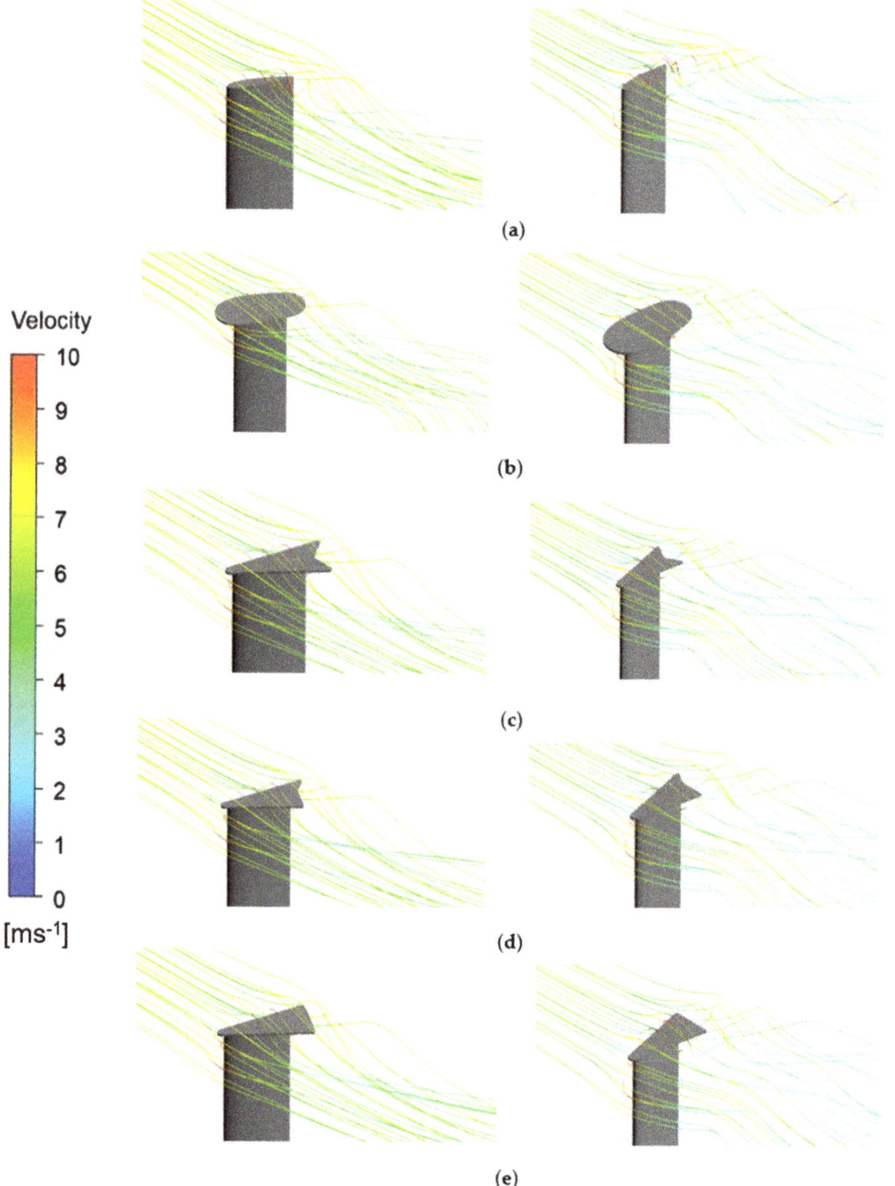

Figure 16. Velocity streamlines across the (**a**) baseline, (**b**) offset, (**c**) symmetric V, (**d**) asymmetric and (**e**) triangular endplates of Blade 1 at $\theta = 60°$ in TSR 1 (left column) and $\theta = 90°$ in TSR 2 (right column).

Although the implementation of endplates is bound to improve the overall performance of a VAWT [16], the additional surface area from the endplates will introduce more drag [9]. Figure 17a,b illustrate the C_P variation of the endplate throughout one revolution in TSRs 1 and 2, respectively. To the best of the authors' knowledge, such results have not been reported in the previous literature. In both cases, the offset endplate encounters the largest C_P reduction. This can be correlated with its surface area, which is at least two times greater than the remaining endplate configurations. However, the C_P penalty is significantly reduced at around $\theta = 190°$. At TSR 1, the offset experiences the largest C_P reduction at $\theta = 345°$, while the rest takes place at $\theta = 30°$. As for TSR 2, the largest drag occurs at $30°$ for all endplate configurations, while minimum drag is observed at $200° < \theta < 240°$. The symmetric V, asymmetric, and triangular endplates perform similarly at TSR 2; however, the symmetric V endplate performs slightly better throughout most of the cycle. A larger discrepancy is observed in TSR 1, and these endplate configurations manage to return a positive C_P at $110° < \theta < 260°$.

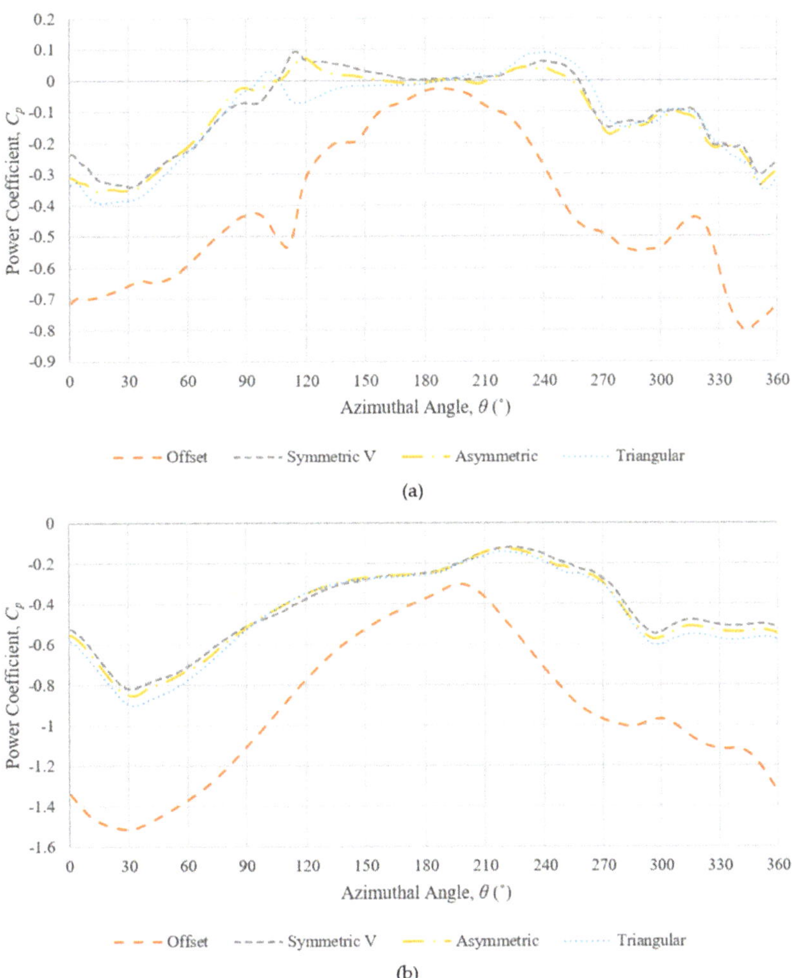

Figure 17. C_P on the endplate surface vs. azimuthal angle at (**a**) TSR 1 and (**b**) TSR 2.

The gain of C_P on the blade itself due to the endplate must overcome the drag introduced in order to achieve overall improvement. Tables 5 and 6 tabulate and compare the $C_{P,ave}$ on the effective blade surface against the net $C_{P,ave}$ generated by Blade 1 overall. The percentage reduction indicates the percentage of $C_{P,ave}$ reduced due to losses from the endplate. At the same time, the percentage improvement and net percentage improvement indicate the improvement over the baseline turbine in the effective blade surface and overall blade, respectively.

Table 5. Breakdown of Blade 1 C_P in TSR 1.

VAWT Configuration	$C_{P,ave}$ on Effective Blade Surface	Net Blade 1 $C_{P,ave}$	Percentage Reduction (%)	Improvement on Effective Blade Surface (%)	Net Percentage Improvement (%)
Baseline	0.0403	0.0403	0	-	-
Offset	0.0364	0.0311	14.56	-9.58	-22.74
Symmetric V	0.0421	0.0409	2.87	4.42	1.43
Asymmetric	0.0425	0.0411	3.22	5.38	1.98
Triangular	0.0415	0.0401	3.43	2.97	-0.56

Table 6. Breakdown of Blade 1 C_P in TSR 2.

VAWT Configuration	$C_{P,ave}$ on Effective Blade Surface	Net Blade 1 $C_{P,ave}$	Percentage Reduction (%)	Improvement on Effective Blade Surface (%)	Net Percentage Improvement (%)
Baseline	0.0939	0.0939	0	-	-
Offset	0.1164	0.0925	20.54	23.97	-1.49
Symmetric V	0.1062	0.0954	10.16	13.13	1.64
Asymmetric	0.1062	0.0950	10.54	13.13	1.21
Triangular	0.1066	0.0949	10.97	13.55	1.09

All endplate configurations on Blade 1 generated negative $C_{P,ave}$ in both TSRs, and the C_P compensated in TSR 2 is greater than TSR 1, as indicated by the larger percentage reduction across the board. Except for the offset endplate in TSR 1, all endplate configurations elevated the $C_{P,ave}$ generated on the effective blade surface. In TSR 1, the offset endplate experiences a 22.74% net performance reduction as it fails to return improvement on the blade in the first place. Although it manages to improve the blade performance by 23.97% in TSR 2, the large drag induces compensates for all the performance enhancement on the turbine blades and, ultimately, leads to a negative value being obtained, contributing to the low overall C_P output. Other endplates manage to maintain a 1.09% to 1.98% improvement after accounting for the endplate losses, except for the triangular endplate at TSR 1. The symmetric V endplate consistently has the lowest drag induced, and the enhancement it brings to the blade surface is in between the asymmetric and triangular endplates. Hence, the symmetric endplate is considered the best design among the four endplates as it performs the best at TSR 2, which is the optimum TSR for this VAWT.

4. Conclusions

In summary, the implementation of well-designed endplate geometry can have a positive effect on the C_P performance of a VAWT. From a CFD simulation, the symmetric V and asymmetric endplates are the best performers in TSRs 1 and 2, respectively, although the overall C_P enhancement is less than 2%. At the blade tip region, endplates can result in a performance boost of up to 163.68% and 1650.82% in TSRs 1 and 2, respectively. Besides that, the endplate effectively blocks the spanwise flow and improves pressure uniformisation near the blade tip. Thus, the loss of pressure difference across the pressure and suction sides of the blade can be minimised. The swirling vortex generated by the baseline turbine at the open-ended blade tip also can be suppressed with endplates. Endplates can effectively improve the C_P performance on the blade surface; however, the induced drag limits the net power improvement. Although the offset endplate offers the best performance enhancement on the blade surface in TSR 2, the large drag induced by the large surface area degrades the VAWT's overall performance. On the other hand, the induced drag from the symmetric V endplate is the lowest among the four endplates tested, and it performs the best at the optimum TSR. Hence, it is considered the best overall design in the current study. Further investigation is still required to explore more endplate geometries as the current study has only focused on four different geometries. This study has only focused on the flow characteristics; other factors such as manufacturing complexity, material strength, and cost are not considered while implementing the endplate geometries for this CFD simulation. An endplate design that can achieve a balance between performance and cost is crucial for minimising the cost of energy (COE) of a VAWT. Therefore, this research contributes to the study of the optimisation of endplate parameters to further enhance the power generation capability of VAWTs.

Author Contributions: Conceptualisation, J.-H.N., K.-H.W. and S.M.; methodology, S.-K.U. and K.-H.W.; validation, S.-K.U. and W.-T.C.; formal analysis, S.-K.U.; investigation, S.-K.U. and K.-H.W.; resources, Y.-H.K. and K.-H.W.; data curation, S.-K.U.; writing—original draft preparation, S.-K.U.; writing—review and editing, Y.-H.K. and K.-H.W.; visualisation, S.-K.U.; supervision, K.-H.W.; project administration, K.-H.W.; funding acquisition, J.-H.N., W.-T.C.; S.M. and K.-H.W. All authors have read and agreed to the published version of the manuscript.

Funding: This research was funded by the Ministry of Higher Education Malaysia under the Fundamental Research Grant Scheme (FRGS), grant number FRGS/1/2020/TK0/USMC/0203. The authors would like to thank Dana Pengukuhan UTM Aerolab.

Acknowledgments: The authors would like to acknowledge all support given by the university and research intern students.

Conflicts of Interest: The authors declare no conflict of interest. The funders had no role in the design of the study; in the collection, analyses, or interpretation of data; in the writing of the manuscript; or in the decision to publish the results.

References

1. GWEC. Global Wind Report 2022. Global Wind Energy Council. 2022. Available online: https://gwec.net/global-wind-report-2022/ (accessed on 10 July 2022).
2. Malla, A.; Han, Z.; Zhou, D. Effect of a winglet on the Power Augmentation of Straight Bladed Darrieus Wind Turbine. *IOP Conf. Ser. Earth Environ. Sci.* **2020**, *505*, 012041. [CrossRef]
3. Tjiu, W.; Marnoto, T.; Mat, S.; Ruslan, M.H.; Sopian, K. Darrieus vertical axis wind turbine for power generation II: Challenges in HAWT and the opportunity of multi-megawatt Darrieus VAWT development. *Renew. Energy* **2015**, *75*, 560–571. [CrossRef]
4. Laín, S.; Taborda, M.A.; López, O.D. Numerical Study of the Effect of Winglets on the Performance of a Straight Blade Darrieus Water Turbine. *Energies* **2018**, *11*, 297. [CrossRef]
5. Sharma, V.; Sharma, S.; Sharma, G. Recent development in the field of wind turbine. *Mater. Today Proc.* **2022**, *64*, 1512–1520. [CrossRef]
6. Arredondo-Galeana, A.; Brennan, F. Floating Offshore Vertical Axis Wind Turbines: Opportunities, Challenges and Way Forward. *Energies* **2021**, *14*, 8000. [CrossRef]
7. Dewan, A.; Gautam, A.; Goyal, R. Savonius wind turbines: A review of recent advances in design and performance enhancements. *Mater. Today Proc.* **2021**, *47*, 2976–2983. [CrossRef]
8. Lam, H.F.; Peng, H.Y. Study of wake characteristics of a vertical axis wind turbine by two- and three-dimensional computational fluid dynamics simulations. *Renew. Energy* **2016**, *90*, 386–398. [CrossRef]
9. Gosselin, R.; Dumas, G.; Boudreau, M. Parametric study of H-Darrieus vertical-axis turbines using CFD simulations. *J. Renew. Sustain. Energy* **2016**, *8*, 053301. [CrossRef]
10. Zhang, T.T.; Elsakka, M.; Huang, W.; Wang, Z.G.; Ingham, D.B.; Ma, L.; Pourkashanian, M. Winglet design for vertical axis wind turbines based on a design of experiment and CFD approach. *Energy Convers. Manag.* **2019**, *195*, 712–726. [CrossRef]
11. Howell, R.; Qin, N.; Edwards, J.; Durrani, N. Wind tunnel and numerical study of a small vertical axis wind turbine. *Renew. Energy* **2010**, *35*, 412–422. [CrossRef]
12. Miao, W.; Liu, Q.; Xu, Z.; Yue, M.; Li, C.; Zhang, W. A comprehensive analysis of blade tip for vertical axis wind turbine: Aerodynamics and the tip loss effect. *Energy Convers. Manag.* **2022**, *253*, 115140. [CrossRef]
13. Jung, J.H.; Kim, M.J.; Yoon, H.S.; Hung, P.A.; Chun, H.H.; Park, D.W. Endplate effect on aerodynamic characteristics of three-dimensional wings in close free surface proximity. *Int. J. Nav. Archit. Ocean. Eng.* **2012**, *4*, 477–487. [CrossRef]
14. Syawitri, T.P.; Yao, Y.; Yao, J.; Chandra, B. A review on the use of passive flow control devices as performance enhancement of lift-type vertical axis wind turbines. *Wiley Interdiscip. Rev. Energy Environ.* **2022**, *11*, e435. [CrossRef]
15. Johansen, J.; Sørensen, N. Numerical Analysis of Winglets on Wind Turbine Blades using CFD. In Proceedings of the in European Wind Energy Conference, Milan, Italy, 7–10 May 2007.

16. Amato, F.; Bedon, G.; Castelli, M.R.; Benini, E. Numerical Analysis of the Influence of Tip Devices on the Power Coefficient of a VAWT. *Int. J. Aerosp. Mech. Eng.* **2013**, *7*, 1053–1060.

17. Mousavi, M.; Masdari, M.; Tahani, M. Power performance enhancement of vertical axis wind turbines by a novel gurney flap design. *Aircr. Eng. Aerosp. Technol.* **2022**, *94*, 482–491. [CrossRef]

18. Jiang, Y.; He, C.; Zhao, P.; Sun, T. Investigation of Blade Tip Shape for Improving VAWT Performance. *J. Mar. Sci. Eng.* **2020**, *8*, 225. [CrossRef]

19. Premkumar, T.M.; Sivamani, S.; Kirthees, E.; Hariram, V.; Mohan, T. Data set on the experimental investigations of a helical Savonius style VAWT with and without end plates. *Data Brief* **2018**, *19*, 1925–1932. [CrossRef]

20. Kassab, S.Z.; Chemengich, S.J.; Lotfy, E.R. The effect of endplate addition on the performance of the savonius wind turbine: A 3-D study. *Proc. Inst. Mech. Eng. Part A J. Power Energy* **2022**, 09576509221098480. [CrossRef]

21. Daróczy, L.; Janiga, G.; Thévenin, D. Optimization of a winglet for improving the performance of an H-Darrieus turbine using CFD. In Proceedings of the 16th International Symposium on Transport Phenomena and Dynamics of Rotating Machinery, Honolulu, HI, USA, 10–15 April 2016.

22. Nathan, K.R.; Thanigaiarasu, S. Effect of different endplates on blade aerodynamic performance and blade loading of VAWT with symmetric airfoil blades. In Proceedings of the 2017 International Conference on Green Energy and Applications (ICGEA), Singapore, 25–27 March 2017; pp. 1–5. [CrossRef]

23. Mishra, N.; Gupta, A.S.; Dawar, J.; Kumar, A.; Mitra, S. Numerical and Experimental Study on Performance Enhancement of Darrieus Vertical Axis Wind Turbine With Wingtip Devices. *J. Energy Resour. Technol.* **2018**, *140*, 121201. [CrossRef]

24. Barnes, A.; Marshall-Cross, D.; Hughes, B.R. Towards a standard approach for future Vertical Axis Wind Turbine aerodynamics research and development. *Renew. Sustain. Energy Rev.* **2021**, *148*, 111221. [CrossRef]

25. Watanabe, K.; Takahashi, S.; Ohya, Y. Application of a Diffuser Structure to Vertical-Axis Wind Turbines. *Energies* **2016**, *9*, 406. [CrossRef]

26. Siddiqui, M.S.; Durrani, N.; Akhtar, I. Quantification of the effects of geometric approximations on the performance of a vertical axis wind turbine. *Renew. Energy* **2015**, *74*, 661–670. [CrossRef]

27. Aihara, A.; Mendoza, V.; Goude, A.; Bernhoff, H. Comparison of Three-Dimensional Numerical Methods for Modeling of Strut Effect on the Performance of a Vertical Axis Wind Turbine. *Energies* **2022**, *15*, 2361. [CrossRef]

28. ANSYS. *ANSYS Fluent—User Guide*, Release 17.2 ed; ANSYS Inc.: Canonsburg, PA, USA, 2016.

29. Menter, F.R. Two-equation eddy-viscosity turbulence models for engineering applications. *AIAA J.* **1994**, *32*, 1598–1605. [CrossRef]

30. Wong, K.H.; Chong, W.T.; Poh, S.C.; Shiah, Y.-C.; Sukiman, N.L.; Wang, C.-T. 3D CFD simulation and parametric study of a flat plate deflector for vertical axis wind turbine. *Renew. Energy* **2018**, *129*, 32–55. [CrossRef]

31. Dol, S.; Khamis, A.; Abdallftah, M.; Fares, M.; Pervaiz, S. CFD Analysis of Vertical Axis Wind Turbine with Winglets. *Renew. Energy Res. Appl.* **2022**, *3*, 51–59.

32. Maître, T.; Amet, E.; Pellone, C. Modeling of the flow in a Darrieus water turbine: Wall grid refinement analysis and comparison with experiments. *Renew. Energy* **2013**, *51*, 497–512. [CrossRef]

33. Yang, Y.; Guo, Z.; Zhang, Y.; Jinyama, H.; Li, Q. Numerical Investigation of the Tip Vortex of a Straight-Bladed Vertical Axis Wind Turbine with Double-Blades. *Energies* **2017**, *10*, 1721. [CrossRef]

Article

Verification of Tilt Effect on the Performance and Wake of a Vertical Axis Wind Turbine by Lifting Line Theory Simulation

Hidetaka Senga [1,*], Hiroki Umemoto [1] and Hiromichi Akimoto [2]

1   Department of Naval Architecture and Ocean Engineering, Graduate School of Engineering, Osaka University, Suita 565-0871, Japan
2   Albatross Technology Inc., Chuo-ku, Tokyo 103-0013, Japan
*   Correspondence: senga@naoe.eng.osaka-u.ac.jp; Tel.: +81-668-797-574

Abstract: Renewable energy has received a lot of attention. In recent years, offshore wind power has received particular attention among renewable energies. Fixed-type offshore wind turbines are now the most popular. However, because of the deep seas surrounding Japan, floating types are more preferable. The floating system is one of the factors that raises the cost of floating wind turbines. Vertical axis wind turbines (VAWT) have a low center of gravity and can tilt their rotors. As a result, a smaller floating body and a lower cost are expected. A mechanism called a floating axis wind turbine (FAWT) is expected to further reduce the cost. FAWT actively employs the features of VAWT in order to specialize itself in the area of offshore floating-type wind turbines. The lifting line theory simulation was used in this study to discuss the performance of the FAWT under the tilted conditions and its wake field. The results show that a tilted VAWT recovers faster than an upright VAWT. This suggests that FAWTs can be deployed in high density and efficiently generate energy as an offshore wind farm using VAWTs.

Keywords: vertical axis wind turbine; floating axis wind turbine; lifting line theory; tilt of rotor; wake

Citation: Senga, H.; Umemoto, H.; Akimoto, H. Verification of Tilt Effect on the Performance and Wake of a Vertical Axis Wind Turbine by Lifting Line Theory Simulation. *Energies* **2022**, *15*, 6939. https://doi.org/10.3390/en15196939

Academic Editors: Francesco Castellani and Davide Astolfi

Received: 29 July 2022
Accepted: 16 September 2022
Published: 22 September 2022

Publisher's Note: MDPI stays neutral with regard to jurisdictional claims in published maps and institutional affiliations.

1. Introduction

In order to achieve the Sustainable Development Goals, various renewable energies have become an active area of research. Offshore wind power's technical potential to meet domestic electricity demands indicates that there is a significant amount of unutilized wind energy [1]. Furthermore, 80% of the world's offshore wind resource potential is located in waters deeper than 60 m [2].

Wind turbines are classified into two types. The first is the horizontal axis wind turbine (HAWT), and the second is the vertical axis wind turbine (VAWT). Each axis type has strong and weak points that arise from structural features, and they are often compared in terms of the power coefficient (C_P) [3,4]. The C_P of VAWT is commonly thought to be lower than that of HAWT, but this is mainly for small wind turbines. In fact, the maximum C_P of the SANDIA 34 m "Test Bed" Darrieus VAWT showed a maximum C_P of 0.41 to 0.42 [5,6]. On the other hand, many studies have been conducted to improve the performance of VAWTs. Daegyoum et al. [7] studied the effect of an upstream deflector plate on the power output. Even though the specific model of wind turbine of high solidity was used for their study, the deflector system could increase the local wind velocity around the turbine by tailoring free-stream flow and the power output was proportional to the cube of the wind velocity. A wind lens with several types of diffuser was experimentally studied by Watanabe et al. [8]. They concluded that a wind lens with a Venturi shape, curved diffuser, and shorter flanges was most effective in producing a greater power augmentation. The shapes of arms or strut also have an effect on the performance of the turbines. Hara et al. [9] numerically investigated the effects of arms with different cross sections, such as an NACA0018 airfoil, rectangular and circular, on the power loss of a small VAWT. They decomposed the tangential forces and resistance torques induced by the arms into pressure- and friction-based components. Their results show that, apart from the manufacturing cost and structural strength, the airfoil cross section is ideal for the arm cross section's shape. Aihara et al. [10] compared several numerical methods to investigate whether these methods could reproduce the strut effect on the performance of a VAWT. Their target was 12 kW H-rotor VAWT and the blade force was simulated using the RANS model, the ALM, and the vortex model. Their results show that the strut influence was significant, especially at a high tip speed ratio, and the RANS model was able to simulate the large influence of the strut better than the other two methods. The effect of the blade's cross-sectional shape was investigated by Hou et al. [11]. They numerically investigated the performance of NACA0012, modified NACA0012, and fish skeleton airfoils. The modified NACA0012 airfoils have cambers which are 3, 5, and 7% of the chord length. The fish skeleton airfoil is passively deformed by the pressure from the fluid and it has a higher lift and lower drag coefficient compared to the NACA0012 airfoil. The blade pitch angle is important for not only HAWTs but also VAWTs. Yang et al. [12] studied, experimentally and numerically, the effect of blade pitch angle on aerodynamic characteristics. Their target was a straight-bladed VAWT with two blades. They obtained an optimum blade pitch angle where the power coefficient of their targeted turbine was the largest. However, they concluded that the blade pitch angle for VAWTs had no significant effect on the power coefficient compared to HAWTs. Because the attack angle of VAWTs' blades continuously varies during one rotor rotation, the effect of active pitch angle control on the power coefficient was investigated numerically by Horb et al. [13]. Their optimized pitch laws could increase the power coefficient by more than 15% in maximum power point tracking mode. Mohammed et al. [14] found that a fixed pitch of -2.5 [deg] could enhance performance of a small-scale straight-bladed Darrieus-type VAWT, although its starting torque capacity could not be improved. Their variable pitch angle of sinusoidal nature could improve the power coefficient despite the low starting capacity.

However, C_P is one of the factors to consider when comparing wind turbine systems, with power generation cost being the most important. If a wind turbine's construction and maintenance are expensive, the power generation cost is high even if a high-efficiency C_P turbine is adapted for the system.

Akimoto et al. proposed a floating offshore wind turbine (FOWT) concept, called the floating axis wind turbine (FAWT), that uses a VAWT to reduce power generation cost [15]. There are several types of FOWT that use vertical axis wind turbines, including semi-submersible [16,17] and spar [18,19]. In Akimoto's concept, the latter was adapted to actively utilize the features of VAWTs, such as the low center of gravity and the fact that the power coefficient of VAWTs is difficult to decrease compared to that of HAWTs if the turbine tilts. A straight-blade VAWT is attached to a spar and they rotate together. Power take-off units are installed on the spar above the water's surface, and the unit is moored so as to keep the position of the FAWT and absorb the reaction torque resulting from power generation. The smaller spar is all that is required for FAWTs because it assumes rotor axis tilt, which reduces construction costs. As a result, it is critical to assess VAWT performance under tilted conditions. Figure 1 depicts the conceptual diagram of the FAWT.

As for constructing a wind farm, the wake behind the turbines is important. Researchers have also concentrated on and studied it experimentally and numerically for HAWTs [20,21] and for VAWTs [22,23]. However, most of the object rotors are upright, and there have been few studies on the wake behind tilted rotors. In a study by Guo et al. [24], the center shaft of the rotor was in an upright condition, and only the blades were inclined by using linkage systems. Meanwhile, in the case of FAWTs, the blades and the center shaft of the rotor incline with the spar.

The performance of the FAWT under the tilted conditions was first discussed by the lifting line theory simulation. Then, its wake field was discussed. The simulation results show that a tilted VAWT recovers faster than both a HAWT and an upright VAWT. This indicates that FAWTs can be deployed in high density at an offshore wind farm.

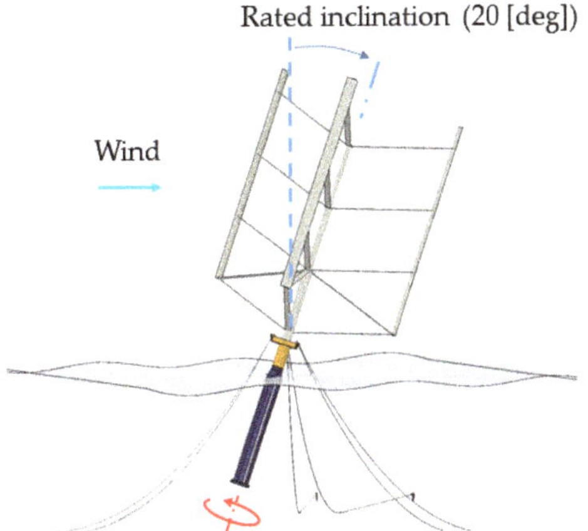

Figure 1. Conceptual diagram of Floating Axis Wind Turbine.

2. Simulation Methods and Turbine Models

In this study, QBlade with v0.963 was used for estimating the performance and wakes of VAWTs under various conditions. It is an open-source wind turbine calculation software that includes 2-dimensional airfoil calculation and the lifting line free-vortex wake (LLFVW) theory simulation which belongs to the vortex method [25]. By using the vortex method, the flow field is modeled as inviscid, incompressible, and irrotational. The effect of fluid viscosity is modeled by the vortices introduced into the flow field from turbine blades, in this case. The rotor is represented by a lifting line which is located at the quarter chord position of the airfoil cross section. The blades are modeled as a lattice of horseshoe vortices using the vortex lattice method. Figure 2 depicts the modeling used in the LLFVW algorithm.

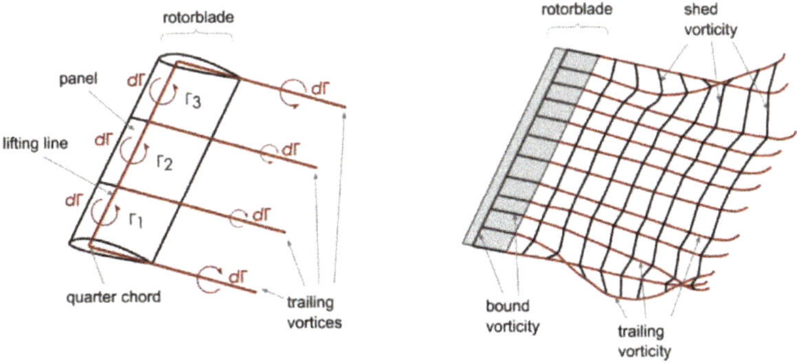

Figure 2. Illustration of blade and wake modeling with the LLFVW algorithm [25].

The circulation of the bound vorticity, which forms the lifting line $(d\Gamma)$, is calculated based on the Kutta–Joukowski theorem as follows.

$$dL = \rho V_{rel} d\Gamma \tag{1}$$

dL is the sectional lift force and V_{rel} is the relative velocity. The relative velocity at an arbitrary position in the analysis field is composed of the free-stream velocity, the velocity of blade motion, and the induced velocity from all vortex elements on the blade and in the wake. The induced velocity from vortex line elements is known as the Biot–Savart law below,

$$V_{ind} = -\frac{1}{4\pi} \int \Gamma \frac{r \times dl}{|r|^3} \tag{2}$$

Meanwhile, the sectional lift force dL can be calculated from the relative velocity and lift coefficient as follows:

$$dL = \frac{1}{2}\rho V_{rel}^2 dA C_L \tag{3}$$

Then, $d\Gamma$ is obtained from Equations (1) and (3) with iteration steps. More details, such as the iteration steps and the convection of vortex elements, can be found in the references [26] and [27].

The authors of [28] compared power coefficients obtained by using QBlade and experimental results, such as SANDIA 34-m "Test Bed" Darrieus VAWT [3] and 1 KW DeepWind turbine test model [29], and validated simulation conditions to obtain sufficient results. Based on it, the simulation conditions in this research were determined. In QBlade, users should design the airfoil first and then simulate its polars, such as lift and drag coefficient, within an adequate attack angle (α) range by using XFOIL, which is integrated into QBlade. The simulated polars can be extrapolated by using the Viterna or Montgomery method. Then, the rotor is designed with the airfoils. The blade forces are calculated by using the lifting line theory. The circulation of the blade is placed at the 1/4 chord positions and its strength is calculated from airfoil data and α, which is calculated from the induced velocity of free wake. The wake is represented by the freely floating vortices which are shed from the trailing edge of the blades during every simulation time step. As the simulation time increases, the total number of vortices shed into the flow field also increase and it results in time-consuming calculation of the induced velocity from the vortices. The wake count can be set as rotor revolutions, time steps, or time. The effect from wake in 8 rotor revolutions was considered in this study.

This study employed two turbine models. One has three straight-blade VAWTs. This model was employed to validate the simulation conditions prior to verifying the tilt effect on the performance of VAWTs. Table 1 summarizes its main features. Despite the fact that NACA0018 was used for the blade profile, the lift (C_L) and drag coefficient (C_D) were defined as the equations that were used in the simulation by Tavernier et al. [30], defined below. Thus, two extrapolated methods described above were not used in this study.

$$C_L = 2\pi \cdot 1.1 \cdot \sin \alpha \tag{4}$$

$$C_D = 0 \tag{5}$$

Table 1. Principal particulars of Model 1 (two straight-blade turbines).

Property	Value
Solidity	0.085
Tip Speed Ratio	3
Radius of Rotor [m]	2.5
Aspect Ratio	0.5, 1, 2, 5
Number of Blades	3
Blade Profile	NACA0018

The other model is a VAWT for testing the tilt effect on its performance and discussing the wake field. The constant tilt of the rotor is assumed to be approximately 20 [deg]. Thus, the angle varied from 0 to 40 by 5 degrees. The C_L and C_D were calculated by using XFOIL and extrapolated into 360 [deg] using Montgomerie's method in the 360 Polar Extrapolation Module. Table 2 summarizes the main characteristics of this mode. The corresponding Reynolds number based on the blade chord length and tip speed is approximately from

1.8×10^4 to 3.4×10^4 for both Model 1 and 2. Flow around the turbine is turbulent under such Reynolds number regions.

Table 2. Principal particulars of Model 2 (VAWT under tilted conditions).

Property	Value
Tilt angle of rotor [deg]	0:5:40
Solidity	0.085
Tip Speed Ratio	3
Radius of Rotor [m]	2.5
Aspect Ratio	1, 2
Number of Blades	3
Blade Profile	NACA0018

In QBlade, the viscous effect is tuned by using the parameter "initial core size". 1/8 of the blade chord length was used in the simulation for both models. This core size is also used in the simulation of CACTUS.

3. Results

3.1. Validation of Simulation Conditions

The power (C_P) and thrust (C_T) coefficients of Model 1, which are defined as the following equations, were compared with Tavernier's work.

$$C_P = \frac{\text{Power}}{\frac{1}{2}\rho V_\infty^3 2RH} \tag{6}$$

$$C_T = \frac{\text{Thrust}}{\frac{1}{2}\rho V_\infty^2 2RH} \tag{7}$$

They compared the results from several simulation methods, including the 2D actuator cylinder model (AC2D) [31], the 2D actuator cylinder with near-wake correction (HAWC2-NW) [32], CACTUS, free- or fixed-wake vortex model [33], and the Actuator Line OpenFOAM model (TurbinesFoam, end effects) [34]. In CACTUS, "fixed-wake" means that the wake convection velocity is kept constant in time, whereas "free-wake" means that the velocity is calculated at each time step based on the induced velocity. Table 3 depicts the comparisons of such simulation results with those of this study. The aspect ratio of the rotor was 1.0 for these calculations. Compared to other calculation methods, appropriate values were estimated by QBlade.

Table 3. Comparisons of Power and Thrust Coefficients.

Method	Power Coefficient (C_P)	Thrust Coefficient (C_T)
AC2D	0.510	0.653
HAWC2-NW	0.400	0.570
CACTUS, fixed-wake	0.509	0.647
CACTUS, free-wake	0.486	0.643
TurbineFoam	0.522	0.660
TurbineFoam, end effects	0.469	0.578
QBlade (this study)	0.503	0.665

Figure 3 depicts the comparison of the tangential force acting on a blade in relation to the azimuth angle. Because its C_P and C_T were closest to those of this study, the result of CACTUS with fixed wake was cited from reference [26]. In Figure 3, "HD" denotes the rotor's aspect ratio and the wind comes from 90 [deg]. Thus, azimuth angles of 0~180 and 180~360 [deg] indicate the upwind and leeward side of the blade, respectively. The same trend with respect to the aspect ratio can be seen in both figures. The blade mainly obtains the energy from wind at around 90 [deg] and also at around 270 [deg]. The results of this study, however, were not smooth on the leeward side. This implies that the initial core size should be adjusted in each case, despite the fact that they were fixed as 8/chord for all aspect ratio cases in this study for comparison.

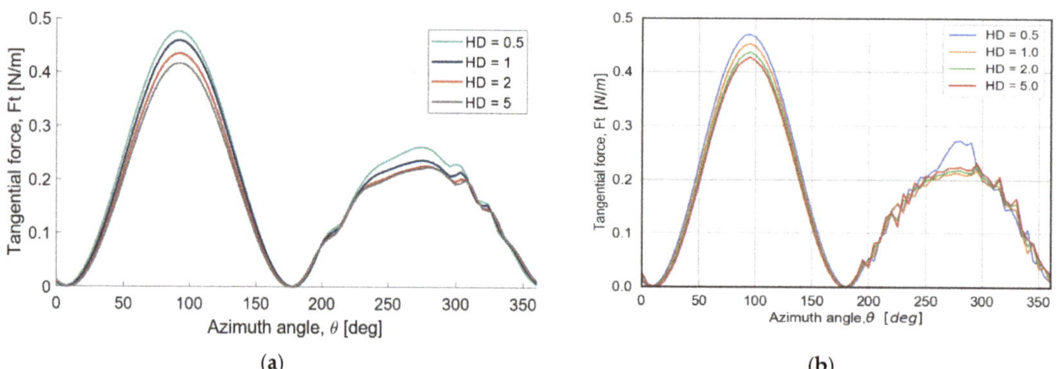

Figure 3. Tangential blade force with respect to the azimuth angle (**a**) CACTUS, fixed-wake [26]; (**b**) this study (QBlade).

Figure 4 depicts tangential blade forces as a contour graph with respect to the azimuth angle and blade height. This turbine is a 1.0. It is the same as Figure 3 in the sense that azimuth angles 0~180 and 180~360 [deg] indicate upwind and leeward sides of the blade's position, respectively, and wind comes from 90 [deg]. The tangential force is symmetric in a blade spanwise direction, and the ground effect is not taken into account in these simulations. There is no discernible difference between Figure 4a,b.

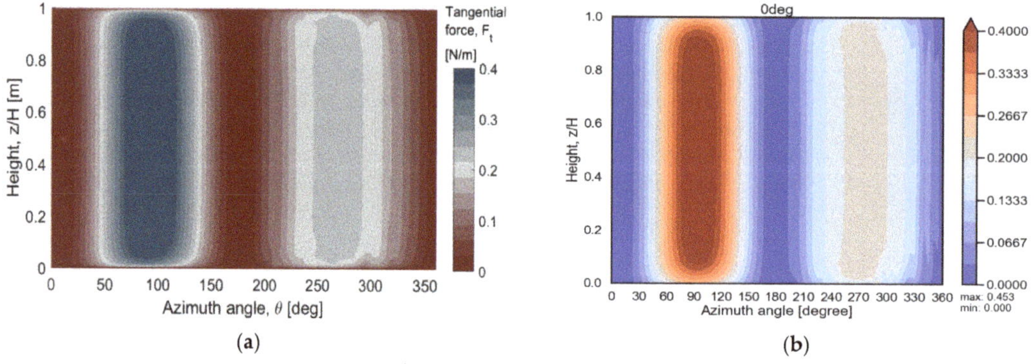

Figure 4. Contour graph of tangential blade force: (**a**) CACTUS, fixed-wake [30]; (**b**) this study (QBlade).

These results demonstrate that the simulation conditions used in this study are appropriate and that such conditions will be used in the following simulations.

3.2. Tilt Effect on the Performance of VAWT

3.2.1. Power and Thrust Coefficient of Tilted VAWT

A FAWT is built with the assumption that its rotor will be at a 20-degree tilt under-rated operation; C_P and C_T for Model 2 were estimated to assess the tilt effect on VAWT performance. The values are shown in Table 4, and Figure 5 depicts the differences in the upright condition. These results show that the power coefficient slightly increases at a 10-degree tilt for a rotor with a 1.0 aspect ratio. The decrease in C_P at a 20-degree tilt is only 4% from the upright condition. This is because the swept area of the VAWT increases when the rotor tilts to some extent [35]. Concerning the rotor with 2.0 aspect ratio, the decrease in performance at a 20-degree tilt is 10% from the upright condition. The decrease in performance for the 2.0 aspect ratio is larger than that for the 1.0 aspect ratio. This is explained by the fact that the rotor radius contributes to the increase in swept area, which becomes relatively small as the aspect ratio increases.

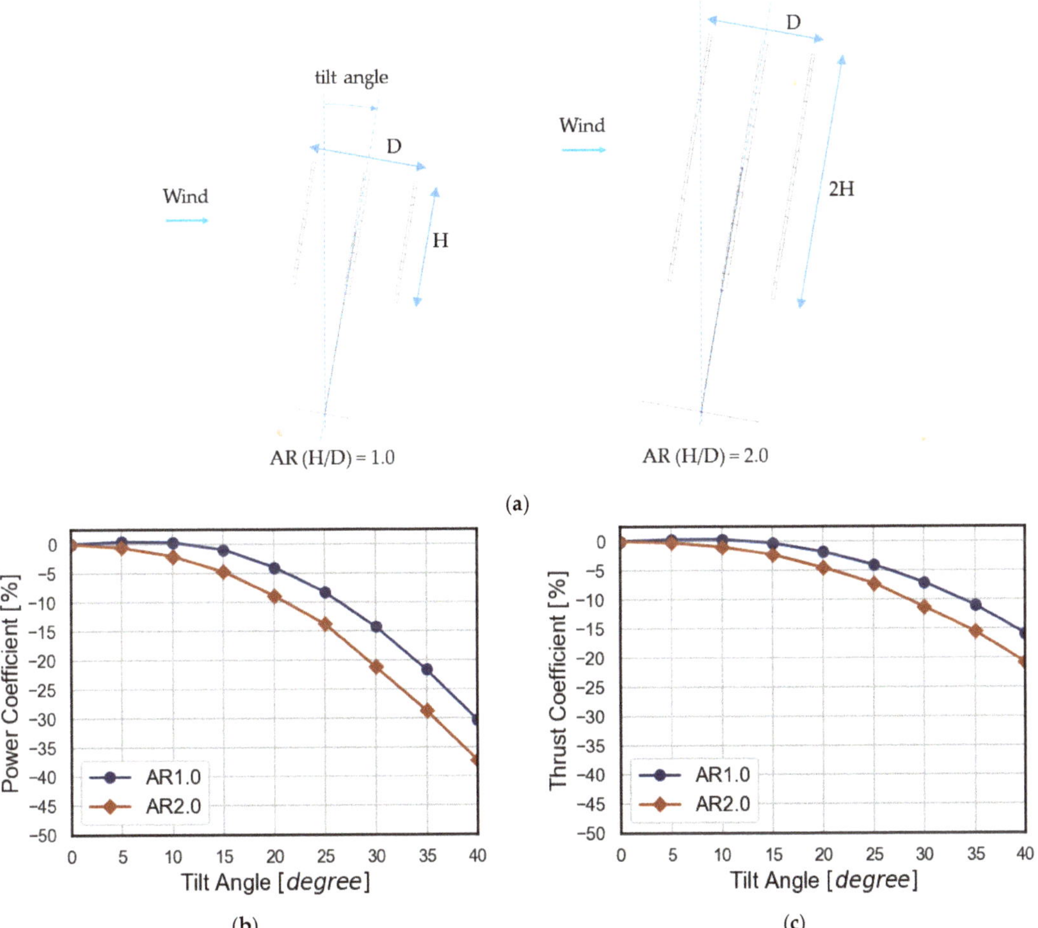

Figure 5. Difference in Power and Thrust Coefficients of a tilted rotor compared to an upright one: (a) configuration of the simulation; (b) power coefficient; (c) thrust coefficient.

Table 4. Power and Thrust coefficients with respect to tilted angle.

Aspect Ratio	1.0		2.0	
Tilt Angle [deg]	C_P	C_T	C_P	C_T
0	0.503	0.665	0.500	0.662
5	0.504	0.667	0.497	0.660
10	0.505	0.667	0.489	0.655
15	0.498	0.662	0.476	0.647
20	0.483	0.652	0.455	0.632
25	0.462	0.638	0.431	0.614
30	0.431	0.618	0.394	0.588
35	0.394	0.592	0.356	0.559
40	0.350	0.559	0.313	0.524

Figure 6 depicts the tangential blade forces as a contour graph in relation to the azimuth angle and blade height. Because the blade height is normalized, the height effect appears in a 2.0 aspect ratio. These figures depict that in the case of a tilted rotor, the lower part of the rotor on the leeward side effectively obtains energy from wind. Under the upright condition, wind velocity on the leeward side decreases from the top to the bottom of the blade height. Meanwhile, when the rotor is in the tilted condition, the wind velocity on the lower part of the rotor on the leeward side stays strong because the blade does not pass the wind on the upwind side. Then, the lower part of the rotor obtains energy from the wind on the leeward side. Even though the smallest part of the rotor can effectively generate the tangential force on the leeward side, the airfoil performance degrades with increasing rotor tilt angle. This may be indicated in the results of the 30- and 40-degree-tilt tests. Based on these findings, it stands to reason that the FAWT rotor is at a 20-degree tilt under the rated operation.

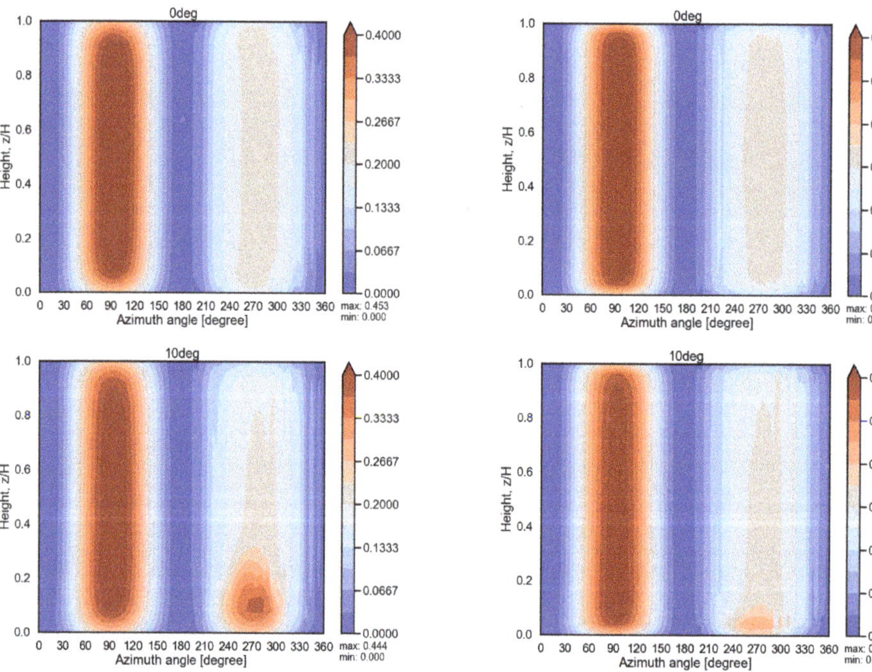

Figure 6. *Cont.*

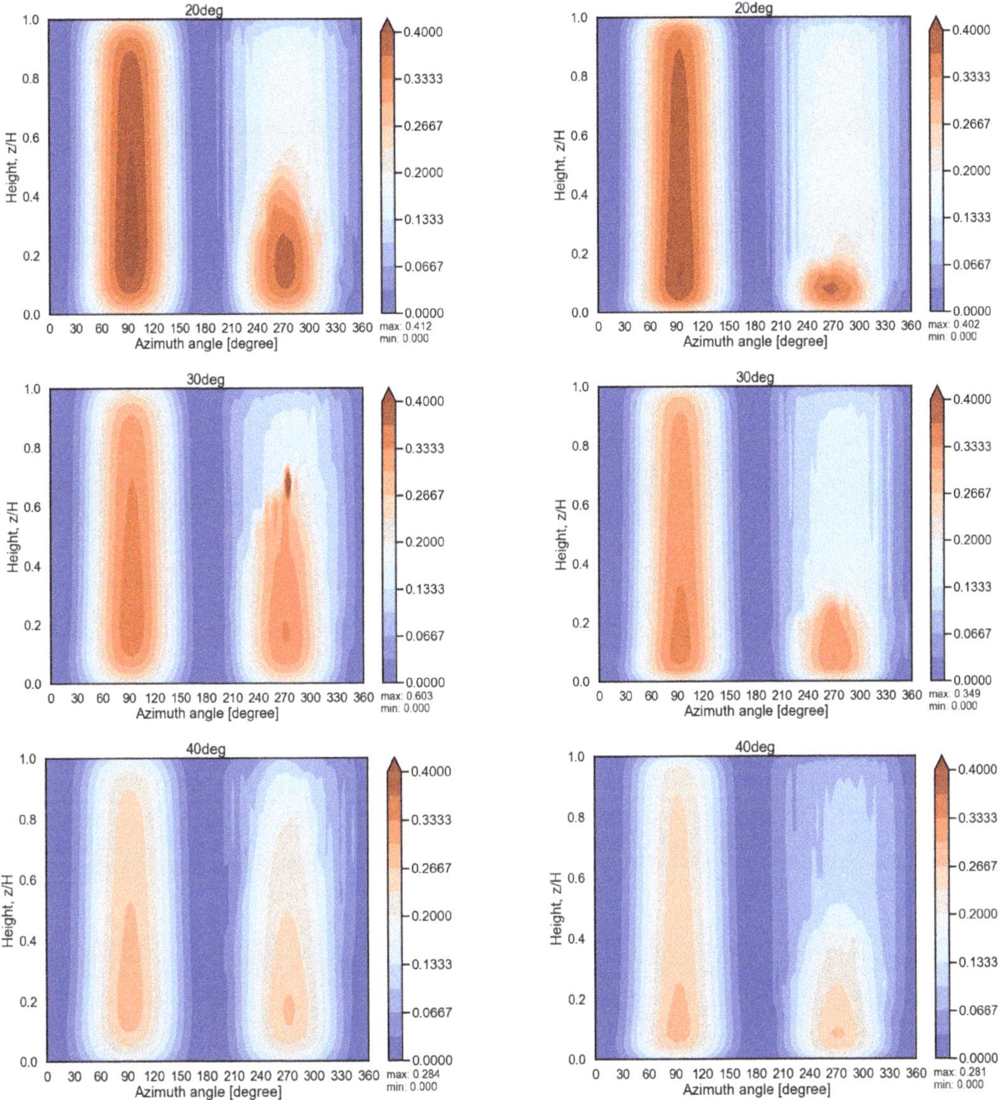

Figure 6. Contour graph of tangential blade force for 0-, 10-, 20-, 30-, and 40-degree-tilted rotors.

3.2.2. The Wake Field of the Tilted VAWT

The installation interval of wind turbines is an important factor to consider when building a wind farm. As a result, the wake field of a VAWT was assessed. The target VAWT is Model 2 with a 1.0 aspect ratio. Figure 7 depicts the instantaneous velocity field center section (XZ-plane) of an upright, 20-degree-tilted VAWT and an upright HAWT. The X-axis is in the direction of the wind, and the coordinate system (O-XYZ) is right-handed. The rotor is represented by a yellow circular cylinder, and the contour color indicates the magnitude of in-plane wind velocity nondimensionalized by uniform flow. The result for the HAWT was simulated by using a sample project of a HAWT contained in QBlade by default. In the case of the upright VAWT and HAWT, the wake goes straight to the leeward side. Meanwhile, in the case of a 20-degree-tilted VAWT, the wake goes upward. The inclination angle is approximately 6 [deg].

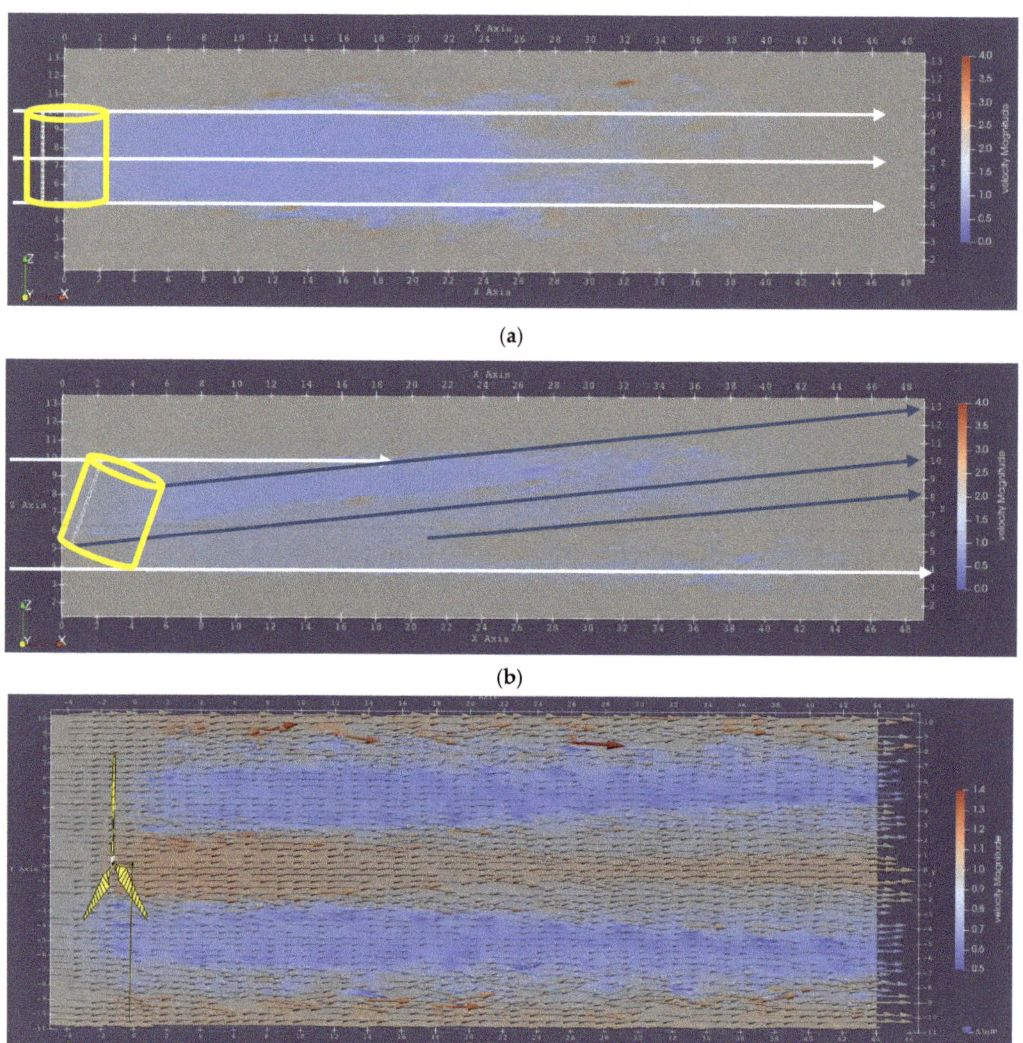

Figure 7. Magnitude of in-plane wind velocity: (**a**) upright VAWT; (**b**) 20-degree-tilted VAWT; (**c**) upright HAWT.

Figure 8 depicts the vertical component of wind velocity corresponding to Figure 7. The colors red and blue represent the upward and downward velocity components, respectively. Compared to the flow field of an upright VAWT, the upward velocity component may be seen just behind the rotor, as opposed to the flow field of an upright VAWT. The orange lines denote the position at four times the rotor diameter's (D) distance from the rotor center, and Figure 9 depicts the wind velocity vector in the YZ-plane at that location. The yellow circular cylinder represents the rotor's position, which is located 4D anteriorly. The color of the contour indicates the magnitude of the velocity in this plane. The colors blue and red represent low and high velocity, respectively. There are upward and downward vectors at the top and bottom of the rotor in both cases. Then, the vectors head into the center region from both sides of the top of the rotor. Figure 10 depicts 3D streamlines of an upright and 20-degree-tilted VAWT. In these figures, the blades of the rotor are shown in black. Under the upright condition, most of the streamlines behind the turbine go straight to the leeward side. On the other hand, the streamlines go obliquely upward under the 20-degree-tilted condition. These results indicate that the momentum is being transferred between the lower- and the higher-velocity region. As a result, the wind velocity at rotor height recovers to its initial velocity faster than that of a HAWT. This allows us to install VAWTs in a dense manner.

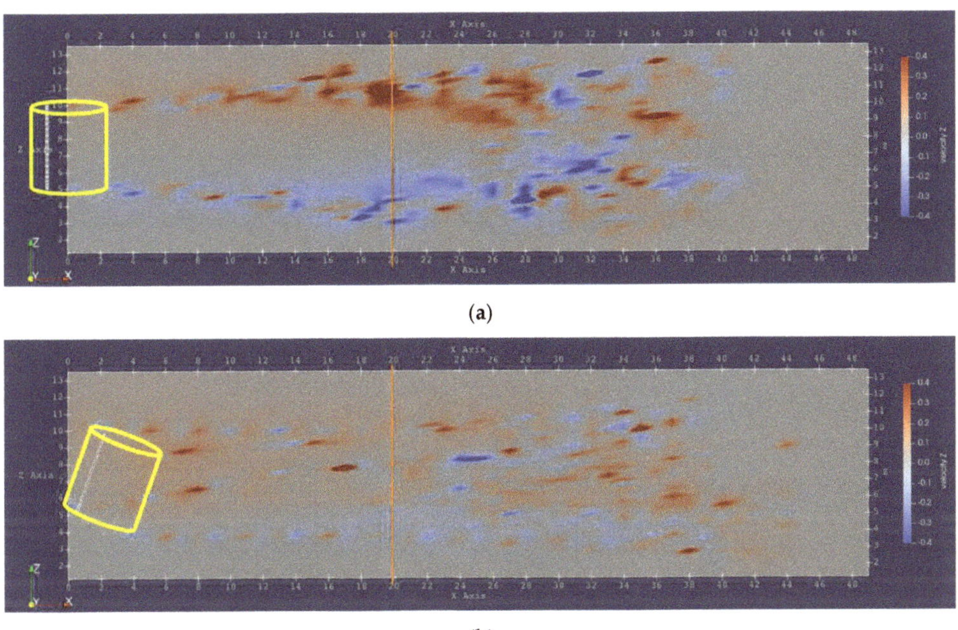

Figure 8. Contour of the vertical component of wind velocity: (**a**) upright; (**b**) 20-degree tilt.

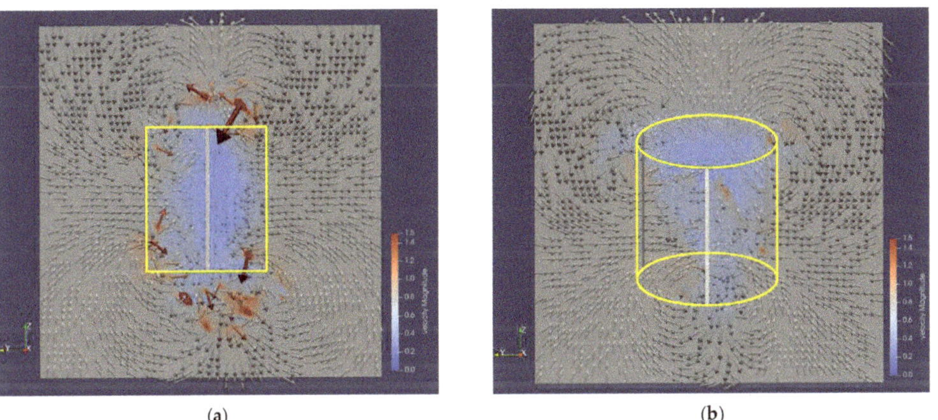

Figure 9. Wind velocity vector in the YZ-plane (**a**) upright; (**b**) 20-degree tilt.

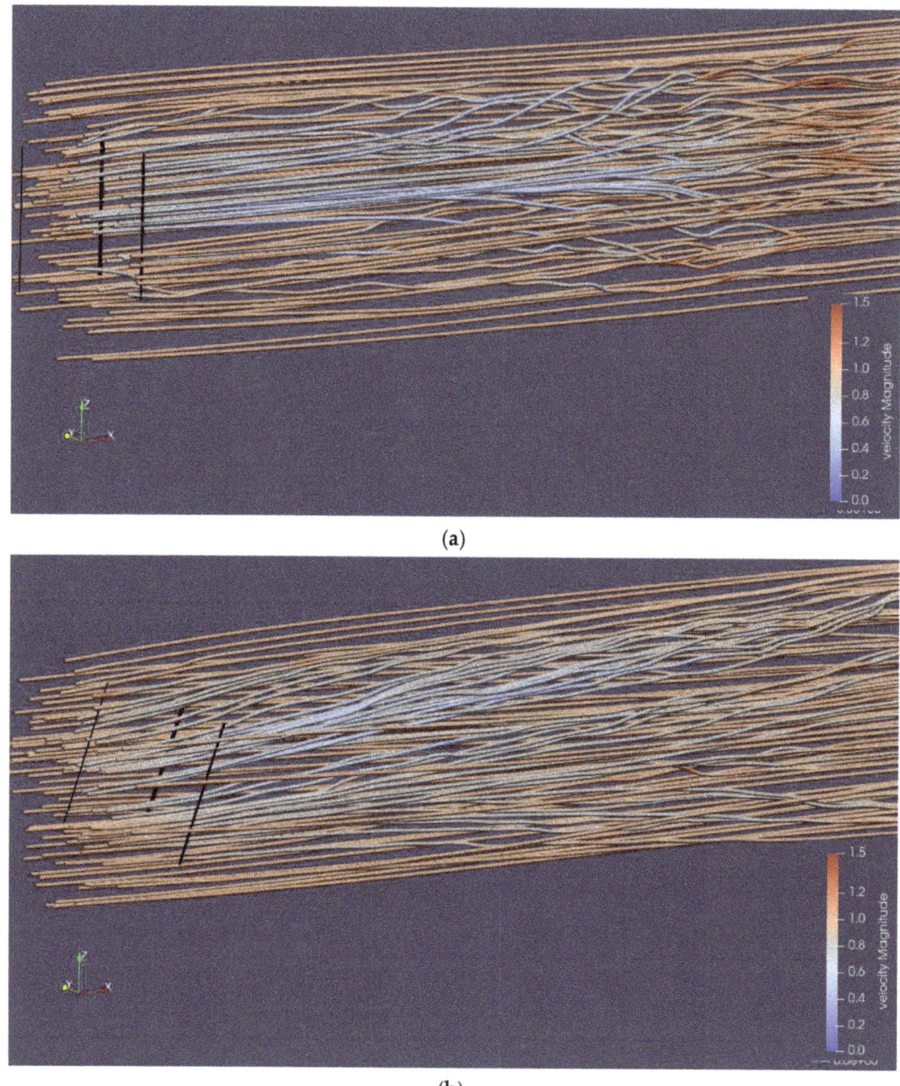

Figure 10. 3D Streamlines (a) upright; (b) 20-degree tilt.

3.2.3. Ground Effect

The ground effect was not considered in the previous section's simulations, so as to verify only the tilt effect on its wake field. Even if the FAWT is properly designed so that the blades do not hit the sea surface when the rotor inclines, the clearance between the rotor blade bottom and the sea surface will affect the wake field when the tilted condition occurs. QBlade implements the ground effect by mirroring the rotor blade and wake vortices at the ground plane. The clearance has been changed from 1.0D, 0.8D, 0.6D, and 0.4D. For all cases, the rotor tilt angle was kept constant at 20 [deg]. The instantaneous velocity field at the center section (XZ-plane) of a 20-degree-tilted VAWT with various ground clearances is depicted in Figure 11. The orange lines represent the position at 4D from the rotor center, and Figure 12 depicts the wind velocity vector in the YZ-plane at that location. In all cases, the upward wind may be seen behind the VAWT, but the smaller the clearance becomes, the greater the distance that velocity recovers.

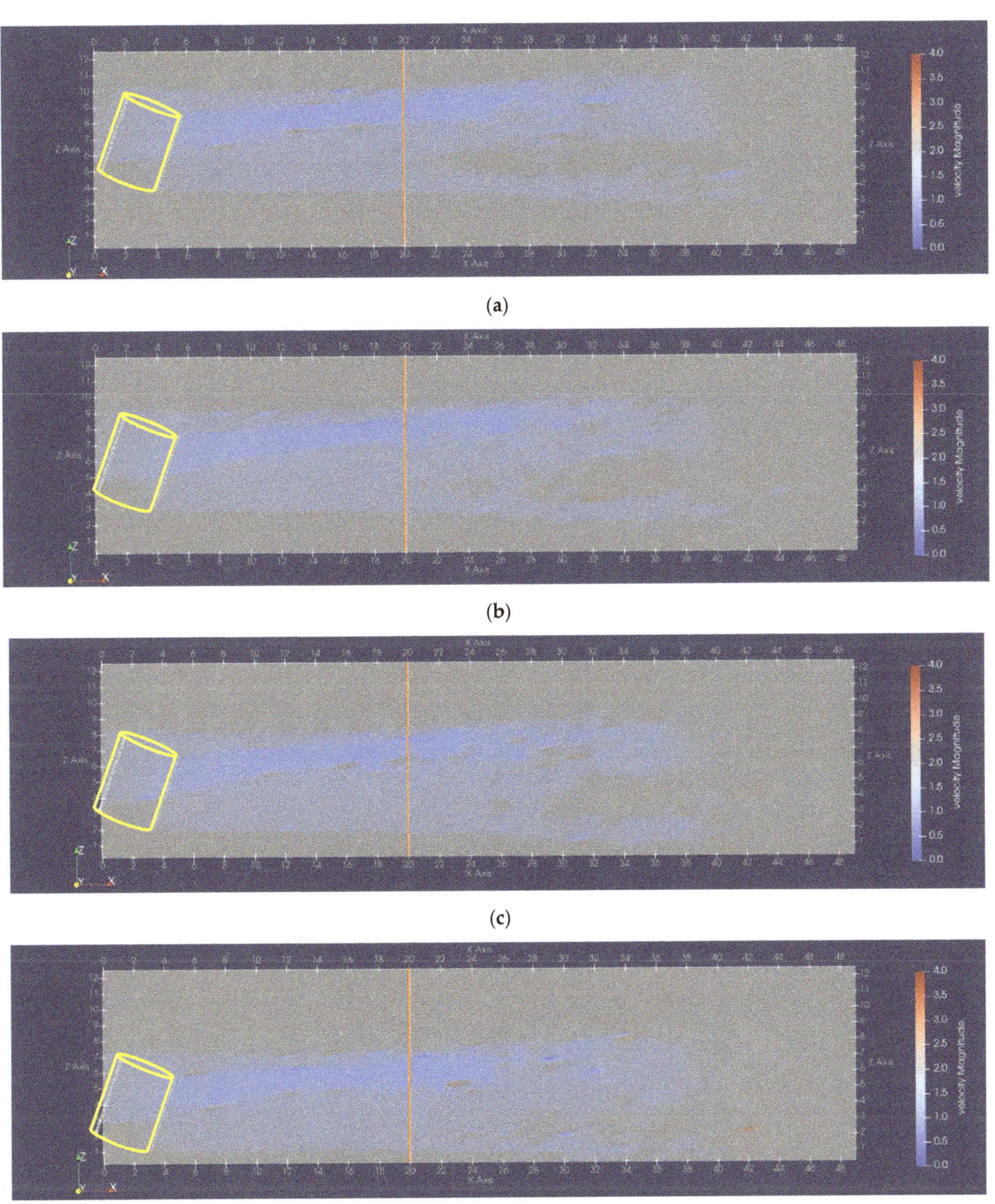

Figure 11. Magnitude of in-plane wind velocity behind a 20-degree-tilted VAWT with different ground clearance: (**a**) 1.0D; (**b**) 0.8D; (**c**) 0.6D; (**d**) 0.4D.

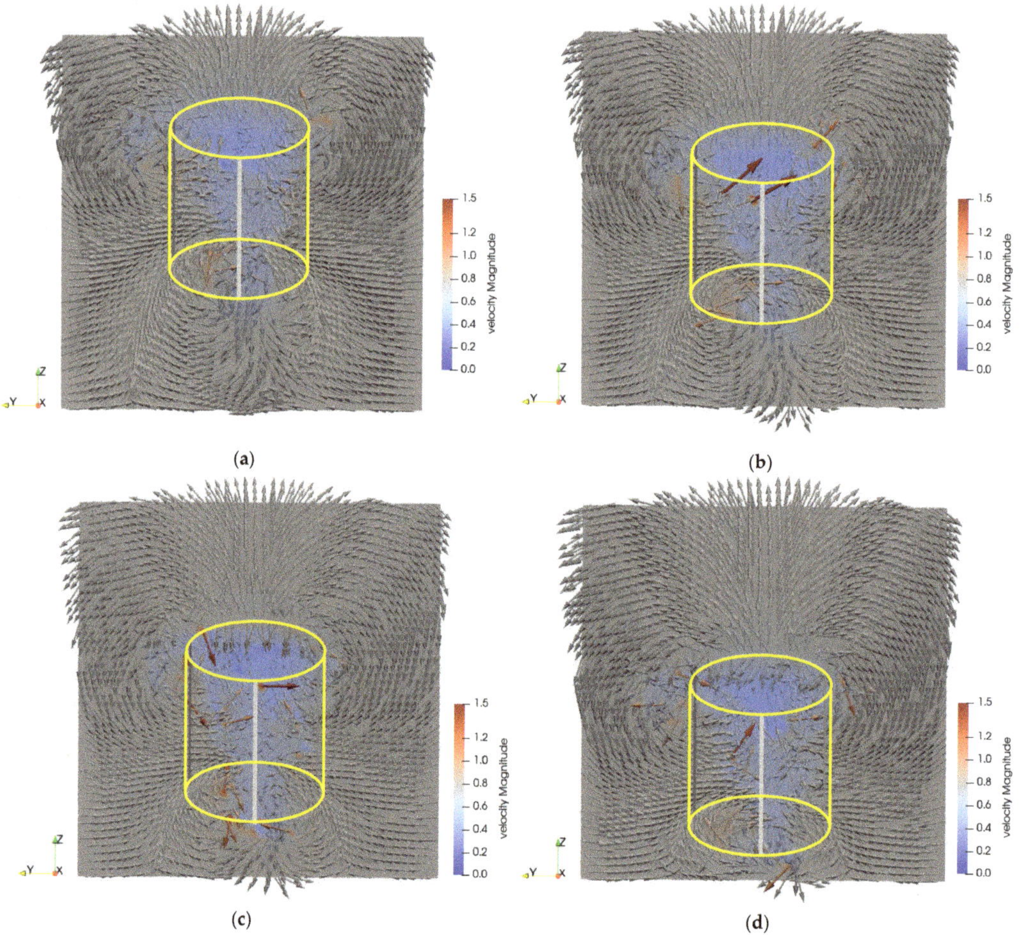

Figure 12. Wind velocity vector in the YZ-plane behind a 20-degree-tilted VAWT with different ground clearance: (**a**) 1.0D; (**b**) 0.8D; (**c**) 0.6D; (**d**) 0.4D.

4. Conclusions

The performance and wakes of VAWTs were simulated in this study using liftingline theory simulation. The simulation conditions were validated using a VAWT model with three straight blades. When compared to the results of the other simulation methods, the simulation conditions used in this study were found to be appropriate. Then, the tilt effect on the performance and wake field of VAWTs were evaluated with the other VAWT model, and the following results were obtained:

At a 10-degree tilt, the power coefficient of a 1.0 aspect-ratio rotor increases slightly. Then, at a 20-degree tilt, the decrease in C_P is only 4% from the upright condition. This increase in the power coefficient with respect to the tilt is in agreement with Balduzzi's work [35], and it is explained by the contour graph of the tangential force on the blade, which shows that in the case of a tilted rotor, the lower part of the rotor at the leeward side effectively obtains the energy from wind. Meanwhile, as the rotor tilt angle increases, so does the performance of the airfoil. Because the increased swept area becomes relatively small as the aspect ratio increases, the performance for the 2.0 aspect ratio decreases more than that for the 1.0 aspect ratio. This, as a result, it is important for VAWTs that the relationship between the increase in swept area and the decrease in airfoil performance be maintained.

The VAWT wake field was then simulated in order to evaluate the installation interval of wind turbines. The upward velocity component appears just behind the rotor, and the wake goes obliquely upward approximately 6 [deg] from the horizontal plane, whereas the wake for the upright VAWT and HAWT goes straight to the leeward side. The velocity field of the plane perpendicular to the wind direction indicates that there are upward and downward velocity components at the top of the rotor's height and the bottom side of the VAWT, respectively. Furthermore, the velocity components at the top of the rotor head into the center region from both sides of the rotor. They result in a momentum exchange. The wind velocity at the top of the rotor of a tilted VAWT recovers to its initial velocity faster than that of a HAWT, allowing us to install VAWTs more densely.

Even though the ground effect was taken into account, the wake goes upward for tilted VAWTs. However, the smaller the clearance between the sea surface and the blades of a tilted VAWT becomes, the further the distance that velocity recovers.

Author Contributions: Conceptualization, H.S., H.U. and H.A.; methodology, H.S., H.U. and H.A.; data curation, H.S. and H.U.; writing—original draft preparation, H.S.; writing—review and editing, H.S.; project administration, H.A. All authors have read and agreed to the published version of the manuscript.

Funding: This research received no external funding.

Data Availability Statement: Not applicable.

Conflicts of Interest: The authors declare no conflict of interest.

Nomenclature

C_P	power coefficient
C_L	lift coefficient
C_D	drag coefficient
C_T	thrust coefficient
D	rotor diameter [m]
dA	cross-sectional area [m^2]
dL	sectional lift force [N]
dl	line element vector [m]
$d\Gamma$	sectional circulation [m^2/s]
H	rotor height [m]
ρ	air density [kg/m^3]
R	rotor radius [m]
r	relative position vector [m]
V_{ind}	induced velocity from vortex line element [m/s]
V_{rel}	relative velocity [m/s]
V_∞	free-stream velocity [m/s]
α	attack angle [deg]

References

1. International Energy Agency. Offshore Wind Outlook. 2019. Available online: https://www.iea.org/reports/offshore-wind-outlook-2019 (accessed on 8 April 2022).
2. Global Wind Energy Council. Global Offshore Wind Report. 2021. Available online: https://gwec.net/global-offshore-wind-report-2021/ (accessed on 8 April 2022).
3. Pope, K.; Dincer, I.; Naterer, G.F. Energy and exergy efficiency comparison of horizontal and vertical axis wind turbines. *Renew. Energy* **2010**, *35*, 2102–2113. [CrossRef]
4. Mendoza, V.; Chaudhari, A.; Goude, A. Performance and wake comparison of horizontal and vertical axis wind turbines under varying surface roughness conditions. *Wind. Energy* **2019**, *22*, 458–472. [CrossRef]

5. Johnston, S.F., Jr. *Proceedings of the Vertical Axis Wind Turbine (VAWT) Design Technology Seminar for Industry*; SANDIA REPORT, SAND80-0984; Sandia National Laboratories: Albuquerque, NM, USA, 1982.

6. Sutherland, H.J.; Berg, D.E.; Ashwill, T.D. *A Retrospective of VAWT Technology*; SANDIA REPORT, SAND2012-0304; Sandia National Laboratories: Albuquerque, NM, USA, 2012.

7. Daegyoum, K.; Morteza, G. Efficiency improvement of straight-bladed vertical–axis wind turbines with an upstream deflector. *J. Wind Eng. Ind. Aerodyn.* **2013**, *115*, 48–52. [CrossRef]

8. Watanabe, K.; Takahashi, S.; Ohya, Y. Application of a diffuser structure to vertical-axis wind turbines. *Energies* **2016**, *9*, 406. [CrossRef]

9. Hou, L.; Shen, S.; Wang, Y. Numerical study on aerodynamic performance of different forms of adaptive blades for vertical axis wind turbines. *Energies* **2021**, *14*, 880. [CrossRef]

10. Hara, Y.; Horita, N.; Yoshida, S.; Akimoto, H.; Sumi, T. Numerical analysis of effects of arms with different cross-sections on straight-bladed vertical axis wind turbine. *Energies* **2019**, *12*, 2106. [CrossRef]

11. Aihara, A.; Mendoza, V.; Goude, A.; Bernhoff, H. Comparison of three-dimensional numerical methods for modeling of strut effect on the performance of a vertical axis wind turbine. *Energies* **2022**, *15*, 2361. [CrossRef]

12. Yang, Y.; Guo, Z.; Song, Q.; Zhang, Y.; Li, Q. Effect of blade pitch angle on the aerodynamic characteristics of a straight-bladed vertical axis wind turbine based on experiments and simulations. *Energies* **2018**, *11*, 1514. [CrossRef]

13. Horb, S.; Fuchs, R.; Immas, A.; Silvert, F.; Deglaire, P. Variable pitch control for vertical-axis wind turbines. *Wind Eng.* **2018**, *42*, 128–135. [CrossRef]

14. Mohammed, A.A.; Sahin, A.Z.; Ouakad, H.M. Numerical investigation of a vertical axis wind turbine performance characterization using new variable pitch control scheme. *J. Energy Resour. Technol.* **2020**, *142*, 031302. [CrossRef]

15. Akimoto, H.; Tanaka, K.; Uzawa, K. Floating axis wind turbines for offshore power generation-a conceptual study. *Environ. Res. Lett.* **2011**, *6*, 044017. [CrossRef]

16. Blusseau, P.; Patel, M.H. Gyroscopic effects on a large vertical axis wind turbine mounted on a floating structure. *Renew. Energy* **2012**, *46*, 31–42. [CrossRef]

17. Ishie, J.; Wang, K.; Ong, M.C. Structural dynamic analysis of semi-submersible floating vertical axis wind turbines. *Energies* **2016**, *9*, 1047. [CrossRef]

18. Paulsen, U.S.; Borg, M.; Madsen, H.A.; Pedersen, T.F.; Hattel, J.; Ritchie, E.; Ferreira, C.S.; Svendsen, H.; Berthelsen, P.A.; Smadja, C. Outcomes of the DeepWind conceptual design. *Energy Procedia* **2015**, *80*, 329–341. [CrossRef]

19. Wen, T.R.; Wang, K.; Cheng, Z.; Ong, M.C. Spar-type vertical-axis wind turbines in moderate water depth: A feasibility study. *Energies* **2018**, *11*, 555. [CrossRef]

20. Lignarolo, L.E.M.; Ragni, D.; Krishnaswami, C.; Chen, Q.; Ferreira, C.J.S.; Bussel, G.J.W. Experimental analysis of the wake of a horizontal-axis wind-turbine model. *Renew. Energy* **2014**, *70*, 31–46. [CrossRef]

21. Rezaeiha, A.; Micallef, D. Wake interactions of two tandem floating offshore wind turbines: CFD analysis using actuator disc model. *Renew. Energy* **2021**, *179*, 859–876. [CrossRef]

22. Tescione, G.; Ragni, D.; He, C.; Ferreira, C.J.S.; Bussel, G.J.W. Near wake flow analysis of a vertical axis wind turbine by stereoscopic particle image velocimetry. *Renew. Energy* **2014**, *70*, 47–61. [CrossRef]

23. Yuan, Z.; Jiang, J.; Zang, J.; Sheng, Q.; Sun, K.; Zhang, X.; Ji, R. A fast two-dimensional numerical method for the wake simulation of a vertical axis wind turbine. *Energies* **2021**, *14*, 49. [CrossRef]

24. Guo, J.; Lei, L. Flow characteristics of a straight-bladed vertical axis wind turbine with inclined pitch axes. *Energies* **2020**, *13*, 6281. [CrossRef]

25. Marten, D. *QBLADE v0.95 Guidelines for Lifting Line Free Vortex Wake Simulations*; Technical University Berlin: Berlin, Germany, 2016. [CrossRef]

26. Van Garrel, A. *Development of a Wind Turbine Aerodynamics Simulation Module*; Technical Report ECN-C-03-079; ECN Wind Energy: Petten, The Netherlands, 2003.

27. Marten, D.; Lennie, M.; Pechlivanoglou, G.; Nayeri, C.N.; Paschereit, C.O. Implementation, Optimization, and Validation of a Nonlinear Lifting Line-Free Vortex Wake Module Within the Wind Turbine Simulation Code QBLADE. *J. Eng. Gas Turbines Power* **2016**, *138*, 072601. [CrossRef]

28. Senga, H.; Akimoto, H.; Fukai, K. Power Coefficient Estimation of Floating Axis Wind Turbine by Lifting Line Theory. *IOP Conf. Ser. Mater. Sci. Eng.* **2021**, *1051*, 012057. [CrossRef]

29. Battisti, L.; Benini, E.; Brighenti, A.; Castelli, M.R.; Dell'Anna, S.; Dossena, V.; Persico, G.; Paulsen, U.S.; Pedersen, T.F. Wind tunnel testing of the DeepWind demonstrator in design and tilted operating conditions. *Energy* **2016**, *111*, 484–497. [CrossRef]

30. Tavernier, D.D.; Sakib, M.; Griffith, T.; Pirrung, G.; Paulsen, U.; Madsen, H.; Keijer, W.; Ferreira, C. Comparison of 3D aerodynamic models for vertical-axis wind turbines: H-rotor and Φ-rotor. *Int. J. Phys. Conf. Ser.* **2020**, *1618*, 052041. [CrossRef]

31. Madsen, H.A. On the Ideal and Real Energy Conversion in a Straight Bladed Vertical Axis Wind Turbine. Ph.D. Thesis, Aalborg University, Aalborg, Denmark, 1983.

32. Beddoes, T.S. A near wake dynamic model. In Proceedings of the AHS National Specialist Meeting on Aerodynamics and Aeroacoustics, Arlington, TX, USA, 25–27 February 1987.

33. Murray, J.C.; Barone, M. The development of CACTUS, a wind and marine turbine performance simulation code. In Proceedings of the 49th AIAA Aerospace Sciences Meeting Including the New Horizons Forum and Aerospace Exposition, Orlando, FL, USA, 4–7 January 2011; AIAA 2011–147. AIAA: Reston, VA, USA, 2011.
34. Bachant, P.; Goude, A.; Wosnik, M. Turbines Foam/Turbinesforam: v0.0.8, Zenodo. 2018. Available online: https://doi.org/10.5281/zenodo.1210366 (accessed on 8 April 2022).
35. Balduzzi, F.; Bianchini, A.; Carnevale, E.A.; Ferrari, L.; Magnani, S. Feasibility analysis of a Darrieus vertical-axis wind turbine installation in the rooftop of a building. *Appl. Energy* **2012**, *97*, 921–929. [CrossRef]

Article

Shades of Green: Life Cycle Assessment of a Novel Small-Scale Vertical Axis Wind Turbine Tree

Duong Minh Ngoc [1,2], **Montri Luengchavanon** [3,4], **Pham Thi Anh** [5], **Kim Humphreys** [6] **and Kuaanan Techato** [7,8,*]

1 Faculty of Environmental Management, Prince of Songkla University, Hat Yai 90110, Thailand
2 Faculty of Economics, Tay Nguyen University, Buon Ma Thuot 630000, Vietnam
3 Sustainable Energy Management Program, Wind Energy and Energy Storage Centre (WEESYC), Faculty of Environmental Management, Prince of Songkla University, Hat Yai 90110, Thailand
4 Cemme Center of Excellence in Metal and Materials Engineering, Prince of Songkla University, Hat Yai 90110, Thailand
5 Institute for Environmental and Transport Studies, Ho Chi Minh City University of Transport, Ho Chi Minh City 72308, Vietnam
6 Independent Researcher, 5561, 5th Line, Alliston ON L9R 1V2, Canada
7 Environmental Assessment and Technology for Hazardous Waste Management Research Center, Faculty of Environmental Management, Prince of Songkla University, Hat Yai 90110, Thailand
8 Program of Sustainable Energy Management, Faculty of Environmental Management, Prince of Songkla University, Hat Yai 90110, Thailand
* Correspondence: kuaanan.t@psu.ac.th; Tel.: +66-74-426843; Fax: +66-74-429758

Abstract: Are small-scale wind turbines green? In this study, we perform a 'cradle to grave' life cycle assessment of a novel domestic-scale 10 kW vertical axis wind turbine tree which uses combined Savonius and H-Darrieus blades. Situated at a test site in Surat Thani, Thailand, SimaPro software was used to evaluate the environmental impact profile of the tree. Comparisons to the Thai grid mix were made, using both with and without end-of-life treatments. Impact profiles were calculated using wind data collected over two years at Surat Thani, and from wind data from a higher capacity factor (C_F) site at Chiang Mai, Thailand. Energy and greenhouse gas payback times were estimated for both locations. The relative magnitudes of impacts were compared with environmental prices protocol, and we investigated reductions in impacts using three mitigative scenarios: changes to design, transportation and materials. The results showed that Chiang Mai had a C_F = 7.58% and Surat Thani had a C_F = 1.68%. A total of 9 out of 11 impacts were less than the grid values at Chiang Mai, but at Surat Thani, 9 of 11 impacts were more than the grid values. End-of-life treatments reduced impacts by an average of 11%. The tower and generator were majority contributors to impacts (average 69%). Greenhouse gas and energy payback times were 28.61 and 54.77 years, and 6.50 and 12.50 years for Surat Thani and Chiang Mai, respectively, with only the Chiang Mai times being less than the turbine's estimated lifetime. Location changes mitigated impacts most, followed by design, transportation, and then materials. We make recommendations to further improve the environmental impact profile of this turbine tree.

Keywords: life cycle assessment; vertical axis wind turbine; turbine tree; environmental impacts; environmental prices; renewable energy; SimaPro; energy payback time; greenhouse gas payback time

Citation: Ngoc, D.M.; Luengchavanon, M.; Anh, P.T.; Humphreys, K.; Techato, K. Shades of Green: Life Cycle Assessment of a Novel Small-Scale Vertical Axis Wind Turbine Tree. *Energies* **2022**, *15*, 7530. https://doi.org/10.3390/en15207530

Academic Editors: Yutaka Hara and Yoshifumi Jodai

Received: 14 September 2022
Accepted: 9 October 2022
Published: 12 October 2022

Publisher's Note: MDPI stays neutral with regard to jurisdictional claims in published maps and institutional affiliations.

1. Introduction

'Green' is a term associated with the reduction of the environmental impacts of technologies and products. Products involving renewable energy are often unquestioningly deemed green and are marketed as such. When measuring the greenness of a product though, a level of focus not only on the finished item is needed, because products are more correctly defined in a broader sense. A product is, in reality, the sum of all the stages of its life, including design, materials acquisition, manufacture, transportation, operation, and end-of-life treatment. Therefore, products may be more comprehensively tested for environmental impacts via a life cycle assessment (LCA). Indeed, there may be aspects of a product's life cycle that make it less meritorious of a 'green' moniker.

LCA is a general methodology that assesses the environmental impacts at all stages of products, processes, or services. Glassbrook et al., Martínez et al., and Wang and Teah are examples of studies in which LCAs have been used to assess environmental impacts, energy payback times and cumulative energy requirements of wind turbines [1–3].

Rising electrical demand, fossil fuel use and greenhouse gas concerns have led to burgeoning use of wind resources. The total global capacity of wind installations increased from 24 GW in 2001 to 591 GW in 2018, a rise of 2363% [4]. Wind energy has been regarded as having more potential than any of the other renewable energy technologies [5].

In Thailand, where our research was carried out, the cumulative wind energy capacity reached 648 MW at the end of 2018 [6], and the Thai government's aim is to generate 3000 MW by 2036 [7].

When LCAs have been used to compare wind with other 'renewable' power generation methods, mixed results have been found. The global warming potential (GWP) for wind energy in Ontario, Canada was found to be intermediate between hydro- and nuclear power [8]. Wang et al. [9] analyzed the environmental impact of hydroelectricity, wind, and nuclear power in China. Wind energy's harmful ecological effects were found most significant, followed by nuclear and hydroelectricity. Asdrubali et al. [10] reviewed 100 case studies concerning renewable energy and identified wind energy as being often lowest in environmental impacts.

Wind turbines can be classified according to their scale: *large-scale* (surpassing 1000 kW), *medium-scale* (100 kW to 1000 kW), *commercial-scale* (16 kW to 100 kW), *domestic-scale* (1.4 kW to 16 kW), *mini-scale* (0.25 kW to 1.4 kW), and *microscale* (0.004 kW to 0.25 kW) [11]. Our LCA in this paper concerns a small prototype vertical axis wind turbine (VAWT) tree which is domestic-scale, rated at 10 kW (see Figure 1).

Figure 1. The prototype VAWT tree.

LCAs of small-scale wind turbines have been hitherto few, with mixed results. A 600 W horizontal wind axis turbine (HAWT) assessed in Taiwan had generally unfavorable LCA results [3]. Kouloumpis et al. [12] analyzed the performance and impacts of a 5 kW VAWT in Poland with varied results. More favorable results were found when Lombardi et al. [5] performed LCAs of small-scale VAWTs in Italy. At a time when small-scale VAWT and HAWT technology is developing rapidly, the environmental impact of these turbines needs to be accurately assessed [3].

In Thailand, with low-to-moderate average wind velocities [13], LCAs for smaller scale turbines are rare. Uddin and Kumar [14] compared the impacts of 300 W VAWTs and 500 W HAWTs. Glassbrook et al. [1] assessed potential economic feasibility and life cycle impacts for Thai 400–20,000 W turbines.

Our research here involves an LCA of a domestic-scale VAWT tree, designed and manufactured by the engineering department at Songkla University, Thailand. The overarching design objective of this tree was to devise a VAWT that could be economically feasible in low-wind, space-limited (urban) situations. This design objective resulted in the following physical attributes, which have implications for an environmental impact assessment (see Table 1 below):

Table 1. VAWT tree design attributes and potential LCA effect.

Design Attribute	Possible Impact on LCA
The design was **small-scale (domestic)** to fit into a space-limited niche.	With decreasing scale of wind turbines and correspondingly lower outputs, it is to be expected that when environmental impacts are expressed in proportion to kWhr, LCA outcomes become less favorable. This has been reported by Uddin and Kumar [14], who found impacts in inverse proportion to capacity factor. Yildiz [15], for example, also noted the inverse relationship between turbine size and energy payback time.
This design was a **vertical axis** wind turbine to utilize fluctuating wind speed and directions in space-limited settings (vertical axis designs typically require less space than horizontal designs) [16].	VAWTs are less efficient than HAWTs due to additional drag created as blades rotate into the prevailing wind [17]. The lower energy output profile, as noted above, means environmental impacts per unit of energy produced may be relatively higher in an LCA analysis.
Since Savonius blades offer lower cut-in speeds but lower efficiency than Darrieus blades [18], our turbine tree was designed with **combined Savonius-Darrieus** turbines (see [19,20]), a design which allows the tree to be used in lower wind situations than Darrieus blades alone, but with higher efficiency than using Savonius blades alone. Additionally, since urban conditions often present complex vortex conditions with varying wind velocity, multiple or stacked turbines can act as out-of-phase generators that reduce moment fluctuations in power [21]. For this reason, a **turbine tree** of 33 paired turbines was implemented.	A turbine tree design, with its repeated use of materials for blades, rotors and generators, may have higher environmental impacts than single turbine designs.
Plastic was used in the design instead of rubber or steel for turbine blades as it was found to have higher performance efficiency [22].	The choice of plastic, rather than a more 'environmentally friendly' material, may result in higher LCA impacts.

The research aims of this study are unique in four ways. First, our research concerns a newly designed VAWT tree, designed for economic feasibility rather than being solely aesthetically pleasing (an economic feasibility study using this design was carried out by Ngoc et al. [23]). Second, few previous LCAs have focused on domestic-scale VAWTs. Third, since LCA research regarding wind turbine feasibility is scarce in Thailand and in most of Southeast Asia, our research augments the LCA literature for this region. Finally, our LCA uses environmental prices [24] to compare impacts with grid values, hitherto rarely carried out for wind systems.

2. Materials and Methods

An LCA may be conducted in four phases [25]:

1. Determination of goals and scope (see Section 2.1 below);
2. Inventory analysis (see Section 2.2 below);
3. Impact assessment throughout the life cycle (see Section 2.3 in Sections 2 and 3);
4. Life cycle interpretation (see Sections 4 and 5).

2.1. Goals and Scope

2.1.1. Goals

Our goals centered on these questions which guided research:

1. How do environmental impacts of the VAWT tree at the lower-wind-speed location of Surat Thani and the higher-wind-speed location of Chiang Mai compare with impacts from the Thai grid mix when assessed with and without an end-of-life option?
2. Which component materials of the VAWT tree contribute most to impacts?
3. What are the energy and greenhouse gas (GHG) payback times at the lower-wind-speed location and the higher-wind-speed location?
4. If the impacts of the VAWT tree are compared with each other using a common basis of comparison, which impacts are most significant?
5. To mitigate important impacts, which change in life cycle aspect would reduce impacts most: transportation, design, or materials? How do these life cycle changes compare with alterations to impacts which result from location change?

2.1.2. Scope

This LCA was 'cradle to grave', using SimaPro 9.3.0.3. software [26].

For the two locations considered, we examined all life cycle stages: raw material acquisition, manufacturing, and transportation of the VAWT components (foundation, tower, generator, inverter, and cabling), installation, operation and maintenance, and end-of-life treatment (see Figure 2 for LCA delineation). The electricity generated enters the local grid (under EGAT—Electricity Generating Authority of Thailand).

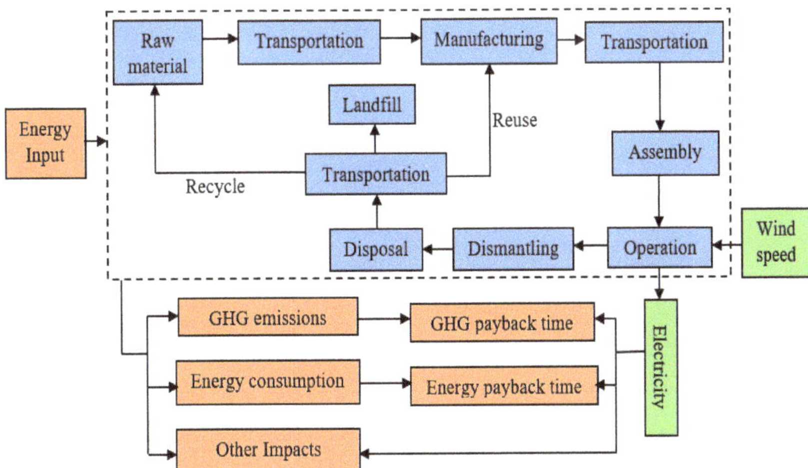

Figure 2. LCA system boundary and impact assessments. The dashed box contains the life cycle components.

2.2. Life Cycle Inventory

The tree consists of thirty-three paired turbines (each pair with a nominal power of 320 W, combining 3 Savonius and 3 Darrieus blades, stacked vertically (see Figure 3)).

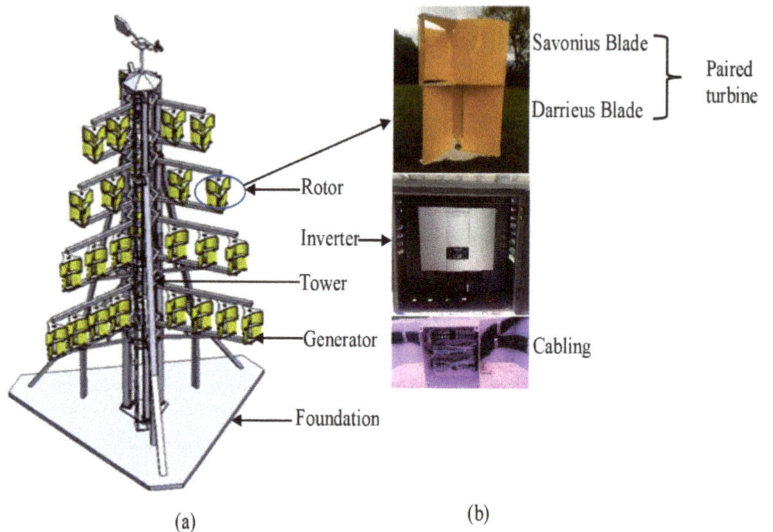

Figure 3. Wind turbine tree (**a**) and turbine structure (**b**).

The dual blade arrangement was designed for low wind sites, having a low cut-in speed (2 m/s). For operational and dimensional details, see Table A1, Appendix A.

System components—material and process inputs for manufacturing and transportation—were matched with relevant background datasets available in Ecoinvent 3 within SimaPro [26].

2.2.1. Raw Material Acquisition and Manufacturing

The VAWT has six material components: the rotor, generator, inverter, cable and controller, tower, and foundation, (see Figure 3). The rotor consists of the blades on the axle. Table A2, Appendix A provides Ecoinvent rotor constituent details. The generator is a synchronous motor with a neodymium magnet (NdFeB). See Table A3 in Appendix A for Ecoinvent inputs. The electrical grid connection comprises an inverter, controller and cables (see Table A6, Appendix A). Composition and proportions for the 5 kW inverter (Shenzhen (China) INVT Electric, Shenzhen, China) were estimated from their website (INVT, 2019) [27]. The tower consists of three 12 m steel H-beam columns reinforced by steel cross struts, supporting four levels of twelve arms hanging 33 turbines. See Table A4 in Appendix A for Ecoinvent inputs. The tree is embedded into a concrete foundation. The foundation contains nine rebar reinforced concrete columns (1.21 × 0.3 × 0.3 m each) and these have thirty-six 0.4 m rebar elements inserted at the top. The equilateral base is rebar-reinforced concrete (sides 6.65 m, 0.15 m thick). See Table A5 in Appendix A for Ecoinvent inputs.

2.2.2. Operation and Maintenance

The turbine does not require any maintenance, lubrication, materials, or energy inputs during its life after commencing operation.

2.2.3. Transport

Transportation was within Thailand (excepting the inverter, which was from China), and included raw material and component delivery and end-of-life disposal. Ecoinvent inputs determined impacts from transport, including fuel extraction, production and use. For calculations, the unit ton-kilometer (tkm) was used. Ecoinvent component details and transportation distances are provided in Table A7 (Appendix A). For Surat Thani, the inverter was assumed to be transported 3015 km by ship from Guangdong to Songkhla

Port, then 360 km to the site by commercial vehicle. Concrete was assumed to be produced in Surat Thani and shipped 80 km by cement mixer truck. For Chiang Mai, the inverter was assumed to be transported 3500 km by ship from Guangdong to Bangkok Port, then 686 km to the site by commercial vehicle. Concrete was assumed to be produced in Chiang Mai and shipped 80 km by cement mixer truck.

2.2.4. Disposal and Recycling

The VAWT was still in operation during this research, so end-of-life forecasting is uncertain. We considered two end-of-life Scenarios for both locations, Surat Thani and Chiang Mai: 'A' (do nothing—our 'base case') and 'B' (reuse, recycle and dispose). For A, the VAWT would be left 'as is'. In B, the foundation would remain in the ground. The permanent magnets are reused [14] since recycling processing is rare [12]. Only glass-fibre-reinforced plastic and paint were assumed 100% disposed of [5]. Other materials were assumed partly disposed of, partly recycled. Aluminum, steel and iron, and copper were considered recyclable at 90% and 95%. Materials and treatments in Scenario B are given in Table A8 (Appendix A).

2.2.5. Turbine Performance and Wind Speed

To compare with other wind turbines, impacts were expressed relative to performance (i.e., power generated) in SimaPro [26], and expressed per kWh. Energy and greenhouse gas payback times are thus functions of local annual wind speed. Power and wind speeds were handled as follows.

Lifetime power generated from the VAWT system was calculated as in [12]:

$$P_{Out} = 8760 \cdot C_F \cdot P_{Rp} \cdot T, \tag{1}$$

with P_{Out} the output power (kWh), and P_{Rp} the rated power (kW), C_F the capacity factor, and T the system lifetime (years). With output power in kWh and lifetime T in years, the constant 8760 h/year was used.

Although uncertain, we conservatively assume the VAWT will have a 20-year minimum lifetime, based on consideration of components. This value has been used in similar studies [5,28], facilitating comparison of results.

The capacity factor is defined as the actual electricity generated by the wind turbine divided by the theoretical maximum amount that can be generated nominally. For our VAWT, based on Equation (1), P_{Out} was 87,600 kWh yearly (P_{Rp} being 10 kW).

Data was collected for the VAWT at Surat Thani for two years (May 2019–April 2021). Monthly electricity generated and average wind speeds were sourced from Ngoc et al. [23].

The above considerations yield power output and capacity factors for Surat Thani and Chang Mai as follows (Table 2):

Table 2. Average annual wind speed and capacity factors for Surat Thani and Chiang Mai.

Location	Average Annual Wind Speed	Reference	Capacity Factor (C_F)
Surat Thani	2.58 m/s	Ngoc et al. [23]	1.65%
Chiang Mai	4.8 m/s	Chaichana and Chaitep [29]	7.58%

2.2.6. Thailand Electrical Grid Mix

We compared impacts from our VAWT with the existing Thailand electrical low-voltage grid (2018 data from SimaPro [26]). The Thai mix sources in 2018 are shown in Figure 4. The large dependency on non-renewable fossil fuels makes sourcing renewable energy desirable in Thailand.

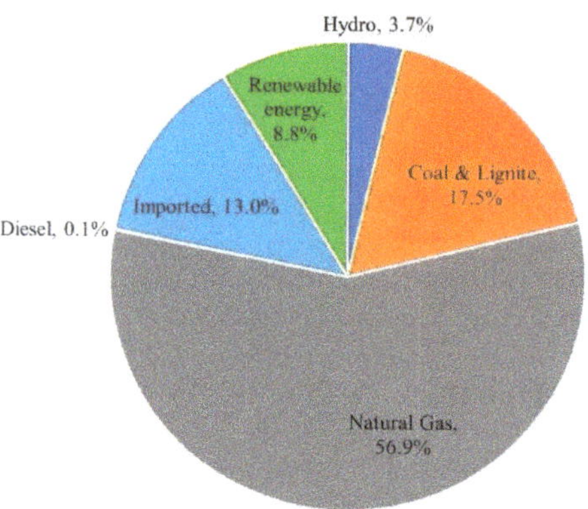

Figure 4. Thailand electrical grid mix [30].

2.3. Life Cycle Impact Assessment (LCIA) Methods

We performed four major analyses for our LCIA, all via SimaPro [26], as listed in Table 3:

Table 3. LCIA Analyses.

Analysis	Goals	Section
Comparison of Scenarios A, B and grid mix (using CML-IA)	1, 2	Section 2.3.1
Energy/GHG payback time (CML-IA) (using Equations (2) and (3) below)	3	Section 2.3.2
Environmental price impacts (using IPCC and ReCIPe 2008, [26]	4	Section 2.3.3.1
Mitigative strategies for impacts of concern (CML-IA)	5	Section 2.3.3.2

2.3.1. CML-IA Baseline 2000 V3.05 Method

We used the CML-IA baseline 2000 V3.05 method when comparing Scenarios A and B for both Surat Thani and Chiang Mai against the Thai grid. Widely used and with clear interpretation, the equal weighting of 11 impact categories also facilitates comparison [12]. CML-IA baseline results are focused on midpoint (unitary environmental problem) indicators [26], and as such, clarity in terms of their cumulative contributions to aggregate impacts (endpoint indicators) is limited [31]. For this reason, we also evaluated impacts using the environmental prices method (see Section 2.3.3.1), which has a common (monetary) impact assessment, facilitating between-category and aggregate category comparison.

2.3.2. Cumulative Energy Demand (CED): Energy, GHG Payback Times and Component Contribution

Two insightful measures add understanding to LCA impacts. One measure, the energy payback time (E_P), compares impacts with the time it takes to generate the same electricity as it took to fabricate, transport, and install the VAWT (see Equation (2) below) [28]. Second, after commencing operation, the VAWT 'replaces' grid mix electricity (significantly fossil fuel-based), and the amount of time it takes to replace GHGs produced by fabricating, transporting, and installing is the greenhouse gas payback time, GHG_P (Equation (3) below) [3]. To calculate GHG_P, we used total greenhouse gas emissions as calculated in SimaPro [26] for all life stage components (GHG_k, in Equation (3)) divided by annual emissions, converting turbine energy output to emissions with conversion factor 0.483 kgCO$_2$eq/kWh for Thailand [32].

From Wang and Teah [3]:

$$E_P = \sum_{k=1}^{n} \frac{E_k}{E_{annual}}, \tag{2}$$

$$GHG_P = \sum_{k=1}^{n} \frac{GHG_k}{0.483\, E_{annual}},\tag{3}$$

E_K is the energy consumed during the VAWT's life cycle. E_{annual} indicates the annual electricity produced by the VAWT.

Using the CED 1.11 method in SimaPro [26], we obtained results for energy payback time, and used CML-IA baseline 2000 V3.05 results to calculate GHG_P. Finally, to examine how payback times might be affected by the use of our prototype in locations with higher wind capacity, we also used a C_F = 7.58% (corresponding to average wind speed = 4.8 m/s) at Chiang Mai [23]. Additionally, when implementing CED, we assessed the proportion of six energy sources (non-renewable fossil, nuclear and biomass, and renewable biomass, water and wind/solar/geothermal) used in the five material components.

The above two analyses allow us to understand impacts with respect to component contributions in life stages with and without end-of-life recycling, and with respect to the Thai grid mix.

2.3.3. Comparative Analyses

In this section, we perform two further comparative analyses. First, we make a life cycle assessment considering the relative severity of impacts, using a common basis of comparison (See Section 2.3.3.1 below). Second, we consider four impacts that might be of most concern to modern society and investigate three mitigative scenarios regarding those impacts (See Section 2.3.3.2 below).

2.3.3.1. Environmental Prices of Impacts

In CML-IA, impact results have different, category-specific units, occluding comparison between different impact categories. An in-common means of comparison between different categories is important when prioritizing alterations to aspects of life cycles that will be most effective in reducing overall impacts. In 2018, Bruyn et al. [24] developed a protocol using shadow prices for monetization of environmental impacts arising in LCAs. The method assesses economic welfare lost when impacts occur, estimating mitigation costs (using midpoint values). We used our Scenario A with environmental prices (based on ReCiPe [33] and IPCC [34] protocol) in SimaPro [26], which expresses prices in USD (the in-common basis of comparison).

Interpretations using environmental prices require caution. First, derivations from shadow prices are necessarily estimates only. Second, Bruyn et al. [24] have hitherto only published data for prices per impact unit in 2015, so estimates after this date are likely too conservative. Third, since prices are derived from European estimates, application to other regions entails further estimation due to differences in income levels and costs of living.

Addressing this last issue, we used a unit transfer, with income adjustment as per Navrud [35] in Equation (4):

$$UP_{THL} = UP_{EU} \left(\frac{GDP_{THL}}{GDP_{EU}} \right)^{\epsilon},\tag{4}$$

UP_{THL} is the unit environmental price in Thailand; UP_{EU} is the unit environmental price in Europe. Europe, ϵ, is the income elasticity. GDP_{THL} and GDP_{EU}, the GDP values for Thailand and Europe, were as assigned in Table 4.

Table 4. GDP per capita (2015, USD) used to estimate environmental prices.

Country	GDP (per Capita)	Source
Europe	25,920	Ghani et al. [36]
Thailand	5840	Macrotrends [37]

2.3.3.2. Comparative Analysis of Three Scenarios on Four Impact Areas

Of the eleven impact categories in CML-IA, among the top globally important impacts of concern today are global warming, ozone depletion, abiotic (resource) depletion, and human toxicity. Considering the likelihood that environmental impacts per kWh were likely to be greater for the low-wind (Surat Thani) location, we sought to assess how these four impacts might be mitigated by changes to life cycle elements. We investigated three different alterations to our Scenario A for this location in SimaPro [26], analyzing with the CML-IA method.

Tremeac and Meunier [28] varied the distance and type of transportation in an LCA for two wind turbines, and found that these could significantly lower impacts in relation to their reference case. In *Scenario 1*, we imagined changing the main method of transport from 'Light Commercial Vehicle' (i.e., by road) to train.

We reasoned, due to construction constraints, that substituting materials for the tower, foundation, and generator would be difficult, but substituting turbine blade material might be possible. Uddin and Kumar [14] found LCA impact reductions when varying materials in turbine construction. Yildiz [15] reported that the choice of steel has been found to have a lower environmental impact than other materials in wind turbine installations. In *Scenario 2*, we investigated changing blade material from glass-fibre-reinforced plastic to stainless sheet steel.

In a life cycle assessment of onshore wind turbine towers by Gkantou, Rebelo and Baniotopoulos [38] involving four- and six-leg hybrid towers, the former was found to result in less environmental impact than the latter. We imagined a design amendment in our VAWT configuration that might accomplish a similar reduction without compromising power output. Although our turbine uses three arms separated by 120 degrees, this configuration may not be optimal when prevailing winds are parallel to one arm. In a case where wind enters an "open V" of two of our wind tree arms, the third arm is mostly in an inefficient wind shadow area [23]. In Surat Thani, where wind speeds are often only slightly greater than the turbine cut-in speed, turbines in a shadow area will contribute minimally to the tree's output. We imagined an altered design where the tree always has an open 'V' optimally facing the wind (i.e., is rotatable). In our *Scenario 3*, the third (wind shadow) arm of the wind tree was eliminated. We can then lower materials used by one third. In this scenario the tree has 22 turbines, keeping the same foundation and infrastructure as necessary to support the 22 turbines. Thus, material weights entered into SimaPro [26] were reduced by a third for the rotor, generator, cable, and tower components. In this scenario we kept the same energy output, assuming the contribution from the 11 eliminated turbines in wind shadow was negligible.

A summary of the specific alterations made in each Scenario above is provided below in Table 5.

Table 5. Changes to CML inputs for Scenarios 1 to 3.

Input/Component Changed		Value in S.T. Base Case Scenario	Value in New Scenario; Scenario Number
Transportation method		Light commercial vehicle	Train; Scenario 1
Material: Turbine blades		Glass fibre-reinforced plastic	Stainless sheet steel; Scenario 2
Design	(a) rotor number	33	22; Scenario 3
	(b) tower weight (kg)	2894.43	2664.012; Scenario 3

After examining the effect of these alterations on impacts over the Surat Thani base case, we then compared their magnitude to the impact changes produced when switching the location of the VAWT tree from Surat Thani to Chiang Mai (see Sections 2.1 and 2.3.1), under Scenario A and with no other change in Scenario.

Finally, to contextualize improvements produced under Scenarios 1 to 3, we compared the eleven new CML-IA numerical impact values of the three Scenarios to the Thai low-voltage grid mix and to our base case Scenario A in Surat Thani. We inspected data for any of the eleven impacts which reduced in value to below those of the base case or Thai grid values.

Since power output is directly related to capacity factor (Equation (1)), and with SimaPro [26] impacts expressed per kWh, we have:

$$\frac{I_a}{I_b} = \frac{C_{Fb}}{C_{Fa}},$$

(5)

where I_a, I_b are impacts for C_{Fa}, C_{Fb}.

Using this relation, we calculated hypothetical C_Fs that would be necessary for turbine impacts to equal grid impacts.

3. Results

3.1. CML Base Case Analysis

3.1.1. CML Impacts per kWh for Scenarios A and B and Thailand Energy Mix

Table 6 presents the impact results from SimaPro [26] for each scenario and the Thailand grid mix for comparison. Two general trends are clear. First, Scenario A had higher impacts in all categories than Scenario B, for both locations. An end-of-life scenario, when included, lowered impacts by an average of about 11% overall for both locations (11.40% and 10.97% for Surat Thani and Chiang Mai, respectively). Second, despite the Thai grid being largely sourced from fossil fuels (Figure 4), both Scenarios A and B for the prototype VAWT at the low-wind-speed location of Surat Thani had impacts that were usually greater than the Thailand grid impacts (9 out of 11 impacts), although, more encouragingly, both Scenarios A and B at Chiang Mai with a higher wind speed had impacts that were usually less than the Thailand grid impacts (only 2 out of 11 impacts were greater than the grid).

Table 6. Scenarios A and B compared with Thailand low-voltage grid impacts (per kWh).

	Impact Category	Unit	Surat Thani		Chiang Mai		Thailand Low-Voltage Grid
			Scenarios A	Scenarios B	Scenarios A	Scenarios B	
1	Abiotic depletion	kg Sb eq	2.44×10^{-5}	2.43×10^{-5}	5.31×10^{-6}	5.30×10^{-6}	5.54×10^{-7}
2	Abiotic depletion (fossil fuels)	MJ	7.71	6.37	1.68	1.39	8.64
3	Global warming (GWP100a)	kg CO_2 eq	6.90×10^{-1}	4.90×10^{-1}	1.50×10^{-1}	1.09×10^{-1}	7.03×10^{-1}
4	Ozone layer depletion (ODP)	kg CFC-11 eq	7.03×10^{-8}	6.41×10^{-8}	1.54×10^{-8}	1.40×10^{-8}	2.98×10^{-8}
5	Human toxicity	kg 1,4-DB eq	2.36	2.29	5.10×10^{-1}	4.99×10^{-1}	2.50×10^{-1}
6	Fresh water aquatic ecotoxicity	kg 1,4-DB eq	8.90×10^{-1}	8.70×10^{-1}	1.90×10^{-1}	1.90×10^{-1}	4.30×10^{-1}
7	Marine aquatic ecotoxicity	kg 1,4-DB eq	3.28×10^{3}	3.16×10^{3}	7.14×10^{2}	6.87×10^{2}	9.37×10^{2}
8	Terrestrial ecotoxicity	kg 1,4-DB eq	6.37×10^{-3}	6.08×10^{-3}	1.38×10^{-3}	1.32×10^{-3}	3.50×10^{-3}
9	Photochemical oxidation	kg C_2H_4 eq	3.21×10^{-4}	2.12×10^{-4}	7.00×10^{-5}	4.62×10^{-5}	8.79×10^{-5}
10	Acidification	kg SO_2 eq	4.57×10^{-3}	3.90×10^{-3}	9.98×10^{-3}	8.50×10^{-3}	2.03×10^{-3}
11	Eutrophication	kg PO_4 eq	2.85×10^{-3}	2.63×10^{-3}	6.21×10^{-4}	5.73×10^{-4}	1.54×10^{-3}

3.1.2. Analysis of Component Contribution

In Figure 5a,b, we present the results of impact contributions from six of the components of the VAWT, those relating to the extraction of raw materials, manufacturing, transportation, and installation and excluding the end-of-life treatment (i.e., Scenario A) for Surat Thani and Chiang Mai. Noteworthy here are the large contributions of the generator and the tower to impacts (69.51%, on average, for Surat Thani and 68.18% for Chiang Mai). These two components usually rank one and two (in 10 out of 11 impact categories for Surat Thani; in 8 out of 11 at Chiang Mai). The other components contributed considerably less (foundation, rotor, cable and controller).

Regarding global warming potential, the tower, the generator, and foundation accounted for most of the impacts: 77.60% for Surat Thani, and 78.54% for Chiang Mai. These components are also the highest contributors toward ozone depletion. For aquatic impacts (eutrophication, acidification, marine aquatic ecotoxicity, fresh water aquatic ecotoxicity), the generator and tower were the major contributors. Some of the components were higher in one particular category (for example, the inverter contributes an anomalously high percentage to abiotic depletion)—see Figure 5a,b and Table 7 for details.

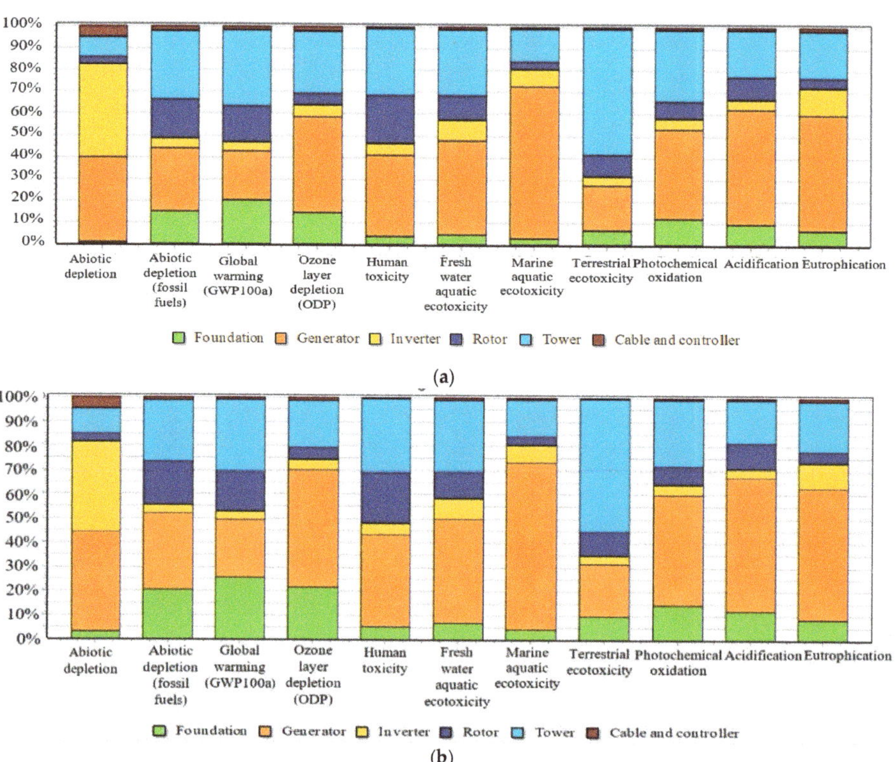

Figure 5. (**a**) Environmental impacts of the VAWT per component (Surat Thani); (**b**) environmental impacts of the VAWT per component (Chiang Mai).

Table 7. Component contribution (%) to impacts.

Impact Category	Foundation		Generator		Inverter		Rotor		Tower		Cable and Controller	
	S. T.	C. M.	S. T.	C. M.	S. T.	C. M.	S. T.	C. M.	S. T.	C. M.	S. T.	C. M.
Abiotic depletion	2.54	3.31	41.30	41.01	37.40	37.12	3.57	3.54	10.20	10.14	4.92	4.88
Abiotic depletion (fossil fuels)	16.20	20.64	33.1	31.37	3.62	3.43	19.20	18.23	26.4	25.01	1.40	1.32
Global warming (GWP100a)	22.10	25.40	24.90	23.84	3.59	3.44	17.50	16.79	30.60	29.30	1.28	1.23
Ozone layer depletion (ODP)	15.90	21.55	52.10	48.58	4.45	4.15	5.62	5.24	20.4	18.99	1.59	1.49
Human toxicity	4.91	5.24	38.00	37.87	4.81	4.79	21.10	21.03	30.60	30.47	0.60	0.60
Fresh water aquatic ecotoxicity	6.31	6.71	43.30	43.11	8.55	8.52	11.00	10.92	29.80	29.71	1.04	1.04
Marine aquatic ecotoxicity	3.88	4.18	69.50	69.24	6.88	6.86	3.92	3.91	15.00	14.93	0.89	0.89
Terrestrial ecotoxicity	8.81	9.49	21.90	21.72	3.74	3.71	9.85	9.78	55.20	54.75	0.55	0.55
Photochemical oxidation	13.20	14.46	45.90	45.26	4.32	4.26	8.10	7.99	27.60	27.25	0.81	0.79
Acidification	11.00	12.23	55.70	54.95	3.71	3.65	10.70	10.51	18.00	17.72	0.96	0.94
Eutrophication	9.76	8.47	54.40	54.07	10.8	10.77	4.69	4.66	20.80	20.65	1.39	1.39

3.1.3. Cumulative Energy Demand (CED): Energy, GHG Payback Times and Component Contribution Results

For Surat Thani and Chiang Mai, the total primary energy consumption was 285,114 MJ and 298,840.6 MJ, respectively, and their annual energy output was 1446.1 kWh and 6640.08 kWh, respectively. From Equations (2) and (3) in Section 2.3.2 above, the energy and GHG payback times were determined to be 54.77 and 28.61 years for Surat Thani and 12.50 and 6.50 years for Chiang Mai (Table 8). When located at the higher wind speed site, Ep and GHGp for the VAWT reduced significantly, by 77.17% and 77.27%, respectively—see Table 8.

Table 8. Energy and GHG payback time for Surat Thani (C_F = 1.65%) and Chiang Mai (C_F = 7.58%).

Location	Capacity Factor	Energy Payback Time (E_P) (Years)	GHG Payback Time (GHG_P) (Years)
Surat Thani	1.65%	54.77	28.61
Chiang Mai	7.58%	12.50	6.50

In Figure 6, component contributions as calculated in SimaPro to the depletion of renewables and non-renewables are shown for the C_F = 1.65% location (results were nearly identical for the C_F = 7.58% case). Again, the generator and tower are the most significant impact sources. The generator impacts non-renewable sources heavily, while the tower affects renewables more than non-renewables.

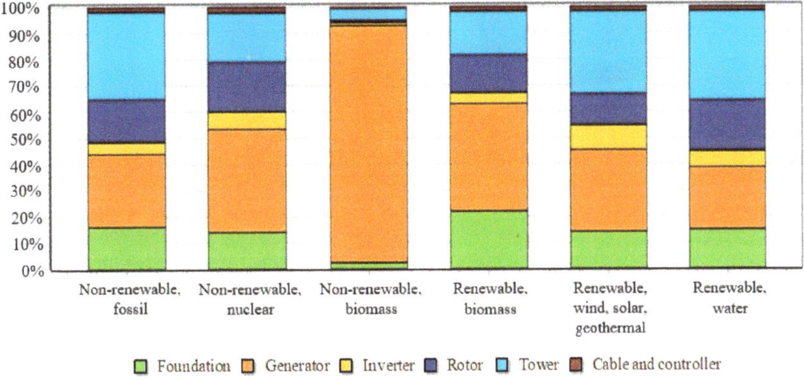

Figure 6. Cumulative energy demand (CED) for renewables and non-renewables per component.

3.2. Environmental Prices Analysis Results

Using our base case scenario for the wind tree prototype and the low-voltage grid, environmental prices were calculated for impacts among 14 categories in the ReCiPe [33] paradigm, and are given in Table 9. These are depicted according to proportion in Figure 7. For the C_F = 1.65% location, impact costs were higher for the VAWT than the Thai grid in every category but one, whereas for the C_F = 7.58% location, costs were lower than the Thai grid in every category but two. The total cost of VAWT impacts at C_F = 1.65% is nearly three times as high as the Thailand low-voltage supply, whereas the total cost of VAWT impacts at C_F = 7.58% is about one third *less* than the Thailand low-voltage supply.

Table 11. Scenarios 1 to 3 vs. Surat Thani, Chiang Mai base cases and Thai low voltage mix impacts (per kWh). Light green indicates values less than the S. T. base case only; dark green indicates values less than both the S. T. base case and the Thai low voltage mix.

	Base Case Surat Thani	Base Case Chiang Mai	Scenario 1: Train	Scenario 2: New Material	Scenario 3: New Design	Thai Low-Voltage Grid
Abiotic depletion	2.44×10^{-3}	5.31×10^{-6}	2.42×10^{-5}	2.45×10^{-5}	1.88×10^{-5}	5.54×10^{-7}
Abiotic depletion (fossil fuels)	7.71	1.683	7.013	7.137	6.162	8.638
Global warming	6.89×10^{-1}	1.5×10^{-1}	6.43×10^{-1}	6.57×10^{-1}	5.71×10^{-1}	7.03×10^{-1}
Ozone layer depletion	7.03×10^{-8}	1.54×10^{-8}	6.18×10^{-8}	7.00×10^{-8}	5.52×10^{-8}	2.98×10^{-8}
Human toxicity	2.36	5.14×10^{-1}	2.35	2.42	1.83	2.45×10^{-1}
Fresh water aquatic ecotoxicity	8.99×10^{-1}	1.95×10^{-1}	8.95×10^{-1}	9.06×10^{-1}	7.05×10^{-1}	4.27×10^{-1}
Marine aquatic ecotoxicity	3.28×10^{3}	7.14×10^{2}	3.27×10^{3}	3.28×10^{3}	2.40×10^{3}	9.37×10^{2}
Terrestrial ecotoxicity	6.40×10^{-3}	1.40×10^{-3}	6.30×10^{-3}	6.40×10^{-3}	5.40×10^{-3}	3.50×10^{-3}
Photochemical oxidation	3.21×10^{-4}	7.00×10^{-5}	3.15×10^{-4}	3.28×10^{-4}	2.56×10^{-4}	8.79×10^{-5}
Acidification	5.00×10^{-3}	9.98×10^{-4}	4.00×10^{-3}	4.00×10^{-3}	3.00×10^{-3}	2.00×10^{-3}
Eutrophication	3.00×10^{-3}	6.21×10^{-4}	2.00×10^{-3}	3.00×10^{-3}	2.00×10^{-3}	1.00×10^{-3}

C_F values allowing the prototype to have impacts equivalent to the grid were calculated (Equation (5)) and are shown in Table 12. Raising the capacity factor to 4% from 1.65% at Surat Thani would change the number of impact categories that are less than the grid from 2 to 7, indicative of the importance of location selection (9 of 11 impacts at Chiang Mai are already less than the grid equivalents at Chiang Mai's C_F of 7.58%). At both of these locations, however, two impacts—abiotic depletion and human toxicity—would still be dramatically higher than grid values even if a very high-capacity factor were attained.

Table 12. C_F values for impacts equivalent to Thai grid.

Impact Category	S.T. (C_F)	C.M. (C_F)
Abiotic depletion	72.62	72.49
Abiotic depletion (fossil fuels)	1.47	1.22
Global warming	1.62	1.17
Ozone layer depletion	3.89	3.57
Human toxicity	15.92	15.42
Fresh water aquatic ecotoxicity	3.48	3.36
Marine aquatic ecotoxicity	5.78	5.56
Terrestrial ecotoxicity	3.01	2.87
Photochemical oxidation	6.03	3.99
Acidification	3.73	3.18
Eutrophication	3.06	2.82

4. Discussion

In Surat Thani, the CML analysis showed Scenario A and Scenario B usually produced *higher* environmental impacts than the Thai low voltage grid. At Chiang Mai, however, the CML analysis showed Scenario A and Scenario B usually produced *lower* environmental impacts than the Thai low voltage grid. These results are concomitant with the fact that the impacts in SimaPro [26] are expressed per kWh (at Surat Thani, our tree had low capacity because of the combination of small size and low wind environment). While scaling up the dimensions of wind turbines can result in proportionally less impacts per kWh due to capacity increases, as well as scaling factors and manufacturing experience [39], scaling up at the Surat Thani site would likely not cause reductions due to consistently low wind speeds.

Environmental impacts can be reduced after decommissioning when there is potential for material reuse and recycling [17], and in our assessment the inclusion of an end-of-life treatment scenario always lowered impacts for both locations. For both scenarios in both regions, however, for with and without an end-of-life treatment, the categories of human toxicity and abiotic depletion remained several times larger than grid values, so mitigation methods targeting these might be investigated.

Our finding that tower and generator components are leading contributors to impacts in both locations is consistent with other LCAs on VAWTs (see, for example [12,14]). These components are energy intensive [14] due to constituent materials and masses. In our case, the generator components are often proportionally higher than that found for LCAs of other *single* wind turbine arrangements, because our design involves a 'tree' of 33 turbines, thereby replicating the generator components 33 times. We therefore consider these components, along with the tower and generators, to be the most important subjects of research into material alternatives and/or reductions from structural redesign, en route to a wind tree that has a greener life cycle.

For Surat Thani, (C_F = 1.65%) energy and GHG payback times exceeded the VAWT's estimated lifetime. While calculated times were less than that found for some studies [3], they were more than others [14]. For Chiang Mai (C_F = 7.58%), both energy and GHG payback times were dramatically reduced to within the estimated VAWT's lifetime, indicating the practicality of using this wind tree design at sites of comparable wind speed. Regarding the cumulative energy demand, we again note the disproportionate contributions of the tower and generator to GHGs and non-renewable fossil fuel depletion, and therefore recommend again that these components be scrutinized for alternatives that mitigate these important impacts.

Comparing impacts using environmental prices, our prototypes in both locations were similar to the Thai grid, in that the four largest impact areas were the same. While all impacts are important, it is perhaps disappointing that these aforementioned largest impacts—human toxicity, particulate matter formation, climate change, and terrestrial acidification—are among those impacts of current global concern.

Comparing our mitigative strategies in Scenarios 1 through 3, it is clear that the proposed design change (Scenario 3—removing an arm in a 'wind shadow' area) was most effective in reducing environmental impacts of concern. Locating the wind tree at the Chiang Mai site with higher capacity would reap even more benefit. The Chiang Mai base case produced less impacts per kWh than all the other changes implemented in Scenarios 1 to 3.

When considering the entirety of impact categories against the C_F = 1.65% base case and Thai grid, design and transportation alterations made the VAWT tree greener, but only location changes made it 'truly green', as only then did it largely outperform the Thai grid. Having half of the impacts of the grid is very feasible (requiring a C_F = 4%, corresponding to a wind speed between 3 to 4 m/s). However, reducing all impacts to under grid values is likely unachievable with this VAWT due to the high corresponding wind speeds necessary to ameliorate human toxicity and abiotic depletion impacts (see Section 3.3).

Other mitigative alterations might be considered. Situating the prototype on the roof of a suitably constructed building could lower foundational component impacts considerably, and increase C_F values as winds increase with height. Ocean and lake shorelines with convectional winds will similarly lead to higher C_Fs.

These findings underscore the importance of planning. Life cycle assessments, as exemplified by this study, reveal that design and site decisions are as important to environmental concerns as they are to engineering concerns [40].

5. Conclusions

We carried out a life cycle assessment on a novel, 10 kW vertical axis wind turbine tree developed by Prince of Songkla University in Thailand, combining Savonius and Darrieus blades for a low cut-in speed. Using calculated capacity factors for Surat Thani [23] and Chiang Mai [29], we comprehensively assessed environmental life cycle impacts using SimaPro [26]. Table 13 summarizes our analyses, findings and recommendations.

Table 13. Summary of life cycle analyses, results and conclusions/recommendations.

Analysis	Results	Conclusions/Recommendations
CML-IA base case analysis, Scenarios A, B (no end-of-life, end-of-life treatment)	In Surat Thani, (C_F = 1.65%), Scenario A and B impacts usually *higher* than Thai grid, only lower for global warming and abiotic depletion (fossil fuels); in Chiang Mai, (C_F = 7.58%), Scenario A and B impacts usually *lower* than Thai grid, only higher for human toxicity and abiotic depletion; recycling and reuse is effective (Scenario B impacts always lower than Scenario A); end-of-life treatment lowers impacts (both locations) by an average of 11%	This VAWT has environmental benefits. Site selection is important. Regarding impacts, the VAWT outperformed the Thai grid at C_F = 7.58% but underperformed the grid at C_F = 1.65%. C_F > 4% recommended to halve impacts. Recycling and reuse strategies should be incorporated into planning. Materials with higher recycling potential should be used.
Component Contribution to Impacts	Tower and generator dominate impacts in 10 of 11 categories.	Research into material alternatives or structural redesign, especially with respect to tower and generator.
Energy and GHG Payback Times	At C_F = 7.58%, energy and GHG payback times are within estimated VAWT lifespan, but exceed lifespan for C_F = 1.65%. CED indicates tower and generator deplete renewables and non-renewables the most.	Installation at locations of higher windspeed (C_F > 4%) is desirable. Research alternate materials for tower and generator.
Environmental Prices	Monetizing impacts with environmental prices indicate four impact categories (human toxicity, particulate matter formation, climate change and terrestrial acidification) account for the majority (94% or 95%) of costs. These percent contributions are independent of C_F.	Research into design, materials and transportation alternatives to mitigate costs from climate change, particulate matter, human toxicity and terrestrial acidification impacts.
Mitigative Alterations to Life Cycle (Scenarios 1 to 3)—effects on GHG, Human Toxicity, Ozone Depletion, Abiotic Resource Depletion; Grid impact C_F equivalents.	The ranking of the potential of mitigative alterations to make the VAWT more environmentally friendly (than grid impact values) is: 1. C_F, 2. design changes, 3. transportation changes, 4. material changes (in that order). C_F = 4% means at least 60% of impact categories are less than grid impact values.	During planning, consider C_F at site with respect to cut-in, rated wind speed, and calculate the C_F necessary to achieve impacts equivalent to or less than grid impacts. Consider prevailing wind direction at site with respect to redesigns. Optimize transportation to reduce fossil fuel consumption.

Space and infrastructure limitations often make a case for the use of small-scale wind turbines, but their application has often been considered limited in terms of economic [41] and environmental [3,12] feasibility. Here, we report favorable results of a life cycle assessment of a novel, domestic-scale hybrid-blade VAWT. When used at a location with C_F = 7.58%, the VAWT has less environmental impacts than grid impacts.

Alternative energy products are sometimes assessed unquestioningly as 'green', or assessed as such based upon a less-than-comprehensive accounting of environmental impacts. Life cycle assessments are an attempt to comprehensively and objectively quantify impacts. In this LCA, we identified aspects of the life cycle of a domestic-scale VAWT that allow it to be 'greener' than the extant electrical grid. Locating in situations of at least 3–4 m/s means a majority of impacts drop below grid values. Design and transportation changes can also improve impacts on the environment, but to a lesser extent.

In terms of further research, optimization of design aspects of this VAWT may lead to environmental suitability of this design in even lower wind speed areas. Future design and material alternatives should focus on tower and generator elements, particularly in relation to the four major contributors to environmental cost that were identified in this research.

Author Contributions: Conceptualization, analyses, methodology, writing (original draft)—D.M.N.; writing (review and editing)—D.M.N., K.H., M.L. and K.T.; supervision—M.L. and K.T.; software—P.T.A. All authors have read and agreed to the published version of the manuscript.

Funding: This research was funded by a Ph.D. research grant [TEH-AC 046/2018], Sustainable Energy Management—Prince of Songkla University, Hat Yai, Thailand. Wind tree construction was subsidized by the Electricity Generating Authority of Thailand (EGAT) (61-F405000-11-IO. SS03F3008347).

Data Availability Statement: Not applicable.

Acknowledgments: The work is supported by the Wind Energy and Energy Storage Centre (WEESYC), the Energy Technology Research Center (ETRC), the Centre of Excellence in Materials Engineering (CEME), and Prince of Songkla University.

Conflicts of Interest: The authors declare no conflict of interest.

Appendix A

Table A1. Dimensions and operational characteristics of the wind tree.

Characteristic	Measure
Rated power (kW)	10
Rotor diameter (m)	0.72
Height of Darrieus blades (m)	0.45
Height of Savonius blades (m)	0.6
Cut-in speed (m/s)	2.0
Cut-out speed (m/s)	15.0
Rated power (kW)	10

Table A2. Rotor inventory and matching Ecoinvent records.

Subcomponents	Raw Material/Ecoinvent Database	Quantity (kg)
Blades	Glass-fibre-reinforced plastic, polyamide, injection moulded {GLO} I market for I APOS, U	172.8
Hub	Glass-fibre-reinforced plastic, polyamide, injection moulded {GLO} I market for I APOS, U	105.6
Bearing	Steel, chromium steel 18/8 {GLO} I market for I APOS, U	6.4
Screw	Steel, chromium steel 18/8 {GLO} I market for I APOS, U	0.26
Shaft	Steel, chromium steel 18/8 {GLO} I market for I APOS, U	160.32
Stick	Steel, chromium steel 18/8 {GLO} I market for I APOS, U	19.2

Table A3. Generator inventory and matching Ecoinvent records.

Subcomponents	Raw Material/Ecoinvent Database	Quantity (kg)
Generator	Permanent magnet, for electric motor {GLO} I production I APOS, U	76.8
Stator	Copper {GLO} I market for I APOS, U	144
	Glass-fibre-reinforced plastic, polyester resin, hand lay-up {GLO} I market for I APOS, U	96

Table A4. Tower inventory and matching Ecoinvent records.

Subcomponents	Raw Material/Ecoinvent Database	Quantity (kg)
Tower	Steel, low-alloyed {GLO} I market for I APOS, U	2849.43
Welding	Welding, arc, steel {GLO} I market for I APOS, U	20
Paint	Acrylic varnish, without water, in 87.5% solution state {GLO} I market for I APOS, U	25

Table A5. Foundation inventory and matching Ecoinvent records.

Subcomponents	Raw Material/Ecoinvent Database	Quantity (kg)
Reinforcement	Reinforcing steel {GLO} I market for I APOS, U	629.83
Concrete base	Concrete block {GLO} I market for I APOS, U	27,600

Table A6. Electrical connection inventory and matching Ecoinvent records.

Subcomponents	Raw Material/Ecoinvent Database	Quantity (kg)
Cable	Copper wire, technology mix, consumption mix, at plant, cross Section 1 mm^2 (duplicate) EU-15 S	138.6
Inverter	Simplified process	15
	Aluminium alloy, AlLi {GLO} I market for I APOS, U	7.64
	Copper {GLO} I market for I APOS, U	3.06
	Steel, low-alloyed {GLO} I market for I APOS, U	1.45
	Electronics	2.82
Controller	Electronics, for control units {GLO} I market for I APOS, U	3

Table A7. Transport methods and distances for components, with Ecoinvent selections.

Journey	Material/Ecoinvent Record	Distance (km) Surat Thani	Distance (km) Chiang Mai
Rotor	Transport, freight, lorry 3.5–7.5 metric ton, euro6 {RER} I market for transport, freight, lorry 3.5–7.5 metric ton, EURO6 I APOS, U	400	600
Generator		600	600
Tower		600	600
Inverter		360	686
	Transport, freight, sea, transoceanic ship {GLO} I market for I APOS, U	3015	3500
Foundation	Transport, freight, lorry 16–32 metric ton, metric ton, euro6 {RoW} I market for transport, freight, lorry 16–32 metric ton, EURO6 I APOS, U	600	600
End of life	Transport, freight, lorry 16–32 metric ton, metric ton, euro6 {RoW} I market for transport, freight, lorry 16–32 metric ton, EURO6 I APOS, U	80	80

Table A8. Scenario B treatments.

Material	Treatment
Aluminum	90% recycled + 10% landfilled
Copper	95% recycled + 5% landfilled
Steel	90% recycled + 10% landfilled
Glass-fibre-reinforced plastic	100% landfilled
Paint	100% landfilled
Electronics	Treatment as hazardous waste mass

References

1. Glassbrook, K.A.; Carr, A.H.; Drosnes, M.L.; Oakley, T.R.; Kamens, R.M.; Gheewala, S.H. Life cycle assessment and feasibility study of small wind power in Thailand. *Energy Sustain. Dev.* **2014**, *22*, 66–73. [CrossRef]
2. Martínez, E.; Sanz, F.; Pellegrini, S.; Jiménez, E.; Blanco, J. Life cycle assessment of a multi-megawatt wind turbine. *Renew. Energy* **2009**, *34*, 667–673. [CrossRef]
3. Wang, W.C.; Teah, H.Y. Life cycle assessment of small-scale horizontal axis wind turbines in Taiwan. *J. Clean. Prod.* **2017**, *141*, 492–501. [CrossRef]
4. Department of Alternative Energy Development and Efficiency. *Annual Report;* Ministry of Energy: Bangkok, Thailand, 2016.
5. Lombardi, L.; Mendecka, B.; Carnevale, E.; Stanek, W. Environmental impacts of electricity production of micro wind turbines with vertical axis. *Renew. Energy* **2018**, *128*, 553–564. [CrossRef]
6. Department of Alternative Energy Development and Efficiency. *Thailand Alternative Energy Situation;* Ministry of Energy: Bangkok, Thailand, 2016. Available online: http://webkc.dede.go.th/testmax/sites/default/files/thailandalternative2016.pdf (accessed on 14 November 2018).
7. Energy Policy and Planning Office Thailand Power Development Plan 2015–2036 (PDP2015). Available online: https://www.egat.co.th/en/images/about-egat/PDP2015_Eng.pdf (accessed on 26 November 2018).
8. Siddiqui, O.; Dincer, I. Comparative assessment of the environmental impacts of nuclear, wind and hydro-electric power plants in Ontario: A life cycle assessment. *J. Clean. Prod.* **2017**, *164*, 848–860. [CrossRef]
9. Wang, L.; Wang, Y.; Du, H.; Zuo, J.; Yi, M.; Li, R.; Zhou, Z.; Bi, F.; Garvlehn, M.P. A comparative life-cycle assessment of hydro-, nuclear and wind power: A China study. *Appl. Energy* **2019**, *249*, 37–45. [CrossRef]
10. Asdrubali, F.; Baldinelli, G.; D'Alessandro, F.; Scrucca, F. Life cycle assessment of electricity production from renewable energies: Review and results harmonization. *Renew. Sustain. Energy Rev.* **2015**, *42*, 1113–1122. [CrossRef]
11. Shah, S.R.; Kumar, R.; Raahemifar, K.; Fung, A.S. Design, modeling and economic performance of a vertical axis wind turbine. *Energy Rep.* **2018**, *4*, 619–623. [CrossRef]
12. Kouloumpis, V.; Sobolewski, R.A.; Yan, X. Performance and life cycle assessment of a small scale vertical axis wind turbine. *J. Clean. Prod.* **2020**, *247*, 119520. [CrossRef]
13. Chingulpitak, S.; Wongwises, S. Critical review of the current status of wind energy in Thailand. *Renew. Sustain. Energy Rev.* **2014**, *31*, 312–318. [CrossRef]
14. Uddin, M.S.; Kumar, S. Energy, emissions and environmental impact analysis of wind turbine using life cycle assessment technique. *J. Clean. Prod.* **2014**, *69*, 153–164. [CrossRef]
15. Yildiz, N.; Hemida, H.; Baniotopoulos, C. Maintenance and End-of-Life Analysis in LCA for Barge-Type Floating Wind Turbine. *Wind* **2022**, *2*, 246–259. [CrossRef]
16. Tasneem, Z.; Al Noman, A.; Das, S.K.; Saha, D.K.; Islam, M.R.; Ali, M.F.; R Badal, M.F.; Ahamed, M.H.; Moyeen, S.I.; Alam, F. An analytical review on the evaluation of wind resource and wind turbine for urban application: Prospect and challenges. *Dev. Built Environ.* **2020**, *4*, 100033. [CrossRef]
17. Casini, M. Small Vertical Axis Wind Turbines for Energy Efficiency of Buildings. *J. Clean Energy Technol.* **2015**, *4*, 56–65. [CrossRef]
18. Jamanun, M.J.; Misaran, M.S.; Rahman, M.; Muzammil, W.K. Performance Investigation of A Mix Wind Turbine Using A Clutch Mechanism at Low Wind Speed Condition. *IOP Conf. Ser. Mater. Sci. Eng.* **2017**, *217*, 012020. [CrossRef]
19. Almotairi, A.M.M.; Mustapha, F.; Ariffin, M.K.A.M.; Zahari, R. Synergy of Savonius and Darrieus types for vertical axis wind turbine. *Int. J. Adv. Appl. Sci.* **2016**, *3*, 25–30. [CrossRef]
20. Kumar, A.; Nikhade, A. Hybrid Kinetic Turbine Rotors: A Review. *Int. J. Eng. Sci. Adv. Technol.* **2014**, *4*, 453–463.
21. Akwa, J.V.; Vielmo, H.A.; Petry, A.P. A review on the performance of Savonius wind turbines. *Renew. Sustain. Energy Rev.* **2012**, *16*, 3054–3064. [CrossRef]
22. Energy Systems Research Institute at Prince of Songkla University. *Project Report; Development of 10 kW Low Speed Vertical Wind Turbine Using Tree Constructor;* Energy Systems Research Institute at Prince of Songkla: Hat Yai, Thailand, 2019.
23. Ngoc, D.M.; Techato, K.; Niem, L.D.; Thi, N.; Yen, H.; Van Dat, N. A Novel 10 kW Vertical Axis Wind Tree Design: Economic Feasibility Assessment. *Sustainability* **2021**, *13*, 12720. [CrossRef]
24. de Bruyn, S.; Bijleveld, M.; de Graaff, L.; Schep, E.; Schrote, A.; Vergeer, R.; Ahdou, S. *Environmental Prices Handbook;* CE Delft: Delft, The Netherlands, 2018.

25. *ISO 14044:2006*; Environmental Management—Life Cycle Assessment—Requirements and Guidelines. International Organization for Standardization: Geneva, Switzerland, 2006; pp. 652–668. [CrossRef]

26. PRé. *Sustainability SimaPro 9.3.0.3*, (Software); PRé: Amersfoort, The Netherlands, 2021.

27. Invt MG 4–5 kW Single Phase On-Grid Solar Inverter—INVT Solar Power. Available online: https://www.invt.com/products/mg-4-5kw-single-phase-on-grid-inverter-100 (accessed on 28 May 2019).

28. Tremeac, B.; Meunier, F. Life cycle analysis of 4.5 MW and 250 W wind turbines. *Renew. Sustain. Energy Rev.* **2009**, *13*, 2104–2110. [CrossRef]

29. Chaichana, T.; Chaitep, S. Wind power potential and characteristic analysis of Chiang Mai, Thailand. *J. Mech. Sci. Technol.* **2010**, *24*, 1475–1479. [CrossRef]

30. Energy Policy and Planning Office. Electricity Statistic. Available online: http://www.eppo.go.th/index.php/en/en-energystatistics/electricity-statistic?orders[publishUp]=publishUp&issearch=1 (accessed on 6 November 2019).

31. RIVM LCIA: The ReCiPe Model—RIVM. Available online: https://www.rivm.nl/en/life-cycle-assessment-lca/recipe%0Ahttps://www.rivm.nl/en/Topics/L/Life_Cycle_Assessment_LCA/ReCiPe (accessed on 3 October 2022).

32. EGAT. *Sustainability Report 2019*; Electricity Generating Authority of Thailand: Bangkok, Thailand, 2019.

33. Goedkoop, M.; Heijungs, R.; Huijbregts, M.; De Schryver, A.; Struijs, J.; Van Zelm, R. *ReCiPe 2008: A Life Cycle Impact Assessment Method Which Comprises Harmonised Category Indicators at the Midpoint and the Endpoint Level*, 1st ed.; Report I: Characterisation; Ministry of Housing, Spatial Planning and the Environment: The Hague, The Netherlands, 2009.

34. Stocker, T.F.; Qin, D.; Plattner, G.K.; Tignor, M.M.; Allen, S.K.; Boschung, J.; Midgley, P.M. *Climate Change 2013: The Physical Science Basis. Contribution of Working Group I to the Fifth Assessment Report of the Intergovernmental Panel on Climate Change*; IPCC: Cambridge, NY, USA, 2013.

35. Navrud, S. *Possibilities and Challenges in Transfer and Generalisation of Monetary Estimates for Environmental and Health Benefits of Regulating Chemicals*; OECD Environment Working Papers; OECD: Paris, France, 2017; Volume 6, p. 10. [CrossRef]

36. Ghani, H.U.; Mahmood, A.; Ullah, A.; Gheewala, S.H. Life cycle environmental and economic performance analysis of bagasse-based electricity in Pakistan. *Sustainability* **2020**, *12*, 10594. [CrossRef]

37. Macrotrends. Thailand GDP per Capita 1960–2021 | MacroTrends. Available online: https://www.macrotrends.net/countries/THA/thailand/gdp-per-capita (accessed on 28 October 2021).

38. Gkantou, M.; Rebelo, C.; Baniotopoulos, C. Life Cycle assessment of tall onshore hybrid steel wind turbine towers. *Energies* **2020**, *13*, 3950. [CrossRef]

39. Caduff, M.; Huijbregts, M.A.J.; Althaus, H.J.; Koehler, A.; Hellweg, S. Wind power electricity: The bigger the turbine, the greener the electricity? *Environ. Sci. Technol.* **2012**, *46*, 4725–4733. [CrossRef]

40. Doerffer, K.; Bałdowska-Witos, P.; Pysz, M.; Doerffer, P.; Tomporowski, A. Manufacturing and recycling impact on environmental life cycle assessment of innovative wind power plant part 1/2. *Materials* **2021**, *14*, 220. [CrossRef]

41. Gough, M.; Lotfi, M.; Castro, R.; Madhlopa, A.; Khan, A.; Catalão, J.P.S. Urban wind resource assessment: A case study on Cape Town. *Energies* **2019**, *12*, 1479. [CrossRef]

Article

Wind-Tunnel Experiments on the Interactions among a Pair/Trio of Closely Spaced Vertical-Axis Wind Turbines

Yoshifumi Jodai [1,*] **and Yutaka Hara** [2]

[1] Department of Mechanical Engineering, National Institute of Technology (KOSEN), Kagawa College, 355 Chokushi, Takamatsu 761-8058, Japan

[2] Faculty of Engineering, Tottori University, 4-101 Koyama-Minami, Tottori 680-8552, Japan

* Correspondence: jodai@t.kagawa-nct.ac.jp; Tel.: +81-87-869-3894

Abstract: To elucidate the wind-direction dependence of the rotor performance in closely spaced vertical-axis wind turbines, wind-tunnel experiments were performed at a uniform wind velocity. In the experiments, a pair/trio of three-dimensional printed model turbines with a diameter of D = 50 mm was used. The experiments were performed systematically by applying incremental adjustments to the rotor gap g and rotational direction of each rotor and by changing the wind direction. For tandem layouts, the rotational speed of the downwind rotor is 75–80% that of an isolated rotor, even at g/D = 10. For the average rotational speed of the rotor pair, an origin-symmetrical and a line-symmetrical distribution are observed in the co-rotating and inverse-rotating configurations, respectively, thereby demonstrating the wind-direction dependence for the rotor pair. The inverse-rotating trio configuration yields a higher average rotational speed than the co-rotating trio configuration for any rotor spacing under the ideal bidirectional wind conditions. The maximum average rotational speed should be obtained for a wind direction of θ = 0° in the inverse-rotating trio configuration. The wind-direction dependence of the rotational speeds of the three turbines was explained via flow visualization using a smoke-wire method and velocity field study using two-dimensional computational fluid dynamics.

Keywords: vertical-axis wind turbine; wind-tunnel experiment; pair of turbines; trio of turbines; closely spaced arrangement; wind-direction dependence; rotational speed; power coefficient

Citation: Jodai, Y.; Hara, Y. Wind-Tunnel Experiments on the Interactions among a Pair/Trio of Closely Spaced Vertical-Axis Wind Turbines. *Energies* **2023**, *16*, 1088. https://doi.org/10.3390/en16031088

Academic Editor: Francesco Castellani

Received: 16 December 2022
Revised: 11 January 2023
Accepted: 16 January 2023
Published: 18 January 2023

1. Introduction

For both onshore and offshore wind farms, studies on the optimal layout of several closely placed vertical-axis wind turbines (VAWTs), using beneficial interactions between turbines in an array [1], are important issues that reflect a great demand for effective usage of abundant wind energy. As a fundamental step in the allocation of VAWTs in a wind farm, this study explores the wind-direction dependence of closely spaced two and three VAWTs by wind-tunnel experiments with miniature model turbines. The aspect ratio of the model turbine, i.e., the ratio of the height H to the diameter D, was set to a widely used value of approximately 1. Shamsoddin et al. [2] reported the effect of the aspect ratio on VAWT wakes.

Dabiri [3] conducted experiments at a field site in Los Angeles; in the experiment, many couples of inverse-rotating rotors were set like a fish schooling. The turbines were a modified version of a commercially available model with H/D = 3.42. Hezaveh et al. [4] showed that *"the wind-farm design with staggered-triangle clusters is the optimal design in terms of cost per unit power produced."* Li et al. [5] showed that the power output is higher in the staggered wind farms (horizontal-axis wind turbines) than in the aligned ones.

From a set of field experiments conducted by Dabiri [3], with a three-VAWT configuration in which the third turbine was placed a distance of $4D$ downstream from the second turbine (i.e., the spacing of the rotors, g, between the surfaces of the second and third rotors is $3D$) in an elbow-like layout, the performance was recovered to within 5% of the single-turbine performance. Zanforlin and Nishino [6] reported the power enhancement in closely arranged VAWTs with two-dimensional ($2D$) numerical simulation assuming fixed rotational speeds.

However, the studies by [3] and [6] did not adopt a co-rotating (CO) configuration. The results of the former include unnatural increases in power in the case near the tandem layout for a rotor gap ratio of $g/D = 0.65$, despite the unpreferred wake interaction between the turbines. The latter assumed that only the upstream turbine is working, regardless of the value of g/D within the range of 0.5–2. Therefore, well-controlled tandem experiments are necessary to clarify these points prior to investigating the wind-direction dependence of VAWT performance.

For the CO configuration, Dessoky et al. [7] performed a numerical simulation for a tandem layout of a pair of VAWTs. They reported that the downwind turbine shows better performance if the upwind turbine is operated at a high tip speed ratio and reported the effect of the turbine spacing on the downwind turbine performance. Recently, Kuang et al. [8] conducted a three-dimensional ($3D$) improved delayed detached-eddy simulation (IDDES) on two tandem CO offshore floating VAWTs. They reported that as the turbine spacing increased, the performance, an increase, and the optimal tip speed ratio of the downwind turbine were enhanced. These results confirm the previous findings reported by Dessoky et al. [7], who explained the improvement in turbine performance as an effect of the reduction in vorticity with the shifting of the downwind rotor in the downstream direction. However, these simulations with fixed rotational speeds still seem to be impractical in variable-speed VAWT operation, which depends on wind direction and flow speed. The importance of the 3D effects of blade geometry on VAWT wakes was reported experimentally (Wei et al. [9]) and numerically (Kuang et al. [8]). Furthermore, Jin et al. [10] reported the $3D$ structure of the wake behind twin VAWTs placed side-by-side and showed that the azimuth angle change has little effect on the rotor performance.

In a study on the wind-direction dependence of a pair of turbines, Sahebzadeh et al. [11] investigated 119 unique turbine arrangements with seven gaps in the range of $g/D = 0.25$–9 and 17 wind directions covering $\pm90°$, and reported the effect of relative distance and angle on the individual and overall power performance of the two rotors.

However, their numerical study was also limited to a CO pair with fixed rotational speed. De Tavernier et al. [12] reported both CO and inverse-rotating (IR) double-rotor configurations in a study using a $2D$ panel/vortex model. From the wind-direction dependence of the power coefficient of two IR turbines, they reported a specific wind direction in which the downwind turbine performs better than the upwind turbine and explained it due to increased mass flow through the rotor because of the induced velocity caused by the first rotor on the second rotor. Regrettably, they intentionally left out the results for the CO configuration, merely stating that *"very similar trends can be observed"*. Therefore, their report does not contain any results on the wind-direction dependence with CO turbines, except for a side-by-side layout.

Although Hezaveh et al. [4] demonstrated an approximate 10% increase in the power generation of a single rotor in well-designed clusters, they reported the gap dependence of the power coefficient only for a wind direction of 60° (one upwind rotor and two downwind rotors) in a triangular-cluster configuration (see their Figures 8 and 9) in a study of a trio of turbines. Among other studies in three-VAWT arrays [13–15], Ahmadi-Baloutaki et al. [15] concluded that *"the optimum range of the streamwise distance of the downstream turbine from the counter-rotating pair and the spacing between the pair was determined to be about three and one rotor diameters, respectively,"* though they only considered a wind direction of 0° (an upwind counter-rotating rotor pair and one downwind rotor) for a nonequilateral-triangular cluster configuration.

Hara et al. [13] performed a $2D$ computational fluid dynamics (hereafter, 2D CFD) study. The strong point of their study was that the rotational speed of each rotor could change according to the interaction between each rotor and flow around it. Yoshino et al. [16] reported the wind-direction dependence of the rotational speed of three VAWTs.

Many studies explored multi-VAWT arrays, such as Zhang et al. [17] with 5 Savonius-type VAWTs (2*D* CFD), Mereu et al. [18] with 16 Savonius-type VAWTs (2*D* CFD), Bangga et al. [19] with 6 VAWTs (2*D* CFD with NACA 0021), Dabiri [3] with 6 VAWTs (field experiment), and Hezaveh et al. [4] with 96 VAWTs (large-eddy simulation).

Therefore, the objectives of this study using two and three miniature VAWTs are to explore the effects of the:

- Rotational directions;
- Gap ratios;
- Wind direction over 360°;

on the arrayed-turbine performance.
The significance of the investigation is:

- Wind-tunnel experiments equivalent to the operation of small variable-speed VAWTs;
- Comparison with the cutting-edge 2*D* CFD with the DFBI method [13,16];
- Well-supported flow patterns obtained by flow visualization.

The rotational speeds of the rotors were simulated in the 2*D* CFD with software STAR-CCM+ by solving an equation of continuity and Reynolds-averaged Navier–Stokes equations in an unsteady incompressible flow, in which the numerical setup was the same as that used in Hara et al. [13] for two rotors and Yoshino et al. [16] for three rotors.

2. Methods

2.1. Configuration of the Flow Field

The wind tunnel used had an outlet area 600 (width) × 350 (height) mm. The speed and direction of the wind were held constant for each experiment. The uniformity of the mean velocity and the turbulent intensity of the flow field are detailed in [20]. Figure 1 shows a model rotor named a butterfly wind turbine (BWT) used in the experiments. Each model rotor was printed on a 3*D* printer. The miniature models were also used in our previous experiments on a pair of VAWTs arranged side-by-side (Jodai and Hara [20]). A BWT is a lift-type VAWT with straight-blade portions, such as a high-performance H-Darrieus wind turbine, which features an armless rotor with looped blades. The height *H* and diameter *D* of the miniature model were 43.4 and 50 mm, respectively. Figure 1b is the cross-section of the model along the equatorial plane.

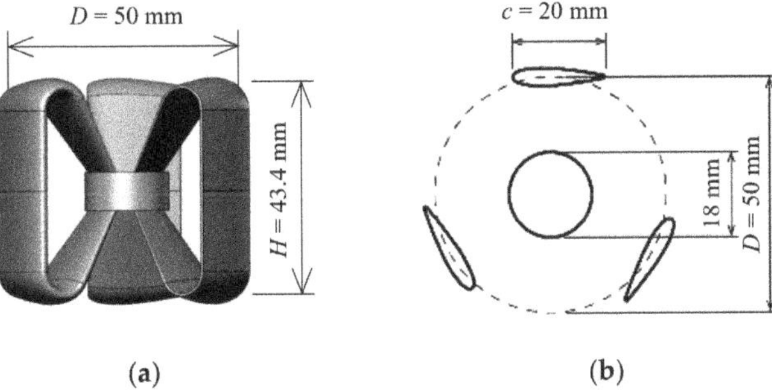

Figure 1. Butterfly wind turbine: (**a**) three-dimensional (3D) image; (**b**) cross-section.

The measurements were conducted on a pair/trio of rotors in an open space beyond the wind-tunnel exit for layouts with 16 or 12 wind directions. An independent tandem experiment was also conducted at a uniform velocity *V* = 10 and 12 m/s prior to the multiple wind-direction experiments. The centers of the two or three turbines were located at the 3*D*-position (i.e., 150 mm) from the wind-tunnel outlet, except for the independent tandem experiments in which the upwind rotor was fixed 50 mm upstream of the center point. The solidity of the turbine model defined by $Bc/(\pi D)$ was $\sigma = 0.382$. The Reynolds number based on D and the tip speed ratio at a rotational speed of $N\sim4000$ rpm were $R_{eD}\sim4.7 \times 10^4$ and $\lambda\sim0.75$, respectively, for the experiments on a pair of turbines with a wind speed of $V = 14$ m/s. On the other hand, in experiments on a trio of turbines with a wind speed of $V = 12$ m/s ($N\sim3400$ rpm), the corresponding values were $R_{eD}\sim4.0 \times 10^4$ and $\lambda\sim0.74$. The Reynolds number using the length of the chord c was $R_{ec}\sim1.9/1.6 \times 10^4$ for the experiments on a pair or trio of turbines. More details about the effect of a Reynolds number on the performance in a VAWT are given in [20] with relevant references. The error in the model rotational speed measurements was ±10 rpm, which corresponds to $\pm0.25\%$ or $\pm0.29\%$ of the rotational speed N (~4000 rpm or ~3400 rpm) of a single rotor with an isolated setting at a uniform wind velocity of $V = 14/12$ m/s. Table 1 provides a comparison of the parameters adopted in related studies with those used in ours, including the cases of multiple rotors with more than three turbines. In Table 1, g_{min}/D is the minimum rotor gap ratio investigated and ω indicates the angular velocity of the rotor.

Table 1. Comparison of the parameters of the vertical-axis turbine. CO and IR represent co-rotating and inverse-rotating configurations, respectively. U is the uniform speed; D is the rotor diameter; R_{ec} is the chord-based Reynolds number; $\lambda = R\omega/V$ is the tip speed ratio; and $\sigma = Bc/(\pi D)$ is the solidity. Values marked with † are examples of nonequilateral-triangular cluster design consisting of three VAWTs.

Study	Layout	U (m/s)	D (m)	$R_{ec}/10^4$ (-)	λ (-)	σ (-)	g_{min}/D (-)
Ahmadi-Baloutaki et al. [15]	0° trio (IR)	6–14	0.30	1.8–4.2	~0.05–0.3	0.239	0.5 †
Bangga et al. [19]	six-parallel (CO&IR)	8.0	2.0	14	1.5–3.0	0.0844	1.0
Dabiri [3]	over 360° overpair (IR) elbow − like trio (IR) six − staggered (IR)	5.7 (7.8)	1.2	4.2 (6)	1.5–3.0	0.102	0.65 0.65 3.0
Dessoky et al. [7]	tandem pair (CO)	8.0	2	14	0.75	0.0844	1.5
De Tavernier et al. [12]	over 360° pair (CO&IR)	1.0	20	6.7	2.5, 3.5	0.032	0.01
Hezaveh et al. [4]	0°, 20°, 40°, 60° trio (CO)	12	1.2	8.8	2.18	0.0875	2.0
Kuang et al. [8]	tandem pair (CO)	8.0	0.8	10.7	0.4–1.5	0.239	1.0
Sahebzadeh et al. [11,21]	over ±90° pair (CO)	9.3	1	15.7	4	0.0191	0.25
Zanforlin and Nishino [6]	over 360° pair (IR)	8.0	1.2	6.8	2.3–3.2	0.102	0.5
Zanforlin [22] (tidal turbines)	over 360° trio (CO)	1.5	1.0	27	1.75	0.175	2.0
Zhang et al. [14]	0°, 60° trio (CO)	4.01	1.48	2	3.7	0.0323	2.0 †
Zheng et al. [23]	0°, 60° trio (IR)	10.6	1.2	9	2.3	0.102	0.6
Present	tandem pair (CO&IR) over 360° pair(CO&IR) over 360° trio(CO&IR)	10, 12 14 12	0.050	1.3, 1.6 1.9 1.6	~0.8	0.382	1.0 0.5 0.5

2.2. Tandem Layouts of a Pair of VAWTs

Figure 2 explains two tandem arrangements using two turbines according to Hara et al. [13] (see Figure 3 in [13]; the rotational direction in the present experiments is opposite to that adopted in their CFD study). In the tandem co-rotating (TCO) layout, the two rotors turn in the same rotational direction (Figure 2a), while in the tandem inverse-rotating (TIR) layout, the two rotors rotate in opposite rotational directions (Figure 2b). Rotor 1 and Rotor 2 are denoted as R1 and R2, respectively. The space between the two rotors is indicated by g. Results were obtained for six gap ratios of $g/D = 1, 2, 4, 6, 8,$ and 10 ($g = 50, 100, 200, 300, 400,$ and 500 mm) in the tandem arrangement. The experiments were performed at a uniform velocity $V = 10$ or 12 m/s. The x-coordinate represents the direction of the wind parallel to the array. The direction normal to x is y. In Figure 2, red arrows show the rotational directions of the rotors.

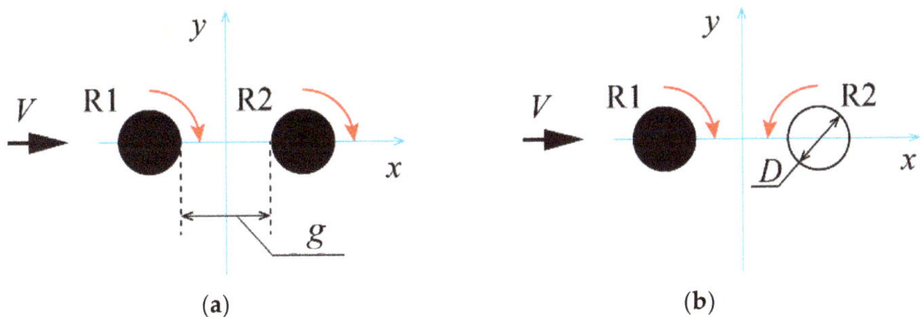

Figure 2. Two tandem layouts against the wind direction in a closely spaced VAWT pair: (**a**) tandem co-rotating (TCO); (**b**) tandem inverse-rotating (TIR).

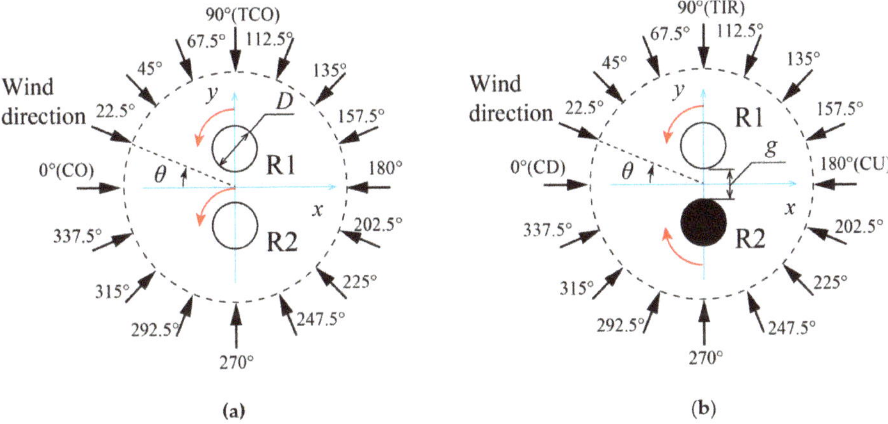

Figure 3. Definition of 16-wind-direction configurations in a closely spaced VAWT pair: (**a**) co-rotation (CO); (**b**) inverse-rotation (IR).

2.3. Wind Directions of a Pair of VAWTs (16 Wind Directions)

Figure 3 shows the definition of 16-wind-direction configurations of a closely spaced VAWT pair according to Sogo et al. [24] (see their Figure 2) or Hara et al. [13] (see their Figure 4). In the co-rotation (CO) configuration, the two rotors rotate in the same rotational direction, as illustrated in Figure 3a. In the inverse-rotation (IR) configuration, the two rotors turn in opposite rotational directions, as shown in Figure 3b. The space between Rotor 1 (R1) and Rotor 2 (R2) is indicated by g. The two gap ratios of $g/D = 0.5$ and 1 ($g = 25$ and 50 mm) were investigated in the configurations. The uniform velocity for the 16-wind-direction experiments was $V = 14$ m/s. θ is the wind-direction angle.

In Figure 3a, the CO layout at $\theta = 0°$ or $180°$ corresponds to the specific layout called CO in a side-by-side arrangement. The tandem co-rotating (TCO) layout at $\theta = 90°$ or $270°$ is introduced in Figure 2a. Although the same layout exists in the symmetric direction with respect to the center of the rotor pair (i.e., *origin symmetry*) in the CO configuration, as seen in Figure 3a, the wind-tunnel experiment was conducted for all cases of the 16-wind-direction configurations to ensure the repeatability of the experiment.

In Figure 3b, the counter-down layout at $\theta = 0°$ and counter-up layout at $\theta = 180°$ are called CD and CU, respectively, in a side-by-side arrangement. The TIR layout at $\theta = 90°$ or $270°$ is shown in Figure 2b. In contrast to the CO configuration in Figure 3a, equivalent cases exist in the symmetrical direction with respect to the *x*-axis (i.e., *line symmetry*) in the IR configuration in Figure 3b. Nevertheless, we still executed the experiment for all 16-wind-direction cases in the IR configuration, since there are no comparable systematic experiments on the wind-direction dependence.

2.4. Wind Directions of a Trio of VAWTs (12 Wind Directions)

Figure 4 shows the definition of 12-wind-direction configurations of a closely spaced VAWT trio. In the co-rotation trio (3CO) configuration, the three rotors turn in the same rotational direction, as illustrated in Figure 4a. In the inverse-rotation trio (3IR) configuration, one of the three rotors (R2) turns in an opposite direction, as shown in Figure 4b. The space between two of the three rotors (R1, R2, and R3) is indicated by *g*. The three gap ratios of $g/D = 0.5, 1,$ and 2 ($g = 25, 50,$ and 100 mm) were investigated in these configurations. The uniform velocity for the 12-wind-direction experiments was $V = 12$ m/s.

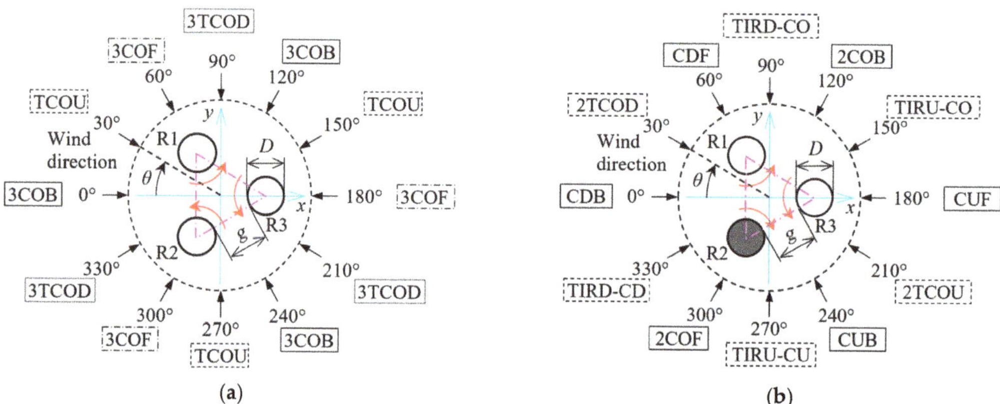

Figure 4. Definition of 12-wind-direction configurations in a closely spaced VAWT trio: (**a**) co-rotation trio (3CO); (**b**) inverse-rotation trio (3IR).

In Figure 4a, the layout at $\theta = 0°$ comprises a pair of CO rotors in a parallel arrangement ($\theta = 0°$ in Figure 3a) and an additional CO rotor (R3). Since the additional rotor R3 is placed behind the other rotors, we define this layout as 3COB. The label of number 3 was added to distinguish this layout from the 2COB layout, which is one of the layouts in the 3IR configuration ($\theta = 120°$ in Figure 4b), as explained later. The layout at $\theta = 90°$ is constituted by a pair of TCO rotors in a tandem arrangement ($\theta = 90°$ in Figure 3a) and an additional CO rotor (R3). We define this layout as 3TCOD, because the blades of the additional rotor R3 move downwind in the gap region (center of three turbines). Again, the label of number 3 was added in order to distinguish this layout from the 2TCOD layout, which is one of the layouts in the 3IR configuration ($\theta = 30°$ in Figure 4b).

Note that the blades of the other rotors, R1 and R2, move upwind in the gap region in the former layout (3TCOD). Similarly, the layouts at $\theta = 30°$ and $60°$ in the 3CO configuration are specific layouts, including a pair of tandem rotors or a pair of parallel rotors, and we define them as TCOU or 3COF, respectively. Since the same layout occurs at every $120°$ in the wind direction, the number of the independent layouts in the 3CO configuration is four (two parallel-like layouts and two tandem-like layouts). However, to obtain reliable experimental results, the wind-tunnel experiment was conducted for all cases of the 12-wind-direction configurations, as shown in Figure 4a. The three rotors are set on the corners of an equilateral triangle shown in lavender dash-dotted line. The length of the sides of the triangle is $g + D$.

In Figure 4b, the layout at $\theta = 0°$ comprises a pair of CD rotors in a parallel arrangement ($\theta = 0°$ in Figure 3b) and an additional rotor (R3). Since the additional rotor R3 is placed behind the other rotors, we define this layout as CDB. The layout at $\theta = 90°$ is comprised of a pair of TIR rotors in a tandem arrangement ($\theta = 90°$ in Figure 3b) and an additional rotor (R3). We define this layout as TIRD-CO, because the blades of the additional rotor R3 turn downwind in the gap region in addition to the fact that R3 rotates in the same direction as R1. This layout differs from the TIRU-CO layout at $\theta = 150°$ in Figure 4b, in which the blades of the additional rotor R1 move upwind in the gap region.

In the same way, the layouts at $\theta = 60°$ (CDF), $120°$ (2COB), $180°$ (CUF), $240°$ (CUB), and $300°$ (2COF) are specific layouts that include a pair of side-by-side rotors. On the other hand, the layouts at $\theta = 30°$ (2TCOD), $210°$ (2TCOU), $270°$ (TIRU-CU), and $330°$ (TIRD-CD) are additional specific layouts that include a pair of tandem rotors. Note that the same layout does not exist in the 3IR configuration. In other words, there are 12 independent layouts in the 3IR configuration (six parallel-like layouts and six tandem-like layouts). Therefore, we also conducted the experiment for all 12-wind-direction cases in the 3IR configuration.

In summary, there are 16 independent layouts (see Table 2 and Figure 4). To our knowledge, the present study is the first comprehensive measurement of the wind-direction dependence on the basis of a wind-tunnel experiment for three closely allocated VAWTs arranged equilaterally without omitting any wind directions, and with not only the 3CO configuration but also the 3IR configuration.

Table 2. Definition of the names of specific layouts in 12-wind-direction configurations (3CO and 3IR) in a closely spaced VAWT trio. Layouts with * are repeated every $120°$.

Wind Direction (°)	3CO (Co-Rotation Trio)	3IR (Inverse-Rotation Trio)
0	3COB	CDB
30	TCOU	2TCOD
60	3COF	CDF
90	3TCOD	TIRD-CO
120	3COB *	2COB
150	TCOU *	TIRU-CO
180	3COF *	CUF
210	3TCOD *	2TCOU
240	3COB *	CUB
270	TCOU *	TIRU-CU
300	3COF *	2COF
330	3TCOD *	TIRD-CD

2.5. Characteristics of a Single Rotor Configuration

The aim of this subsection is to briefly describe the properties of the miniature turbine. Details of the torque measurement for the power calculation have been shown in Jodai and Hara [20]. Figure 5a shows the variation in the power coefficient C_p with the tip speed ratio λ of a wind turbine (see Table 1). The relationship between the rotational speed N and the motor power (power P of the model turbine at equilibrium) is shown in Figure 5b, where the red circles are points interpolated from the experiment. The approximate black curve (P vs. N) in Figure 5b is expressed by Equation (1).

$$P[\text{mW}] = 0.2047\left(\frac{N[\text{rpm}]}{1000}\right)^3 + 0.0442\left(\frac{N[\text{rpm}]}{1000}\right)^2 + 21.042\left(\frac{N[\text{rpm}]}{1000}\right) - 0.0851 \quad (1)$$

In the present experiments, a direct current motor was used only to start the rotation of a turbine in a uniform wind. The model rotational speed N [rpm] was measured using a noncontact-type digital tachometer with an accuracy of ± 1 rpm, from a distance of 250–300 mm from the rotors (see [20] for details).

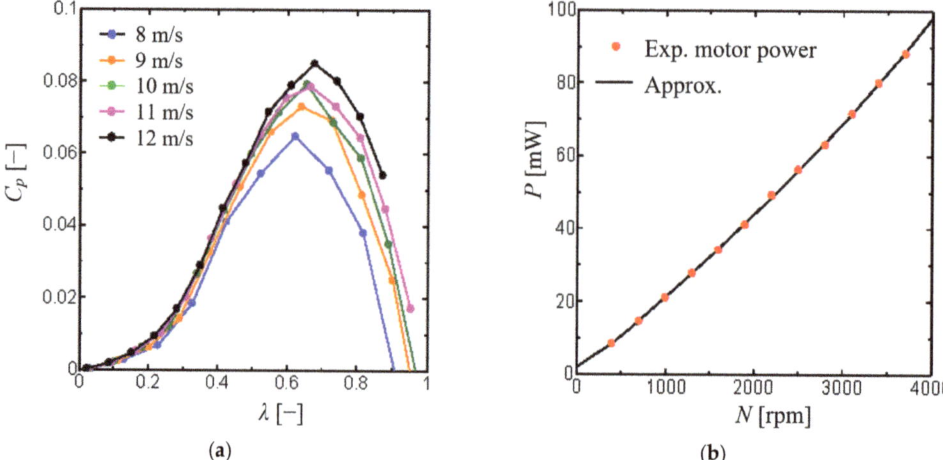

(a)

(b)

Figure 5. Experimental data of a 3D-printed miniature rotor: (**a**) power coefficient as a function of tip speed ratio; (**b**) power as a function of rotational speed (reproduced with permission from Jodai and Hara [20]).

2.6. Experimental Setup of a Pair of VAWTs and a Trio of VAWTs

Figure 6 shows the normalized rotational speed in the case of the isolated turbine N_{SI}/N_{SI0} placed $y = 0$ mm (see Figures 2–4) at a uniform velocity $V = 10$ or 12 m/s. Here, N_{SI} is the rotational speed of the single rotor and N_{SI0} is that located at $x/D = 0$. Hereafter, the rotational speed in the experimental result means the value measured in the case without a power supply to the startup motor, i.e., the free rotational speed (see Figure 7). The origins of the x- and y-axes correspond to the centers of the two/three rotors. Note that the upstream rotor (R1) is fixed at $x/D = -1$ and only the streamwise position of the downstream rotor (R2) was adjusted according to the rotor gap g in the tandem experiment (Figure 2). We confirmed the constant rotational speed within the error of $\pm 2\%$ in the range of $0 \leq x/D \leq 12$ (Figure 6). This 600 mm streamwise range covers the full length of the tandem experiment, including a pair of two rotors at the maximum gap of $10D = 500$ mm.

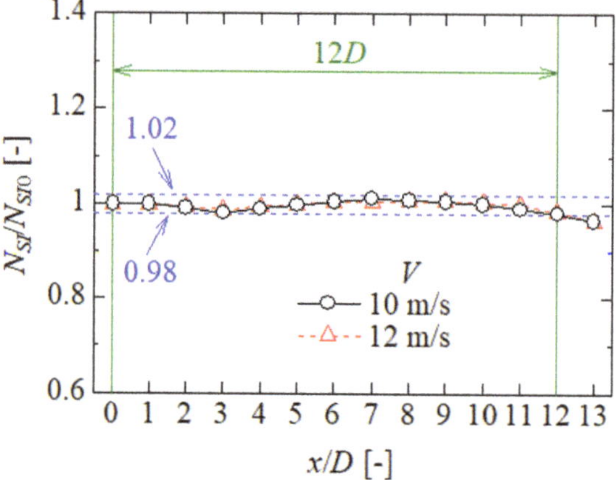

Figure 6. Free rotational speed of the single rotor along the streamwise direction at $y = 0$ mm.

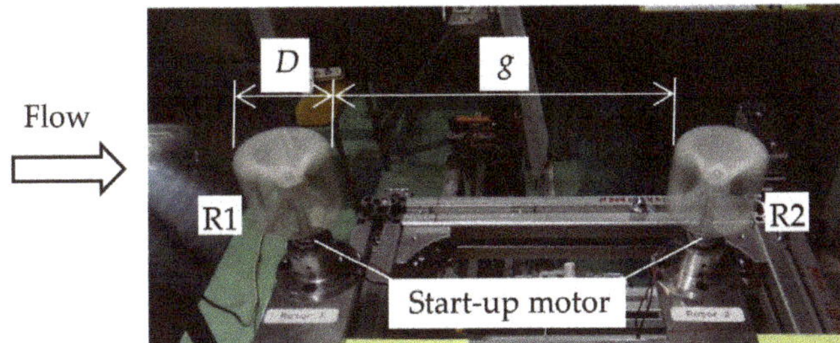

Figure 7. Setup of two VAWT models with a gap of $g/D = 4$ arranged in tandem.

Figure 7 shows the setup of two VAWT models with a 200 mm gap ($g/D = 4$) arranged in tandem. Rotor 1 and Rotor 2 are aligned with the wind direction, as seen in Figure 2. The rotors can easily move along the rail in the $\pm x$ direction to realize the required spacing between the two rotors.

Figure 8 is the setup of two VAWT models with a 50 mm gap ($g/D = 1$) in IR configuration at $\theta = 112.5°$. The figure shows the view from the top (see Figure 3b). The wind direction can be adjusted by using a rotating stage.

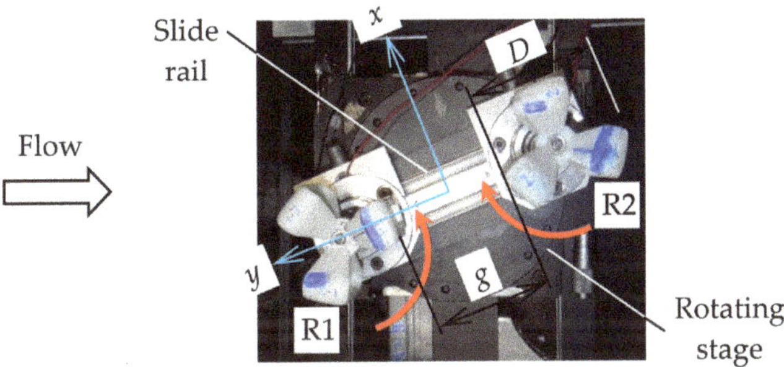

Figure 8. Setup of two VAWT models with a gap of $g/D = 1$ in IR configuration at $\theta = 112.5°$.

Figure 9 shows the experimental setup of a trio of 3D-printed VAWT models with a 50 mm gap ($g/D = 1$) in 3IR configuration at $\theta = 120°$, viewed from the top (see Figure 4b).

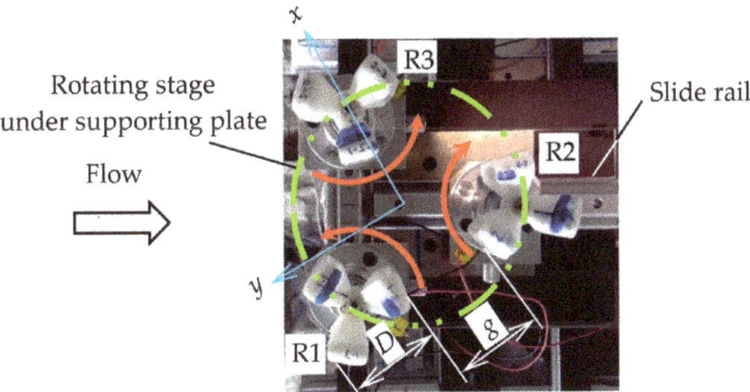

Figure 9. Experimental setup of a trio of 3D-printed VAWT models with a gap of $g/D = 1$ in the 3IR configuration at $\theta = 120°$ (2COB).

3. Results and Discussion

3.1. Rotational Speeds and Power of Closely Spaced VAWTs in Tandem Layouts

Figure 10 shows the variations in the rotational speed N [rpm] with the gap g between the two rotors. The results for the TCO layout (Figure 2a) and for the TIR layout (Figure 2b) are shown in Figures 10a and 10b, respectively. The rotational speed of the single turbine obtained in the experiment was 2900 rpm at $V = 10$ m/s.

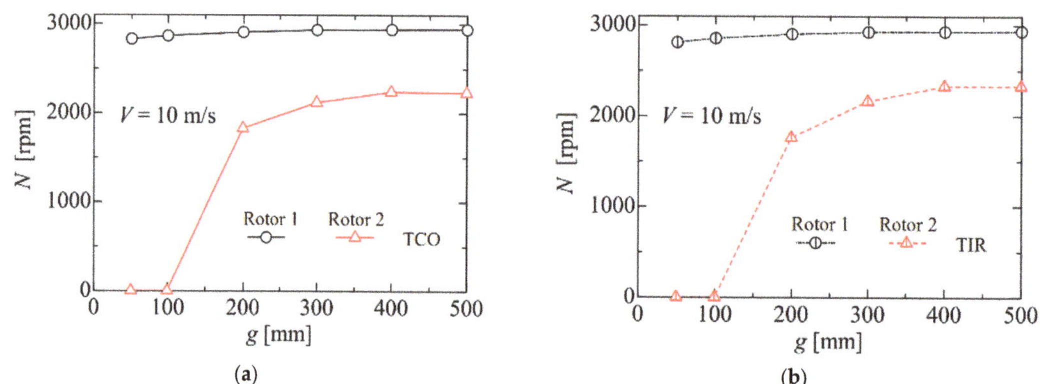

Figure 10. Variations in rotational speed with the gap between the rotors in tandem layouts at $V = 10$ m/s: (**a**) TCO; (**b**) TIR.

In the TCO layout in Figure 10a, the rotational speeds N of Rotor 1 in the upwind location and Rotor 2 in the downwind location decreased as the gap decreased because of the interaction between the rotors. Similarly, in the TIR layout (Figure 10b), the rotational speeds of Rotor 1 and Rotor 2 decreased as the gap decreased. In particular, the decreasing tendency in the case of Rotor 2 for smaller gaps is remarkable. As seen in Figure 10, the gap dependence of the rotational speed is almost the same for the TCO and TIR layouts: (1) at the smallest gap of $g = 50$ mm ($g/D = 1$), the value of the rotational speed of Rotor 1 is 97% of that of an isolated rotor; (2) even at $g = 500$ mm ($g/D = 10$), the value of the rotational speed of Rotor 2 is 75–80% of that of an isolated rotor. Note that for smaller gaps of $g = 50$ and 100 mm ($g/D = 1$ and 2), downwind Rotor 2 cannot continue to rotate due to the decelerated wake flow behind the upwind Rotor 1 at $V = 10$ m/s. Zanforlin and Nishino [6] presented the reduction of the power of the upstream turbine under the assumption that only this turbine is working in the case of the TIR layout at $g/D = 0.5$, 1, 1.5, and 2, using 2D CFD. However, their results showing the same power coefficient as that of an isolated turbine seem to be unexpected (see 90° in their Figures 15, 17, and 18). De Tavernier et al. [12] showed a power reduction of both turbines in the case of the TIR layout at $g/D = 0.2$, based on a 2D simulation with a panel/vortex model. According to their Figure 9, the power values of Rotor 1 and Rotor 2 are approximately 80% and 30% of that of an isolated rotor, respectively.

To investigate the decreasing tendency in smaller gaps, we also experimented with the TCO layout at a higher wind speed, as shown in Figure 11a. At the smallest gap of $g = 50$ mm, the value of N of Rotor 1 is 97% of that of an isolated rotor (3700 rpm for $V = 12$ m/s) and at $g = 500$ mm, the value of N of Rotor 2 is approximately 80% of that of an isolated rotor, as in the case of $V = 10$ m/s. At this higher uniform wind speed, the downwind Rotor 2 continued to rotate at 54% and 62% of rotational speed of an isolated rotor, even at smaller gaps of $g = 50$ and 100 mm, respectively. Figure 11b shows the gap dependence of the power P obtained using Equation (1) explained in Section 2.5. The main results are as follows: (1) At $g = 50$ mm ($g/D = 1$), the power values of Rotor 1 and Rotor 2 are 97% and 49% of that of an isolated rotor (88.7 mW for $V = 12$ m/s); (2) at $g = 500$ mm ($g/D = 10$), the power of Rotor 2 is 78% of an isolated rotor. At the largest gap ratio, the power of Rotor 1 recovers to the value of a single rotor. According to Figure 6b,c in Sahebzadeh et al. [21], in the TCO layout for $V = 9.3$ m/s, the power values of Rotor 1 and Rotor 2 are 92% and 44%, respectively, of that of an isolated rotor at $g/D = 0.5$; these values changed to 95% and 40%, respectively, at $g/D = 1.25$. These values are qualitatively consistent with our experimental values for $g/D = 1$. Since the gap dependence of N is similar to that of P, as shown in Figure 11, hereafter, we will only use N for the discussion on the gap/wind-direction dependence for all configurations.

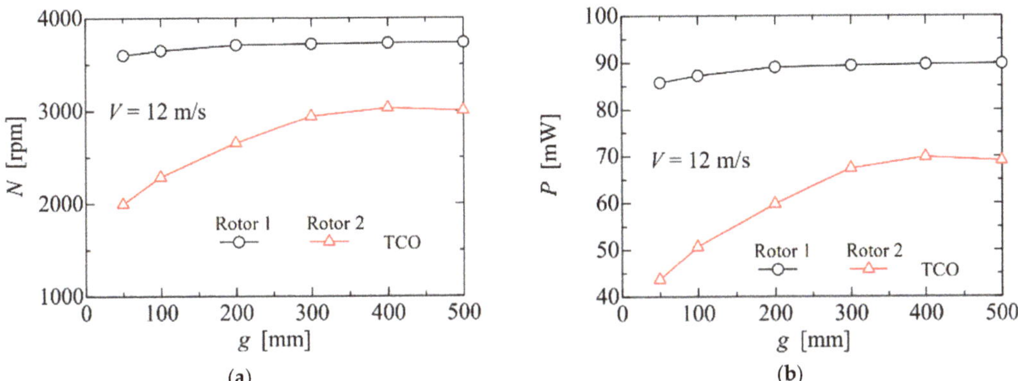

Figure 11. Gap dependence of (**a**) rotational speed and (**b**) power in the TCO layout at $V = 12$ m/s.

Next, we examine the normalized rotational speed N_{norm}, defined in Equation (2).

$$N_{norm} = \frac{N}{N_{SI}} \tag{2}$$

Figure 12 compares the results of N_{norm} obtained from the wind-tunnel experiment (Figure 10) with those obtained by Hara et al. [13] via 2D CFD analysis using the dynamic fluid body interaction (DFBI) method. The gap between the two rotors (g on the abscissa) is also nondimensionalized using the diameter of each rotor D. Regardless of the layout type (TCO and TIR), the normalized rotational speed decreased as the gap decreased in the experimental and CFD results.

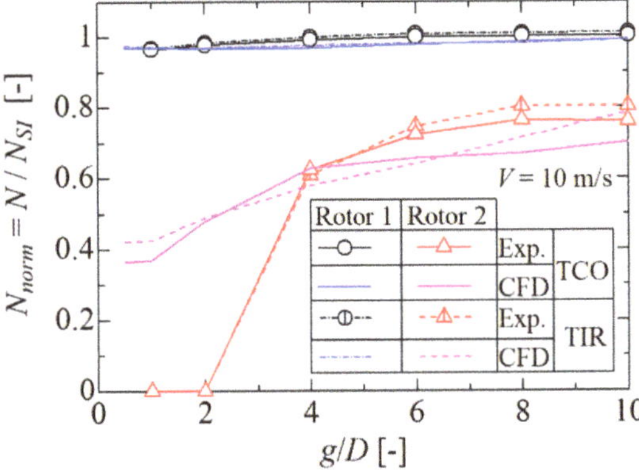

Figure 12. Comparison between the wind-tunnel experiment and CFD (reproduced with permission from Hara et al. [13]) on VAWTs in tandem layouts at $V = 10$ m/s.

Dessoky et al. [7] have reported the importance of the rotor spacing on the pair performance. This is based on the CFD code developed at their institute for wind turbine applications using 2 m diameter Darrieus turbine rotors of the two-bladed NACA 0021 airfoil. Sahebzadeh et al. [11] showed that the turbine wake is broken at a downstream distance of approximately $8D$ for the TCO layout based on the unsteady Reynolds-averaged Navier–Stokes (URANS) simulations with one-bladed turbines of an NACA 0018 airfoil. Their results also show a decreasing tendency of power for both rotors with decreasing g/D. They stated that the drop was caused by the upstream induction of the downstream turbine. However, their downstream rotor starts generating higher power at $0.25 \leq g/D \leq 1.25$ in an unexpected manner; the minimum power is obtained at $g/D = 2$ and 4 in their simulation (see their Figure 6a). The parameter g in our work corresponds to their R-d (R is the center-to-center distance of two rotors and d is the diameter of the rotor in Sahebzadeh et al. [11]). Recently, Kuang et al. [8] reported that *"the gap ratio of $g/D = 5$ can appropriately balance the power and space cost"* only in the case of the TCO layout. Our experimental result (Figure 12), showing a slight change in the N_{norm} with the gap at $g/D > 5$, substantiates their appropriate gap ratio.

3.2. Wind-Direction Dependence of a Pair of VAWTs (16 Wind Directions)

Figure 13 shows the 16-wind-direction dependence of the rotational speed in the CO configuration. The surrounding numbers indicate the wind direction θ (°). At wind direction $\theta = 90°$, called the tandem co-rotating (TCO) layout, the rotational speed of the downwind rotor R2 decreases considerably (Figure 13b) or R2 stops (Figure 13a). This happens because of the wake of upstream rotor R1, as in the cases in Figure 11a ($V = 12$ m/s) and Figure 10a ($V = 10$ m/s). Interestingly, R2 also stops at $\theta = 112.5°$. This can be explained by the existence of a deflected wake flow of R1 toward R2 by the Magnus effect [25]. Huang et al. [26] reported *"wake shape deformation and deflection"* of a VAWT using advanced robotic PIV. Strom et al. [1] have shown coherent structures in the wake of a vertical-axis turbine in a water channel experiment, in addition to the wake deflection (see their Figure 5). Furthermore, they have explained the mechanisms for the wake asymmetry using forces acting on the blade (see their Figure 6).

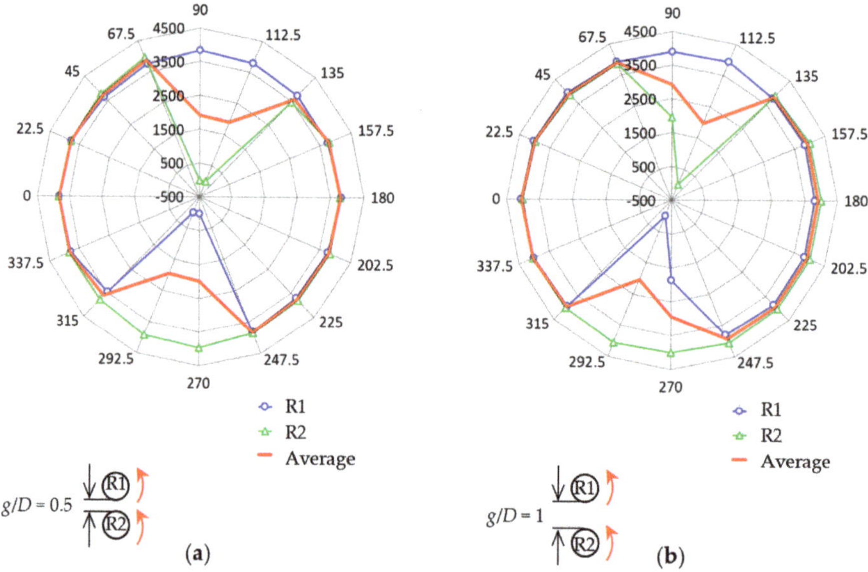

Figure 13. CO configuration of 16-wind-direction dependence on two VAWTs at $V = 14$ m/s: (a) $g/D = 0.5$; (b) $g/D = 1$.

Sogo et al. [24] have reported this deflected wake in the cases of TCO and TIR layouts by visualizing the streak lines with a smoke-wire method in their Figures 4 and 5. In Figure 13, the same phenomena (decrease in rotational speed of R1 or stoppage of R1) can be confirmed at $\theta = 270°$ and $292.5°$, which are in origin symmetry for $\theta = 90°$ and $112.5°$, respectively. Therefore, these phenomena ensure the repeatability of the experiment explained in Section 2.3. Sahebzadeh et al. [21] have defined the wind direction near the TCO layout ($\theta = 90–120°$ or $\theta = 270–300°$ and $g/D < 4$ in our coordinate) as the "*wake-interaction regime*" with low total performance in the TCO layout. This concurs well with our findings showing an origin-symmetrical distribution in the CO configuration.

Figure 14 shows the 16-wind-direction dependence of the rotational speed in the IR configuration. At wind direction $\theta = 90°$, called the tandem inverse-rotating (TIR) layout, the rotational speed of the downwind rotor R2 decreases considerably (Figure 14a) or R2 stops (Figure 14b). This is the result of a wake interference, caused by the upwind rotor R1, with downwind rotor R2, as in the case in Figure 10b ($V = 10$ m/s). At $\theta = 112.5°$, although R2 stops in the case of $g/D = 0.5$, it continues to rotate in the case of $g/D = 1$. In Figure 14, a slowdown in rotational speed or stopping of R1 can also be confirmed at $\theta = 270°$ and $247.5°$, which are in line symmetry for $\theta = 90°$ and $112.5°$, respectively.

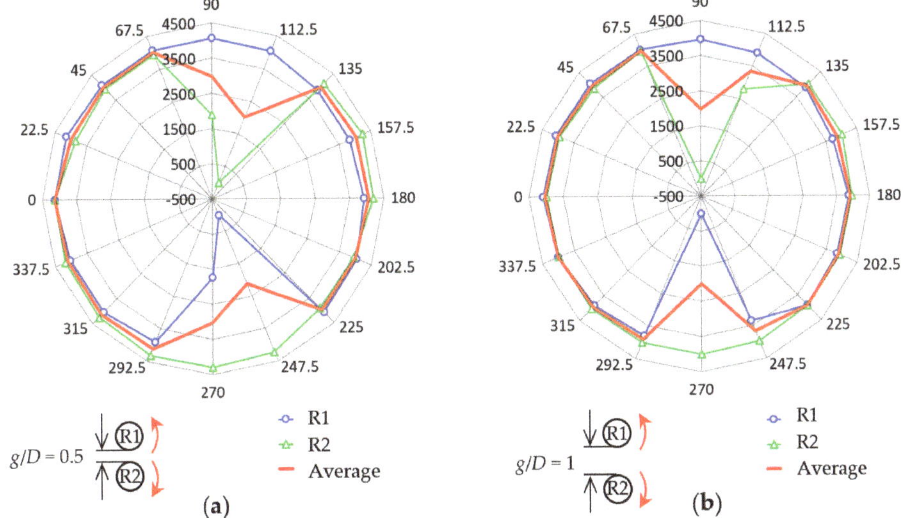

Figure 14. IR configuration of 16-wind-direction dependence on two VAWTs at $V = 14$ m/s: (a) $g/D = 0.5$; (b) $g/D = 1$.

This line symmetry is seen with respect to the line connecting the wind direction of $\theta = 0°$ and $180°$. Consequently, the size of the left half area surrounded by red lines (indicating the average rotor speed of R1 and R2) and the line connecting the wind direction of $\theta = 90°$ and $270°$ is larger than the size of the corresponding right half area. This imbalance is emphasized with the decrease in g/D, as predicted by the numerical simulation by Hara et al. [13] (see their Figure 13b,d,f). In contrast, unnatural increases in power in the case of tandem layout are reported by Dabiri [3], based on field measurements (see their Figure 4), and Zanforlin and Nishino [6] (see their Figure 15), conducted by 2D numerical simulation. The former is seen in a wind direction of approximately $\theta = 110°$ and $290°$ (approximately $20°$ and $200°$ in their coordinates). The latter is seen in a wind direction of approximately $\theta = 90°$ (approximately $\gamma = 90°$ in their coordinate). The line-symmetrical distribution of the wind-direction dependence in the IR configuration was also seen in the results of De Tavernier et al. [12], obtained using the 2D panel/vortex method (see their Figure 9), though they do not mention it. However, increases seen in their results for approximately $\theta = 45°, 135°, 225°$, or $315°$ are unexplained.

Next, we compare the experimental results with those obtained by 2D CFD [13] conducted by our group. Figure 15 depicts the comparison between the experimental and CFD results on 16-wind-direction dependence of the average normalized rotational speed in the CO configuration. In each figure, the red line represents the experimental results and the green dotted line shows the CFD results. Decreases in the average rotational speed on the TCO layout ($\theta = 90°$ and $270°$) and near the tandem layout ($\theta = 112.5°$ and $292.5°$) are seen in both the experimental and CFD results. This origin-symmetrical distribution on 16-wind-direction dependence in the CO configuration can be confirmed for $g/D = 0.5$ and 1. At $\theta = 112.5°$ and $292.5°$, the decrease in rotational speed occurs more explicitly in the experimental results compared to the CFD results. This implies that the wake deflection, accompanied by the stopping of a downwind rotor, in the 3D experiment is stronger than that of the 2D CFD.

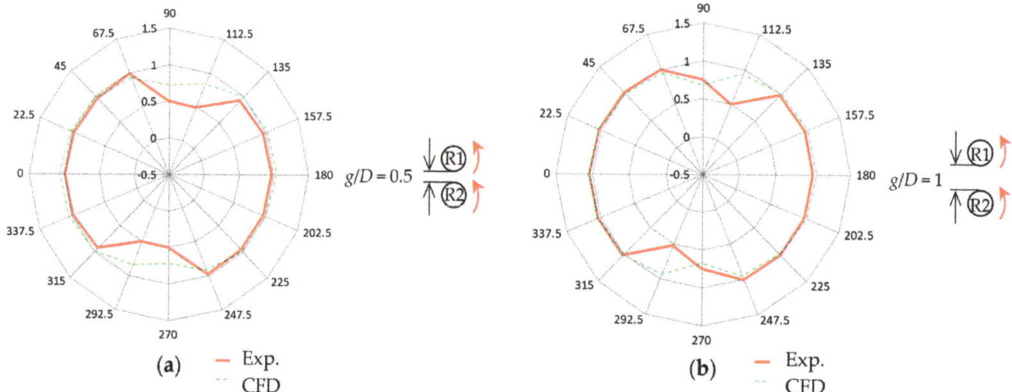

Figure 15. Comparison between the experimental and CFD results on 16-wind-direction dependence of average normalized rotational speed in a pair of VAWT models in CO configuration at $V = 14$ m/s: (a) $g/D = 0.5$; (b) $g/D = 1$.

Figure 16 presents a comparison between the experimental and CFD results on the 16-wind-direction dependence of the average normalized rotational speed in the IR configuration. The CFD results also show an obvious slowdown in the average rotational speed in the TIR layout ($\theta = 90°$ and $270°$) and near the tandem layout ($\theta = 112.5°$ and $247.5°$). This supports the experimental results showing the line-symmetrical distribution on 16-wind-direction dependence in the IR configuration.

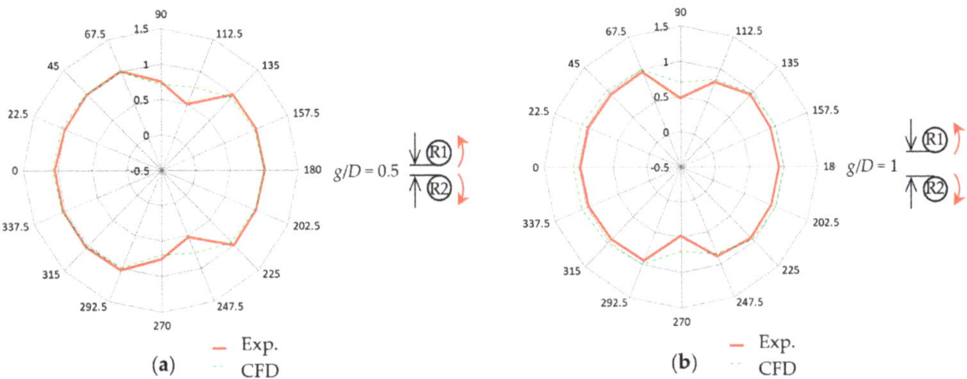

Figure 16. Comparison between the experimental and CFD results on 16-wind-direction dependence of average normalized rotational speed in a pair of VAWT models in IR configuration at $V = 14$ m/s: (**a**) $g/D = 0.5$; (**b**) $g/D = 1$.

3.3. Wind-Direction Dependence of a Trio of VAWTs (12 Wind Directions)

Figure 17 shows the 12-wind-direction dependence of the normalized rotational speed N_{norm} in the 3CO and 3IR configurations. In each figure, the blue circles indicate the rotational speed N of Rotor 1 (R1), the green crosses plot the N of Rotor 2 (R2), and the black triangles show the N of Rotor 3 (R3). The red squares represent the average rotor speed of R1, R2, and R3, defined as $N_{norm,\ ave}$.

First, we investigate the 3CO configuration in Figure 17a,c,e. At wind direction $\theta = 0°$, called 3COB (see Figure 4a), the downwind rotor R3 stops at the smallest gap of $g/D = 0.5$ (Figure 17a) but R3 rotates faster with increasing the gap (Figure 17c,e). At wind direction $\theta = 30°$, called TCOU, all rotors rotate at a relatively high speed, with the average rotational speed of the three rotors ($N_{norm,ave}$) reaching a maximum value at $g/D = 0.5$ and 1 (Figure 17a,c). At $\theta = 60°$, called 3COF, one of the downwind rotors, R3, stops rotating at $g/D = 0.5$ and 1 (Figure 17a,c). Similarly, at $\theta = 90°$, called 3TCOD, the downwind rotor R2 stops at $g/D = 0.5$ and 1 (Figure 17a,c). In this layout, a significant decrease in the rotational speed of R2 occurs, even in the largest gap of $g/D = 2$ (Figure 17e). As a result, the average rotational speed $N_{norm,ave}$ is remarkably low in the 3TCOD. As explained in Section 2.4, the same layout occurs in every 120° in the 3CO configuration. In fact, the experimental results (Figure 17a,c,e) show the excellent *rotational symmetry* for the average normalized rotational speed $N_{norm,ave}$.

Second, we examine the 3IR configuration in Figure 17b,d,f. At wind direction $\theta = 0°$ (CDB; see Figure 4b), the downwind rotor R3, in addition to the upwind rotors (R1 and R2), continues to rotate even at the smallest gap of $g/D = 0.5$ (Figure 17b). This completely differs from the result in 3COB ($\theta = 0°$ in the 3CO layout). The average rotational speed of the three rotors, $N_{norm,ave}$, in the 3IR configuration reaches its maximum value at $\theta = 0°$ (i.e., CDB) at $g/D = 0.5$, 1, and 2 (Figure 17b,d,f). Hence, $N_{norm,ave}$ in other layouts tends to have a much smaller value than it does in CDB, especially in the cases with a smaller gap distance. Although we do not explain all the layouts in detail here, the experimental results shown in Figure 17b,d,f prove that there is no rotationally symmetric distribution in $N_{norm,ave}$, as expected by the 12 independent layouts in the 3IR configuration explained in Section 2.4.

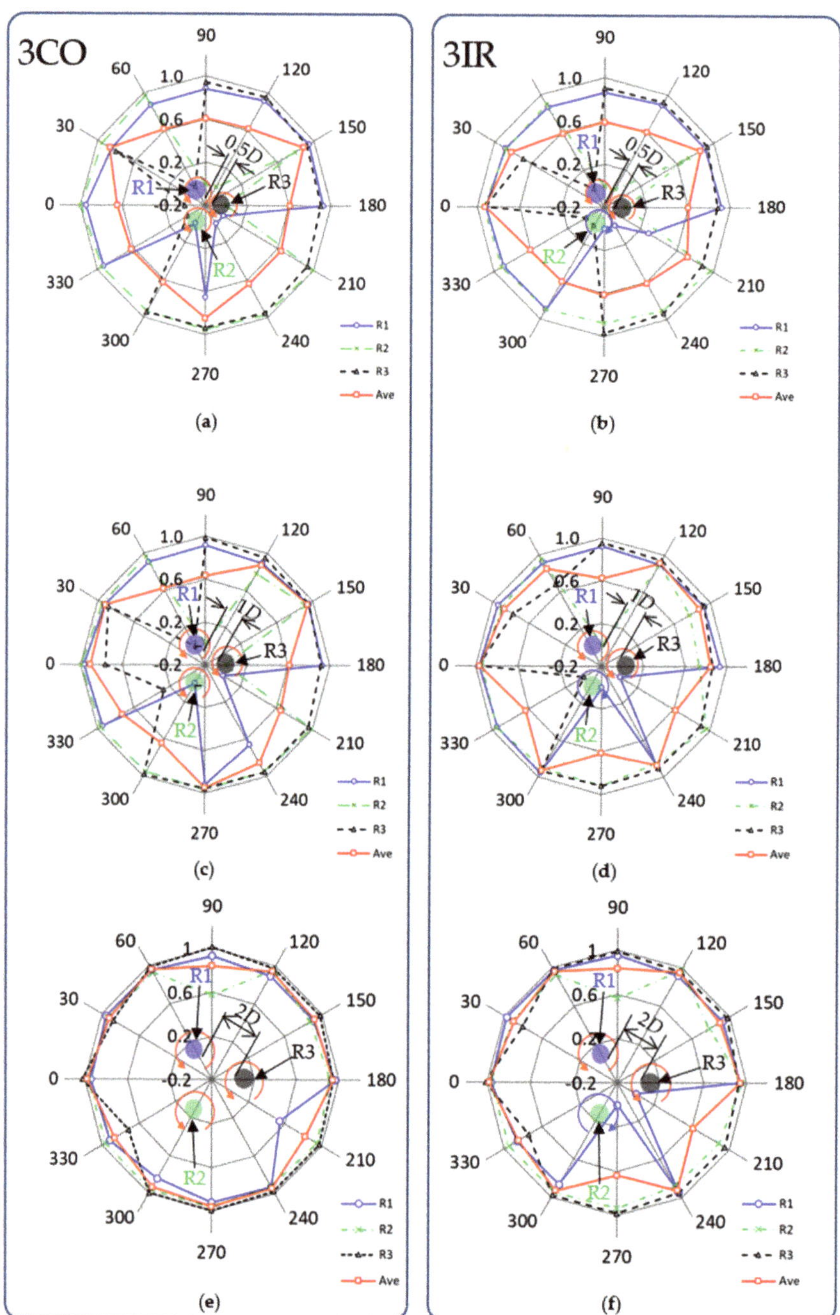

Figure 17. Configurations of 12-wind-direction dependence of normalized rotational speed in a trio of VAWT models at $V = 12$ m/s: (**a**) 3CO configuration, $g/D = 0.5$; (**b**) 3IR configuration, $g/D = 0.5$; (**c**) 3CO configuration, $g/D = 1$; (**d**) 3IR configuration, $g/D = 1$; (**e**) 3CO configuration, $g/D = 2$; (**f**) 3IR configuration, $g/D = 2$.

Hezaveh et al. [4] termed the mean value of three turbines as *"cluster-averaged power."* For $\theta = 60°$, they suggested the best gap ratios of 2, 3, and 4 with the highest cluster-averaged power value (see their Figure 8). These values support our results that the best gap ratio is $g/D = 2$ with maximum $N_{norm,ave}$ value (Figure 17).

Here, we define the footprint radius R_{foot} and the footprint area A_{foot} of a trio of rotors, as in Equations (3) and (4). R_{foot} is the radius of the circular region that connotes the three rotors with a gap distance of g. The second term of the right hand of Equation (3) represents a distance from the center of the rotor trio to the corner of an equilateral triangle (lavender dash-dotted line in Figure 4). Since the diameter of each rotor D is 50 mm, the values of R_{foot} at $g = 25$, 50, and 100 mm are 68.30, 82.74, and 111.60 mm, respectively. The A_{foot} values at $g = 25$, 50, and 100 mm are then 0.01466, 0.02150, and 0.03913 m^2, respectively. We define the simple average of the $N_{norm,ave}$ of the 12 wind directions as the $N_{norm,\,12\text{-}wind}$. Table 3 lists the values of $N_{norm,\,12\text{-}wind}$ based on the wind-tunnel experiments and those multiplied by 3 (i.e., the number of rotors) divided by the corresponding A_{foot}.

Table 3. Comparison of average normalized rotational speed $N_{norm,\,12\text{-}wind}$ and $3N_{norm,\,12\text{-}wind}/A_{foot}$ in a trio of VAWT models in an isotropic 12-directional wind speed.

Average Speed	$g/D = 0.5$		$g/D = 1$		$g/D = 2$	
	3CO	3IR	3CO	3IR	3CO	3IR
$N_{norm,\,12\text{-}wind}$ (-)	0.686	0.677	0.778	0.801	0.918	0.869
$3N_{norm,\,12\text{-}wind}/A_{foot}$ (-/m^2)	140.4	138.6	108.5	111.7	70.4	66.6

By referring to Table 3, the value of $N_{norm,\,12\text{-}wind}$ increases with an increase in g/D. This gap dependence is emphasized by the plots in Figure 18, in which $N_{norm,ave}$ of the different rotor spacings are presented in one radar chart for 3CO and 3IR, respectively. In each figure, the blue circles indicate the average normalized rotational speed $N_{norm,ave}$ at $g/D = 0.5$, the green triangles plot the $N_{norm,ave}$ at $g/D = 1$, and the red squares show the $N_{norm,ave}$ at $g/D = 2$.

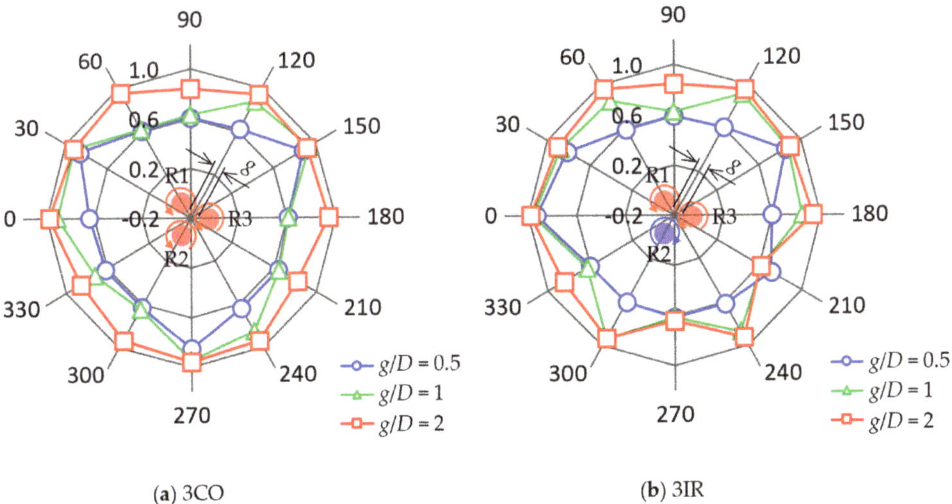

(a) 3CO   **(b)** 3IR

Figure 18. Experimental results of the gap dependence of the 12-wind-direction for the average normalized rotational speed in a trio of VAWT models at $V = 12$ m/s in: **(a)** 3CO configuration; **(b)** 3IR configuration.

In contrast, the performance indicated by $3N_{norm,\,12\text{-}wind}/A_{foot}$ in a unit footprint area shows the opposite tendency, i.e., the advantage of a smaller rotor spacing (see Table 3). However, it is crucial to contemplate not only the performance (wind-direction dependence in $N_{norm,\,12\text{-}wind}$ or $3N_{norm,\,12\text{-}wind}/A_{foot}$) of a trio of turbines, but also the velocity deficit of the wake flow.

This interference between a trio and the other trios is essential for future research on the design of an optimal layout of the three-turbine array or multi rotor systems with very many rotors (e.g., [27]).

$$R_{foot} = \frac{D}{2} + \frac{g+D}{\sqrt{3}} \tag{3}$$

$$A_{foot} = \pi R_{foot}^{\,2} \tag{4}$$

Now, we define another normalized rotational speed, $N_{norm,bi\text{-}wind}$, in Equation (5). Here, $N_{norm,ave,\,0°}$ and $N_{norm,ave,\,180°}$ are average normalized rotational speeds of parallel-like layouts at $\theta = 0°$ and $\theta = 180°$, respectively. Therefore, $N_{norm,bi\text{-}wind}$ indicates the performance of a trio of turbines in an isotropic bidirectional wind speed. An example of the bi-wind in nature is a daytime sea breeze or a nighttime land breeze.

Table 4 shows the values of $N_{norm,bi\text{-}wind}$ for different gap ratios. Table 4 also contains the values of $3N_{norm,bi\text{-}wind}/A_{foot}$, which indicates the advantage of a smaller rotor spacing. As shown in Table 4, the 3IR configuration yielded a higher average rotational speed than the 3CO arrangement at any rotor spacing in the ideal bidirectional wind conditions. It is interesting that the change in wind from 12-wind-direction (Table 3) to 2-wind-direction (Table 4) acts negatively in the 3CO configuration but positively in the 3IR configuration, on VAWT performance.

$$N_{norm,\,bi-wind} = \frac{N_{norm,\,ave,\,0°} + N_{norm,\,ave,\,180°}}{2} \tag{5}$$

Table 4. Comparison of normalized rotational speed $N_{norm,bi\text{-}wind}$ and $3N_{norm,bi\text{-}wind}/A_{foot}$ in a trio of VAWT models in an isotropic bidirectional wind speed.

Average Speed	$g/D = 0.5$		$g/D = 1$		$g/D = 2$	
	3CO	3IR	3CO	3IR	3CO	3IR
$N_{norm,bi\text{-}wind}$ (-)	0.629	0.763	0.762	0.903	0.960	0.961
$3N_{norm,bi\text{-}wind}/A_{foot}$ (-/m^2)	128.8	156.1	106.3	126.0	73.6	73.7

Figures 19 and 20 compare the experimental results with corresponding 2D CFD results with the DFBI method [16] for the 3CO configuration and the 3IR configuration, respectively. As for the rotational direction and the gap length ($g/D = 1$), refer to the inset in Figures 19 and 20. Broadly speaking, the speed-down tendencies of the downwind rotor are confirmed in both the experimental and the CFD results.

Next, we examine the difference between the 3CO and 3IR configurations. Figure 21 plots average normalized rotational speed in a trio of VAWT models at $g/D = 1$. The green circle and the purple triangle indicate the 3CO and 3IR, respectively. Figure 21a shows the experimental results with solid lines, while Figure 21b presents the CFD results with broken lines. It is worth noting that $\theta = 60°$ in 3IR (CDF) with a higher $N_{norm,ave}$ value is an advantageous wind direction against $\theta = 60°$ in the 3CO (3COF).

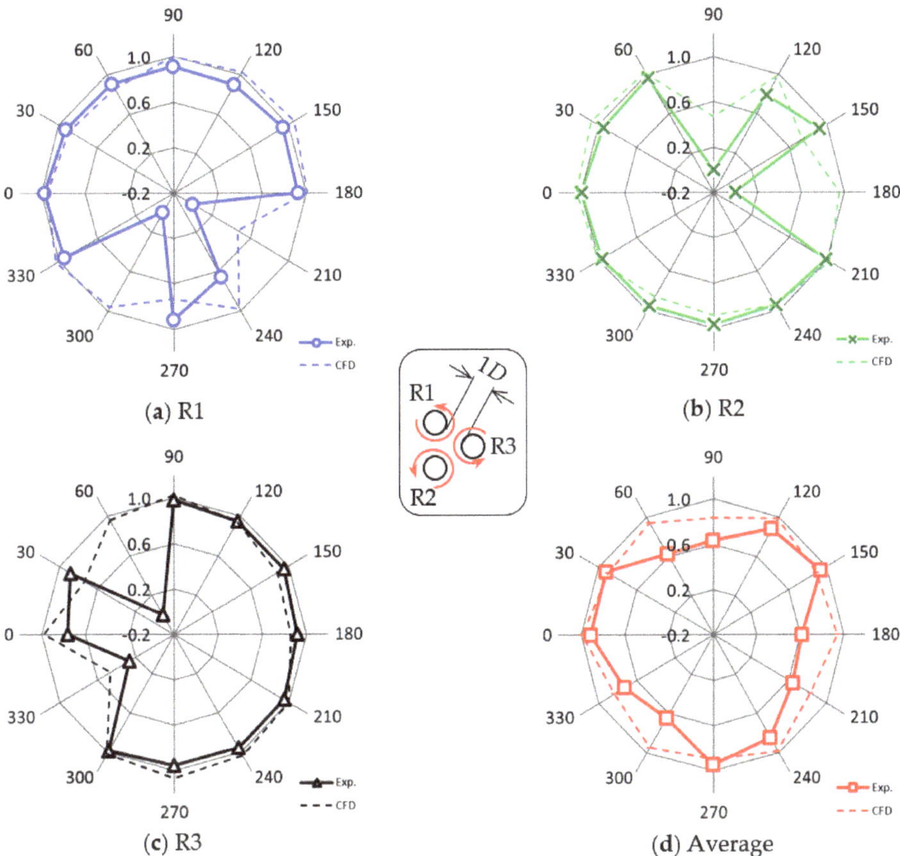

Figure 19. Comparison between the experimental and CFD results on 12-wind-direction dependence of normalized rotational speed in a trio of VAWT models at $g/D = 1$ in the 3CO configuration: (**a**) Rotor 1; (**b**) Rotor 2; (**c**) Rotor 3; (**d**) Average.

On the other hand, $\theta = 270°$ in 3IR (TIRU-CU) with a lower $N_{norm,ave}$ value is a disadvantageous wind direction against $\theta = 270°$ in 3CO (TCOF). In total, the value of $N_{norm,\,12\text{-}wind}$ (the simple average of the $N_{norm,ave}$ of the 12 wind directions) of 0.801 of the 3IR is 3% larger than 0.778 of the 3CO (see Table 3).

Figures 22 and 23 show the streak lines observed using a smoke-wire method in the cases of the 3CO configuration and 3IR configuration, respectively. A stainless-steel wire, which is enclosed in the yellow dotted frame in Figure 22a, was horizontally set 2.7D upstream of the center of a trio of rotors. The apparatus is the same as that used in [20]. The critical parameter in the visualization of the flow is the rotational speed N of each rotor in order to maintain the tip speed ratio $\lambda = \pi DN/(60V)$. Since the ratio of a uniform velocity in visualization to rotational speed measurements at 12 m/s is 1/6, each rotor speed is precisely adjusted one-sixth of the rotational speed presented in Figure 17 by using a variable external resistance. Electric cables located far below the trio of turbines are also seen in the visualization, as shown in Figure 22a.

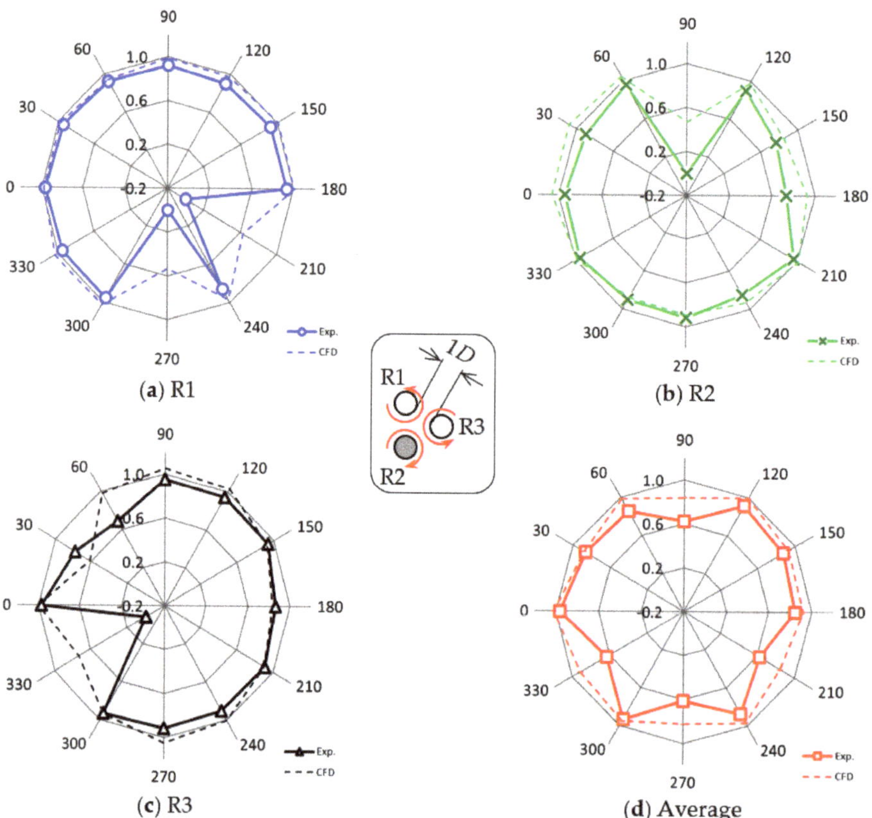

Figure 20. Comparison between the experimental and CFD results on 12-wind-direction dependence of normalized rotational speed in a trio of VAWT models at $g/D = 1$ in the 3IR configuration: (a) Rotor 1; (b) Rotor 2; (c) Rotor 3; (d) Average.

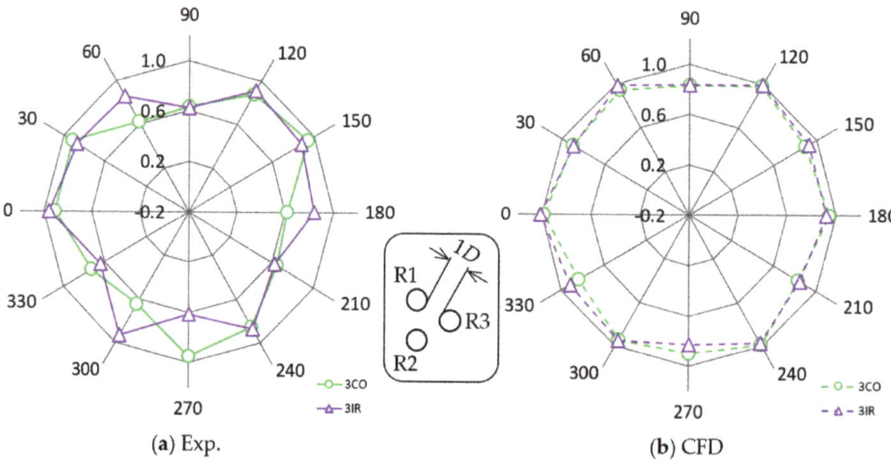

Figure 21. Comparison between 3CO and 3IR configurations on the average normalized rotational speed in a trio of VAWT models at $g/D = 1$: (a) experimental results; (b) CFD results.

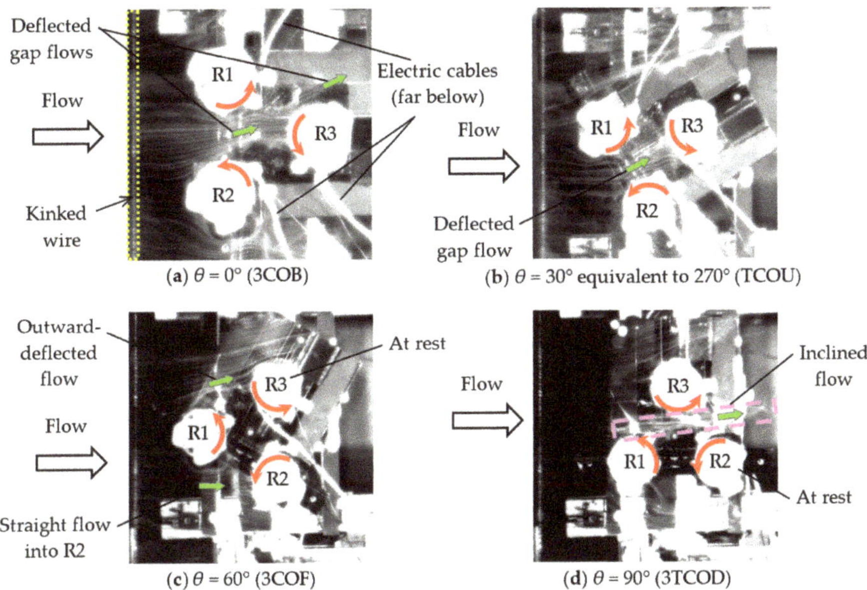

Figure 22. Photographs of smoke flow through three turbines at $g/D = 1$ under $V = 2$ m/s in the 3CO configuration: (**a**) $\theta = 0°$; (**b**) $\theta = 30°$; (**c**) $\theta = 60°$; (**d**) $\theta = 90°$.

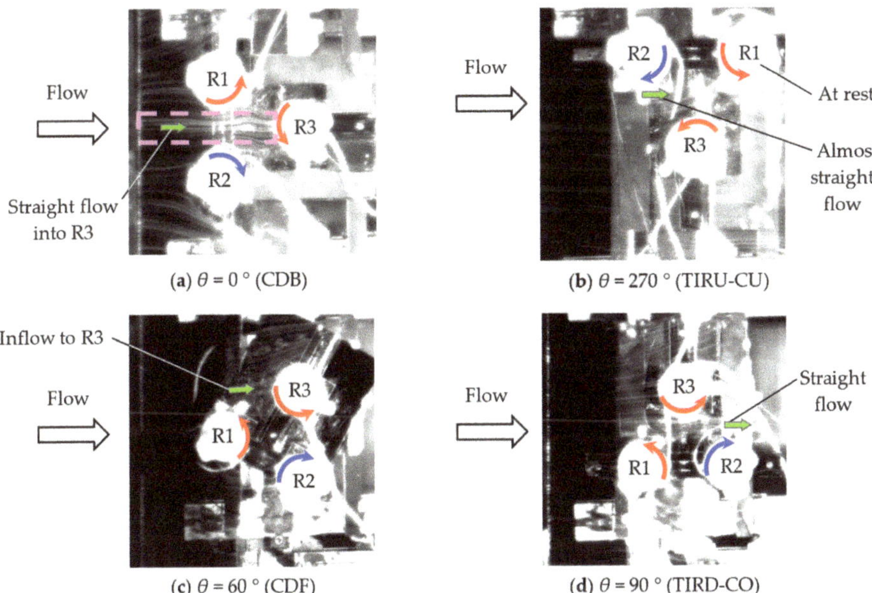

Figure 23. Photographs of smoke flow through three turbines at $g/D = 1$ under $V = 2$ m/s in the 3IR configuration: (**a**) $\theta = 0°$; (**b**) $\theta = 270°$; (**c**) $\theta = 60°$; (**d**) $\theta = 90°$.

Figures 24 and 25 present the distributions of the streamwise component of flow velocity conducted by the corresponding 2D CFD with the DFBI method [16] in the cases of the 3CO and 3IR configurations, respectively. The color in Figures 24 and 25 indicates the magnitude of the velocity.

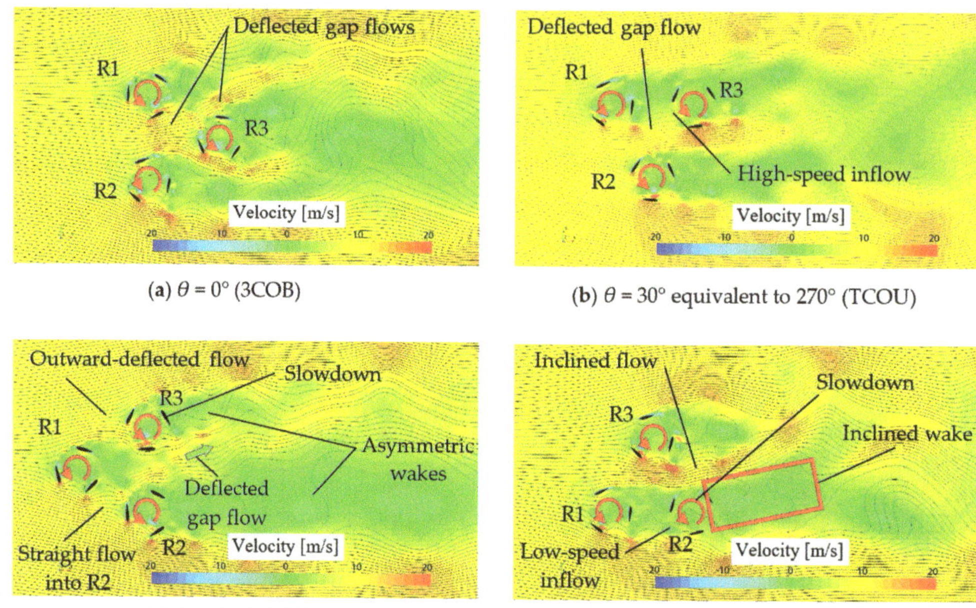

(**a**) $\theta = 0°$ (3COB)

(**b**) $\theta = 30°$ equivalent to 270° (TCOU)

(**c**) $\theta = 60°$ (3COF)

(**d**) $\theta = 90°$ (3TCOD)

Figure 24. Color contours of flow velocity at $g/D = 1$ under $V = 10$ m/s in the 3CO configuration: (**a**) $\theta = 0°$; (**b**) $\theta = 30°$; (**c**) $\theta = 60°$; (**d**) $\theta = 90°$.

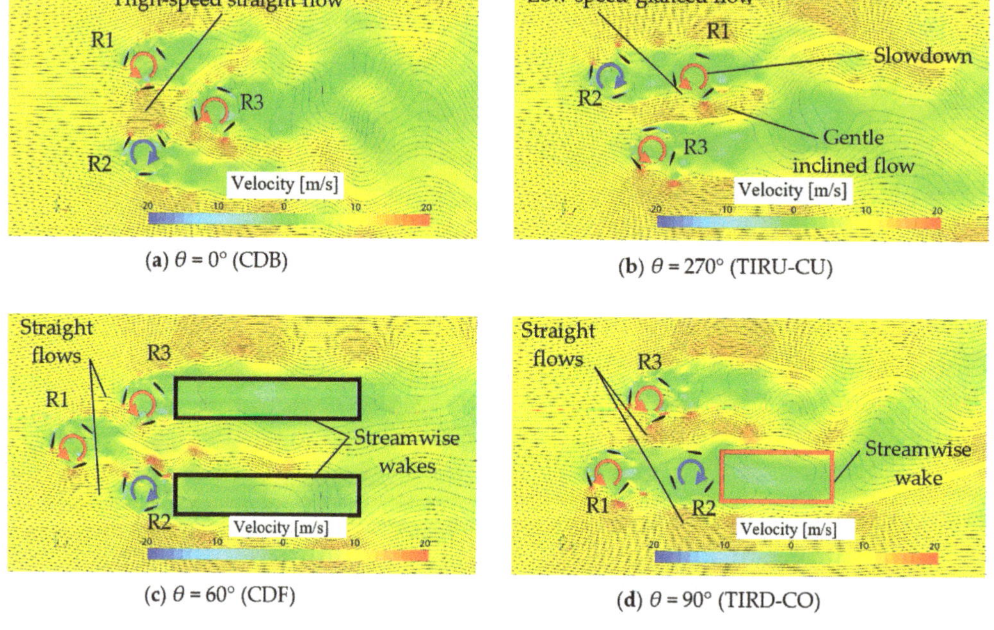

(**a**) $\theta = 0°$ (CDB)

(**b**) $\theta = 270°$ (TIRU-CU)

(**c**) $\theta = 60°$ (CDF)

(**d**) $\theta = 90°$ (TIRD-CO)

Figure 25. Color contours of flow velocity at $g/D = 1$ under $V = 10$ m/s in the 3IR configuration: (**a**) $\theta = 0°$; (**b**) $\theta = 270°$; (**c**) $\theta = 60°$; (**d**) $\theta = 90°$.

First, we explore the flow patterns in the 3CO configuration in Figure 22. At wind direction $\theta = 0°$, called 3COB (see Figure 22a), the distances of the adjacent streak lines contract between the upwind rotors (R1 and R2), which implies the flow acceleration between R1 and R2. This is also confirmed in the CFD results in Figure 24a. Subsequently, the accelerated gap flow deflects toward the R1 side (upper-side) and flows partially into the downwind rotor (R3).

At wind direction $\theta = 30°$, called TCOU, in which all rotors rotate relatively fast, the accelerated flow between R1 and R2 deflects toward the center of R3 in Figure 22b (see the relatively high-speed inflow shown in yellow just in front of R3 in Figure 24b).

At $\theta = 60°$, called 3COF, the outward-deflected flow in front of R3 glanced off the outer edge (upper-side) of R3, as shown in Figures 22c and 24c. Furthermore, the flow in front of R3 receives a negative effect from R2-driven induced velocity. This outward deflection is caused by R1- and R2-driven induced velocities, resulting in deceleration in front of R3. On the other hand, R1- and R3-driven induced velocities have an accelerating effect on the flow in front of R2. These results imply that much momentum flows into R2; subsequently, asymmetric wakes (accompanied by deflected gap flow between them) are generated behind R2 and R3, as shown in Figure 24c. These are the reasons for the stoppage of R3 (black triangle in Figure 17c).

The deflected gap flow from R2 to R3 can also be seen in a 3CO-like layout (not in an equilateral triangle) using three Savonius turbines conducted by Shaheen et al. [28] (see their Figure 19).

Similarly, at $\theta = 90°$, called 3TCOD, the downwind rotor R2 stops, indicated by the green cross in Figure 17c (a remarkable decrease in the speed in CFD in Figure 19b), since the flow between R2 and R3 diverges toward the opposite side (upper-side) of R2 enclosed by the lavender broken frame in Figure 22d. This can be seen in the inclined high-speed zone shown in the red contour in the CFD results in Figure 24d. Note that the low-speed inflow shown in green just in front of R2 in Figure 24d contrasts to the high-speed inflow in front of R3 seen in Figure 24b. Furthermore, the remarkably inclined wake zone enclosed by a red solid frame behind R2 is also contrastive to the straight wake zone, which is investigated later in Figure 25d in the opposite rotational direction of R2.

The above-mentioned flow deflection occurs through the combinational results of the Magnus effect and induced-velocity effect, originating from the three rotating turbines. Regarding the two parallel-like layouts in the 3CO configuration, the average rotational speed performance in the 3COB layout ($\theta = 0°$) is greater than that in the 3COF layout ($\theta = 60°$). Regarding the two tandem-like layouts in the 3CO configuration, the TCOU layout ($\theta = 30°$) shows better performance compared with that in the 3TCOD layout ($\theta = 90°$). These two relationships are common results obtained by the experiments and the CFD.

The former relationship (3COB > 3COF) is also confirmed in the normalized overall power coefficient by Zanforlin [22] for $gap/D = 2$ (see her Figure 10b), by Shaheen et al. [28] (see their Figures 22 (3COF-like layout) and 26 (3COB-like layout)), and by Silva and Danao [29] for $gap/D = 1$ (see their Table 6), with all based on 2D CFD.

Regarding the order of rotational speed $N_{norm,\ ave}$ at $\theta = 60°$ in the 3CO configuration (i.e., 3COF), R2 rotates at the fastest speed for the $gap/D = 0.5$ and 1. This is the same result in three Savonius turbines in a 3COF-like layout independent of rotor gap ratio conducted by Shaheen et al. [27] (see their Figure 21) and on three Darrieus turbines in a 3COF layout by Silva and Danao [28] for $gap/D = 1$.

Next, we examine the flow patterns in the 3IR configuration in Figure 23. At wind direction $\theta = 0°$, called CDB (see Figure 23a), the contracted streak lines between the upwind rotors (R1 and R2) directly flow into the downwind rotor (R3). The large high-speed zone surrounded by the three rotors is seen in the CFD results in Figure 25a. Therefore, the value of $N_{norm,ave}$ in this layout takes the maximum value of 0.9618 among the whole 16 independent layouts (see Table 2). This value is 106% of 0.9087 obtained in the case of 3COB ($\theta = 0°$ in the 3CO layout). The corresponding 2D CFD shows 103% in this comparison (CDB to 3COB).

The order of rotational speed $N_{norm,\ ave}$ at $\theta = 0°$ in the 3IR configuration (i.e., CDB) is R1 > Ave. > R2/R3, as seen in Figure 17d or Figure 17f for $gap/D = 1$ or 2. This agrees with the order of the power coefficient for CDB obtained by Zheng et al. [23] (see their Figure 19) based on 2D CFD.

At wind direction $\theta = 60°$, called CDF, although the induced velocity of the upwind rotor (R1) causes negative effects on the flow before the downwind rotor (R3), the opposite rotation of R2 produces positive effects on the flow in front of R3. The wakes behind R2 and R3 show a relatively symmetric straight flow pattern enclosed by black solid frames in Figure 25c.

This indicates that the upper-side flow of R1 gives its streamwise momentum relatively directly into R3. This flow pattern is clearly different from that in the largely outward-deflected wake flow in the 3CO configuration at the same wind direction of $\theta = 60°$ (3COF). Hence, R3 continues to rotate in the case of CDF, which is in contrast to the result (R3 stopped) in the case of 3COF. This leads to a significant advantage in the $N_{norm,ave}$ at the CDF layout ($\theta = 60°$ in the 3IR configuration) over $N_{norm,ave}$ at 3COF, as seen in Figure 21 (indicated by CDF > 3COF).

At $\theta = 90°$, called TIRD-CO, both flows of the upper and lower sides of R2 straightly move downstream, generating the above-mentioned streamwise wake zone behind R2 (Figure 25d).

At $\theta = 270°$, called TIRU-CU (see Figure 23b), since the gap flow between R2 and R3 is almost straight in contrast to the largely deflected flow in Figure 22b, the gap flow in the TIRU-CU layout does not flow fully into the downwind rotor R1 but glances off the inner edge (lower-side) of R1.

Consequently, R1 stops to rotate in the experiment, or R1 takes the minimum $N_{norm,ave}$ value in the CFD (see Figure 20a). Remember that the TCOU layout in Figure 22b for $\theta = 30°$ is equivalent to that in $\theta = 150°$ and $270°$ only for the 3CO configuration, as explained in Table 2. There is a remarkable difference in $N_{norm,ave}$ between the TIRU-CU layout and the TCOU layout in the same wind direction (see Figure 21).

Finally, it is interesting to note that among the six parallel-like layouts ($\theta = 0°$, $60°$, $120°$, $180°$, $240°$, and $300°$), the $N_{norm,ave}$ value in the 3IR configuration always exceeds that in the 3CO configuration in the experiments (see Figure 21a). In contrast, the $N_{norm,ave}$ value in the 3IR configuration falls below that in the 3CO configuration regarding the six tandem-like layouts ($\theta = 30°$, $90°$, $150°$, $210°$, $270°$, and $330°$).

4. Conclusions

Our wind-tunnel experiments on the interaction between a pair/trio of closely spaced VAWTs have revealed a wind-direction dependence at a uniform wind speed.

For both the tandem co-rotating (TCO) and the tandem inverse-rotating (TIR) pair:

- The decrease in the rotational speed and rotor power of a pair of turbines arranged in tandem was demonstrated;
- The amount of decrease depended on the g/D ratio, with the value of the rotational speed of the downwind rotor 75–80% of that of an isolated rotor even at $g/D = 10$, although the value of the upstream rotor was 100%;
- The corresponding power value of the downwind rotor was approximately 80%.

In the 16-wind-direction experiments on a pair of VAWTs:

- The *"origin-symmetrical"* distribution of the average rotational speed of two rotors in the CO pair configuration and the *"line-symmetrical"* distribution in the IR pair configuration were demonstrated;

- The existence of a deflected wake flow accounted for the decreasing tendency of the rotational speed at $\theta = 90°, 112.5°, 270°$, and $292.5°$ (showing the origin symmetry) in the CO pair configuration;
- A wake interaction caused a slowdown tendency at $\theta = 90°, 112.5°, 270°$, and $247.5°$ (showing the line symmetry) in the IR pair configuration.

We examined 16 independent layouts (eight parallel-like layouts and eight tandem-like layouts) in the experiments on a trio of VAWTs:

- The performance in a unit footprint area demonstrated the advantage of a smaller rotor spacing for not only the average of 12-directional wind (Table 3) but also the average of bidirectional wind (Table 4);
- The inverse-rotating trio (3IR) configuration takes a higher average rotational speed than the co-rotating trio (3CO) configuration at any rotor gap under the ideal bidirectional wind conditions;
- The maximum average rotational speed can be obtained at a wind direction of $\theta = 0°$ in the 3IR configuration, which is 6% faster than that in the 3CO configuration;
- The average rotor speed of the three rotors $N_{norm, \, ave}$ is remarkably low at $\theta = 90°, 210°$, and $330°$ in the 3CO configuration (3TCOD). The 3CO configuration demonstrates the *"rotational symmetry"* for $N_{norm, \, ave}$ in every $120°$;
- The 3IR configuration does not show the rotational symmetry for $N_{norm, \, ave}$, as expected.

The corresponding flow patterns and velocity fields have been discussed in detail, along with flow visualization and 2D CFD results obtained by adopting the DFBI method:

- The relationship 3COB > 3COF for $N_{norm, \, ave}$ is explained by accelerated gap flows at $\theta = 0°$ and the decelerating effect on the flow at $\theta = 60°$, in the 3CO configuration;
- The relationship CDF > 3COF for $N_{norm, \, ave}$ is explained by straight wakes at $\theta = 60°$ in the 3IR configuration and asymmetric wakes at $\theta = 60°$ in the 3CO configuration.

For future work:

- We are confident that our research will serve as a base for future studies on designing a wind farm consisting of sets of these turbine pairs and trios. Further investigation on the wake interference between them is essential for future research on the design. We are currently making preparations for the wind-tunnel experiments with 12 BWTs to determine the optimal arrangement, supported by JSPS KAKENHI below.

Author Contributions: Conceptualization, Y.J. and Y.H.; methodology, Y.J.; software, Y.J. and Y.H.; validation, Y.J.; formal analysis, Y.J. and Y.H.; investigation, Y.J.; resources, Y.J. and Y.H.; data curation, Y.J. and Y.H.; writing—original draft preparation, Y.J.; writing—review and editing, Y.J. and Y.H.; visualization, Y.J. and Y.H.; supervision, Y.H.; project administration, Y.H.; funding acquisition, Y.J. and Y.H. All authors have read and agreed to the published version of the manuscript.

Funding: This research was supported by JSPS KAKENHI Grant Numbers JP 18K05013 and JP 22K12456 and the International Platform for Dryland Research and Education (IPDRE), Tottori University.

Data Availability Statement: The data supporting the findings of this study are available from the corresponding author, Y.J., upon reasonable request.

Acknowledgments: We gratefully acknowledge the help provided by Y. Sogo, T. Kitoro, K. Marusasa, T. Okinaga, and K. Yoshino. In addition, we would like to thank the anonymous referees for their constructive comments.

Conflicts of Interest: The authors declare no conflict of interest.

References

1. Strom, B.; Polagye, B.; Brunton, S.L. Near-wake dynamics of a vertical-axis turbiners. *J. Fluid Mech.* **2022**, *935*, A6. [CrossRef]
2. Shamsoddin, S.; Porté-Agel, F. Effect of aspect ratio on vertical-axis wind turbine wakes. *J. Fluid Mech.* **2020**, *889*, R1. [CrossRef]
3. Dabiri, J.O. Potential order-of-magnitude enhancement of wind farm power density via counter-rotating vertical-axis wind turbine arrays. *J. Renew. Sustain. Energy* **2011**, *3*, 043104. [CrossRef]
4. Hezaveh, S.H.; Bou-Zeid, E.; Dabiri, J.; Kinzel, M.; Cortina, G.; Martinelli, L. Increasing the power production of vertical-axis wind-turbine farms using synergistic clustering. *Bound.-Layer Meteorol.* **2018**, *169*, 275–296. [CrossRef]
5. Li, C.; Liu, L.; Lu, X.; Stevens, R.J.A.M. Analytical model of fully developed wind farms in conventionally neutral atmospheric boundary layers. *J. Fluid Mech.* **2022**, *948*, A43. [CrossRef]
6. Zanforlin, S.; Nishino, T. Fluid dynamic mechanisms of enhanced power generation by closely spaced vertical axis wind turbines. *Renew. Energy* **2016**, *99*, 1213–1226. [CrossRef]
7. Dessoky, A.; Zamre, P.; Lutz, T.; Krämer, E. Numerical investigations of two Darrieus turbine rotors placed one behind the other with respect to wind direction. In Proceedings of the 13th International Conference of Fluid Dynamics, ICFD13-EC-6023, Cairo, Egypt, 21–22 December 2018.
8. Kuang, L.; Lu, Q.; Huang, X.; Song, L.; Chen, Y.; Su, J.; Han, Z.; Zhou, D.; Zhao, Y.; Xu, Y.; et al. Characterization of wake interference between two tandem offshore floating vertical-axis wind turbines: Effect of platform pitch motion. *Energy Convers. Manag.* **2022**, *265*, 115769. [CrossRef]
9. Wei, N.J.; Brownstein, I.D.; Cardona, J.L.; Howland, M.F.; Dabiri, J.O. Near-wake structure of full-scale vertical-axis wind turbines. *J. Fluid Mech.* **2021**, *914*, A17. [CrossRef]
10. Jin, G.; Zong, Z.; Jiang, Y.; Zou, L. Aerodynamic analysis of side-by-side placed twin vertical-axis wind turbines. *Ocean Eng.* **2020**, *209*, 107296. [CrossRef]
11. Sahebzadeh, S.; Rezaeiha, A.; Montazeri, H. Towards optimal layout design of vertical-axis wind-turbine farms: Double-rotor arrangements. *Energy Convers. Manag.* **2020**, *226*, 113527. [CrossRef]
12. De Tavernier, D.; Ferreira, C.; Li, A.; Paulsen, U.S.; Madsen, H.A. Towards the understanding of vertical-axis wind turbines in double-rotor configuration. *J. Phys. Conf. Ser.* **2018**, *1037*, 022015. [CrossRef]
13. Hara, Y.; Jodai, Y.; Okinaga, T.; Furukawa, M. Numerical analysis of the dynamic interaction between two closely spaced vertical-axis wind turbines. *Energies* **2021**, *14*, 2286. [CrossRef]
14. Zhang, J.H.; Lien, F.-S.; Yee, E. Investigations of vertical-axis wind-turbine group synergy using an actuator line model. *Energies* **2022**, *15*, 6211. [CrossRef]
15. Ahmadi-Baloutaki, M.; Carriveau, R.; Ting, D.S.-K. A wind tunnel study on the aerodynamic interaction of vertical axis wind turbines in array configurations. *Renew. Energy* **2016**, *96*, 904–913. [CrossRef]
16. Yoshino, K.; Hara, Y.; Okinaga, T.; Kitoro, T.; Jodai, Y. Comparison between numerical analysis and experiment on rotational states of a trio of vertical-axis wind turbines closely arranged in the corners of atriangle with sides of two-times rotor diameter. In Proceedings of the 52nd Student Member Presentation of Graduation Theses, Chugoku-Shikoku Branch Meeting, JSME, Kochi, Japan, (Online), 3 March 2022.
17. Zhang, B.; Song, B.; Mao, Z.; Tian, W. A novel wake energy reuse method to optimize the layout for Savonius-type vertical axis wind turbines. *Energy* **2017**, *121*, 341–355. [CrossRef]
18. Mereu, R.; Federici, D.; Ferrari, G.; Schito, P.; Inzoli, F. Parametric numerical study of Savonius wind turbine interaction in a linear array. *Renew. Energy* **2017**, *113*, 1320–1332. [CrossRef]
19. Bangga, G.; Lutz, T.; Krämer, E. Energy assessment of two vertical axis wind turbines in side-by-side arrangement. *J. Renew. Sustain. Energy* **2018**, *10*, 033303. [CrossRef]
20. Jodai, Y.; Hara, Y. Wind tunnel experiments on interaction between two closely spaced vertical-axis wind turbines in side-by-side arrangement. *Energies* **2021**, *14*, 7874. [CrossRef]
21. Sahebzadeh, S.; Rezaeiha, A.; Montazeri, H. Impact of relative spacing of two adjacent vertical axis wind turbines on their aerodynamics. *J. Phys. Conf. Ser.* **2020**, *1618*, 042002. [CrossRef]
22. Zanforlin, S. Advantages of vertical axis tidal turbines set in close proximity: A comparative CFD investigation in the English channel. *Ocean Eng.* **2018**, *156*, 358–372. [CrossRef]
23. Zheng, H.-D.; Zheng, X.Y.; Zhao, S.X. Arrangement of clustered straight-bladed wind turbines. *Energy* **2020**, *200*, 117563. [CrossRef]
24. Sogo, Y.; Jodai, Y.; Hara, Y.; Kitoro, T.; Okinaga, T. Wind tunnel experiments on the rotational speed of a vertical axis wind turbine pair against 16-direction wind distributions. In Proceedings of the 59th Chugoku-Shikoku Branch Meeting, JSME, Okayama, Japan, (Online), 5 March 2021.
25. Maruyama, Y. Study on the physical mechanism of the Magnus effect. *Trans. JSASS* **2011**, *54*, 173–181. [CrossRef]
26. Huang, M.; Ferreira, C.; Sciacchitano, A.; Scarano, F. Experimental comparison of the wake of a vertical axis wind turbine and planar actuator surfaces. *J. Phys. Conf. Ser.* **2020**, *1618*, 052063. [CrossRef]
27. Jamieson, P.; Ferreira, C.S.; Dalhoff, P.; Störtenbecker, S.; Collu, M.; Salo, E.; McMillan, D.; McMorland, J.; Morgan, L.; Buck, A. Development of a multi rotor floating offshore system based on vertical axis wind turbines. *J. Phys. Conf. Ser.* **2022**, *2257*, 012002. [CrossRef]

28. Shaheen, M.; El-Sayed, M.; Abdallah, S. Numerical study of two-bucket Savonius wind turbine cluster. *J. Wind Eng. Ind. Aerod.* **2015**, *137*, 78–89. [CrossRef]
29. Silva, J.E.; Danao, L.A.M. Varying vawt cluster configuration and the effect on individual rotor and overall cluster performance. *Energies* **2021**, *14*, 1567. [CrossRef]

Article

Numerical Simulation of the Effects of Blade–Arm Connection Gap on Vertical–Axis Wind Turbine Performance

Yutaka Hara [1,*], Ayato Miyashita [2] and Shigeo Yoshida [3,4]

[1] Advanced Mechanical and Electronic System Research Center (AMES), Faculty of Engineering, Tottori University, 4-101 Koyama-Minami, Tottori 680-8552, Japan

[2] Department of Mechanical and Aerospace Engineering, Tottori University, 4-101 Koyama-Minami, Tottori 680-8552, Japan; miyashita.ayato@ma.mee.co.jp

[3] Institute of Ocean Energy (IOES), Saga University, Saga 840-8502, Japan; yoshidas@ioes.saga-u.ac.jp

[4] Research Institute for Applied Mechanics (RIAM), Kyushu University, Fukuoka 816-8580, Japan

[*] Correspondence: hara@tottori-u.ac.jp; Tel.: +81-857-31-6758

Abstract: Many vertical-axis wind turbines (VAWTs) require arms, which generally provide aerodynamic resistance, to connect the main blades to the rotating shaft. Three–dimensional numerical simulations were conducted to clarify the effects of a gap placed at the blade–arm connection portion on VAWT performance. A VAWT with two straight blades (diameter: 0.75 m, height: 0.5 m) was used as the calculation model. Two horizontal arms were assumed to be connected to the blade of the model with or without a gap. A cylindrical rod with a diameter of 1 or 5 mm was installed in the gap, and its length varied from 10 to 30 mm. The arm cross section has the same airfoil shape (NACA 0018) as the main blade; however, the chord length is half (0.04 m) that of the blade. The simulation shows that the power of the VAWT with gaps is higher than that of the gapless VAWT. The longer gap length tends to decrease the power, and increasing the diameter of the connecting rod amplifies this decreasing tendency. Providing a short gap at the blade–arm connection and decreasing the cross–sectional area of the connecting member is effective in increasing VAWT power.

Keywords: vertical–axis wind turbine; arm; gap; computational fluid dynamics; three–dimensional effects; drag; surface pressure; wall shear stress

Citation: Hara, Y.; Miyashita, A.; Yoshida, S. Numerical Simulation of the Effects of Blade–Arm Connection Gap on Vertical–Axis Wind Turbine Performance. *Energies* **2023**, *16*, 6925. https://doi.org/10.3390/en16196925

Academic Editor: Frede Blaabjerg

Received: 30 August 2023
Revised: 22 September 2023
Accepted: 27 September 2023
Published: 2 October 2023

1. Introduction

Most of the wind turbines currently used for wind power generation are propeller–type horizontal–axis wind turbines (HAWTs) [1,2]. A different wind turbine rotor structure is the vertical–axis wind turbine (VAWT) [3–5], which has features such as having no wind direction dependence. Unlike HAWTs, many lift–driven VAWTs (Darrieus type) structurally require arms (or struts) to connect the main blades to the rotating shaft. If the cross section of the arm is airfoil–shaped, the slant–installed arm generates a rotational force using lift, which improves the starting performance. However, it becomes the main cause of aerodynamic resistance in a high–rotation–speed state, regardless of the cross section, when installed horizontally. Several previous studies have investigated the effects of arms on the performance of lift–type VAWTs.

Li and Calisal [6] investigated the three–dimensional (3D) effects of a vertical–axis tidal turbine, but not wind turbines, through numerical analysis using the vortex method. However, their numerical simulation did not include the effects of the arm, which were analytically modeled based on separate towing experiments. In their research, two types of arm structures (types A and B) were analyzed. In type B, the arm with the airfoil section was directly connected to the blade. Conversely, in type A, the arm had a bluff body–shaped cross section, which was attached to the blade via a clamping mechanism to adjust the angle of attack. As a result, type A caused larger arm loss than type B.

Qin et al. [7] conducted a computational fluid dynamics (CFD) analysis based on the Reynolds–averaged Navier–Stokes (RANS) equations of an H–type VAWT with a rooftop (diameter: 2.5 m, blade length: 2 m, chord length: $c = 0.2$ m) consisting of three blades. The 3D model was equipped with a flat cylindrical rotor hub (top disk) with a diameter of 0.5 m. Although the cross–sectional shape of the arm is unknown, the length of the arm connecting the top disk and a blade was set to 1 m. A comparison of the 2D and 3D calculation results showed that the average torque decreased by 40% because of the 3D effects. The authors considered that the observed negative torque, which is larger downstream than upstream, must be attributed to the interaction between the wake generated by the blade passing upwind half and the blade downstream. In Figure 14(a) of their paper, the vorticity generated from the arms was visualized to demonstrate its significant effects. However, the azimuth dependency of the arm effects and the importance of the connection part were not mentioned.

De Marco et al. [8] conducted a CFD analysis on the characteristics of a straight–bladed VAWT (diameter: 2.6 m, blade length: 2.5 m, blade chord length: 20 mm, blade cross section: NACA 0018) equipped with slant arms. The cross–sectional shape of the upper and lower slant arms mounted at 30 degrees on a horizontal arm placed on the equatorial plane was similar to the main blade. The torque generated by the horizontal arms was not considered in the CFD analysis. Three cases in which the number of blades changed from one to three were examined, and in each case, models of the main blade only, arm set only, and full element (both main blades and arms) were calculated. In each case, the full–element characteristics were shown to be inferior to the simple sum of the blade–only and arm set–only models, and the complex effects of wake and blockage with an increasing number of blades were discussed. Figure 14(b) of their paper showed the improvement in rotor power when the chord length of the slant blade was reduced. However, their slant arms were directly attached to the blades, and the effects of the connection portion on the turbine rotor were not mentioned.

Marsh et al. [9] performed an unsteady RANS (URANS) analysis on a vertical–axis tidal current turbine with three straight blades (radius: 0.457 m, blade length: 0.686 m, blade chord length: 65 mm, blade cross section: NACA 0012) using k–ω SST as the turbulence model. Three types of turbines were analyzed. Turbine A was equipped with horizontal struts (section NACA 0012) directly connected to the blade tips (upper and lower). The other two turbines (turbines B and C) were structured with a horizontal strut attached to the quarter position of the blade length from each upper and lower blade tip. The cross section of the strut of turbine B had a bluff body shape, and the strut was connected to a blade with a connection tab. In contrast, the strut section of turbine C was airfoil–shaped (NACA 0012), and the strut was directly connected to the blade (faired joint). Turbine A clearly showed a smaller power coefficient (C_p) than turbines B and C. Turbine B generated a smaller rotational torque than turbine C; however, it is not clear whether the shape of the strut or the connecting tab had a greater effect, with the authors stating that the distinction is difficult to determine.

Li et al. [10] conducted a wind tunnel experiment using an experimental rotor with a diameter, blade length, and chord length of 20 m, 1.2 m, and 0.265 m, respectively, to clarify the aerodynamic characteristics of a straight–blade VAWT. The cross section of the main blade was NACA 0021. The number of blades was changed from two to five to investigate the effects of solidity and Reynolds number. The shapes of the arms were not described in detail. In addition to pressure measurements using 32 pressure taps at the mid–span cross section of the blade, torque meter measurements and six–component force balance measurements were performed. The C_p obtained by the pressure measurement was larger than the C_p measured by the torque meter or force balance. This difference was attributed to the fact that the pressure measurement did not include the blade tip or arm effects. Their study also presented CFD results showing the azimuth dependence of the torque at different positions in the span direction of one blade. The results illustrated that at the upstream position (azimuth 90°), where the maximum torque was generated, the torque generated near the equatorial level was approximately 40% higher than that near the blade tip.

Aihara et al. [11] performed 3D–CFD (RANS, turbulence model: realizable k–ε) on a 12 kW three–bladed VAWT with a diameter and blade length of 6.5 m and 5 m, respectively. The azimuth dependency of the normal and tangential forces of one blade and one inclined strut (upper and lower) was calculated for three cases: blade only, with strut, and with tower. The C_p for the blade–only model was 0.277 and that with struts decreased by 43% to become 0.158. The influence of the tower was smaller than that of the struts. The tangential force was more strongly influenced by the strut than the normal force and decreased at the height of the connection between the strut and the blade. Furthermore, at 90° azimuth (upstream), the reduction in the tangential force owing to the struts was large. Note that there was no gap between the strut and the blade in their model.

In most vertical–axis turbines actually installed in the field, such as the aforementioned 12 kW VAWT, an arm and a blade are directly connected without creating any gap or changing the cross–sectional area of the arm (see Figure 1 of reference [12] and the conceptual video or images shown in the web page [13]).

Hara et al. [14] performed a CFD analysis to elucidate the effects of the arms on the performance of a VAWT. The analysis objects (diameter: 0.75 m, blade length: 0.5 m, blade chord length: 80 mm, blade cross section: NACA 0018) consisted of two straight blades without arms (armless rotor) or those with arms having three different cross sections (airfoil, rectangular, and circular). The study showed that the model with airfoil cross–sectional arms reduced the power by approximately 50% under the maximum power condition (tip speed ratio: $\lambda = 3$) compared to the model without arms. The drag caused by the surface pressure of the arm tended to increase near the connection between the arm and blade, regardless of the arm cross–sectional shape. Importantly, the rotational force caused by the pressure distribution on the main blade was significantly affected by the arm mounting portion, and the influence largely depended on the cross–sectional area of the arm, that is, the contacting area between the arm and the blade. In this study, the main blade and arm were connected without a gap.

The present study targeted a VAWT model of the same size as that used in a previous study [14]. However, a gap was placed between the main blade and arm to investigate the effects of the connection portion on the rotor performance. In addition, the effects of the gap length and the size of the connecting members installed in the gap were clarified.

2. Methods

2.1. Object Model for Calculation

Similarly to a previous study [14], a two–bladed H–type Darrieus wind turbine was selected as the calculation object, which had the same size as the experimental VAWT (DU–H2–5075) [15] at the Delft University of Technology (TU Delft) (Figure 1). The cross section of the straight blade (main blade) is NACA 0018 (symmetrical airfoil), with chord length $c = 0.08$ m, rotor diameter D or $2R = 0.75$ m, and blade span $2H = 0.5$ m. The blade–mounting position is 40%c (0.032 m). The computational model did not consider the hub or the rotating shaft (or tower) of the wind turbine rotor. Two arms were attached horizontally to each of the main blades, and the arm–mounting position was at one–quarter of the blade span from the top and bottom tips of the main blade. The cross–sectional shape of the arm is NACA 0018, and the chord length of the arm cross section is $c_{\text{arm}} = 0.04$ m. However, in this research, a gap is put in between the arm and the main blade. Three gap lengths—$L_{\text{gap}} = 10$ mm, 20 mm, and 30 mm—were assumed for comparison. A model without the gap (gapless), corresponding to the case of $L_{\text{gap}} = 0$ mm, was analyzed in addition to the other opposite extreme–case model (i.e., armless rotor) that corresponds to $L_{\text{gap}} = \infty$. In this study, the gap length was defined as the distance between the neutral line of the blade cross section and the outer end of the arm, as shown in Figure 1b. The inner end of the arm is $r_{\text{root}} = 20$ mm away from the rotor center, and the arm length L_{arm} varies between 355 mm and 325 mm depending on the gap length. A cylindrical rod with a diameter of $d = 1$ mm or 5 mm was installed in the gap as a member connecting the main blade and the arm. Table 1 summarizes the calculation cases, that is, the specifications of each model. In the present CFD analysis, the upstream wind speed was set to $U_\infty = 7$ m/s, and the tip speed ratio was set to λ or $R\omega/U_\infty = 3.0$, which almost matches the maximum C_p state of the object turbine. Therefore, the rotation speed of the wind turbine was set to 535 rpm. The Reynolds number based on the chord length c is $Re = 1.1 \times 10^5$. Figure 1 also shows an absolute coordinate system (x–y–z) with the origin at the center of the rotor. The wind turbine is assumed to begin the rotation from the azimuth origin ($\psi = 0$), defined as the positive y–axis direction, and rotate counterclockwise as viewed from above.

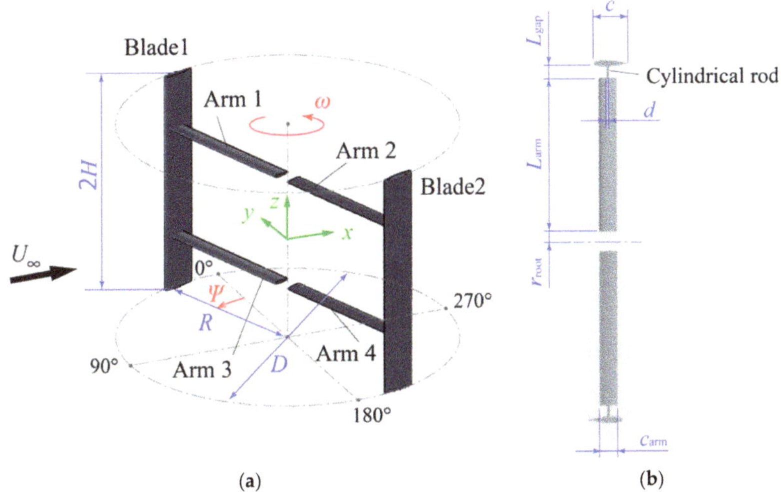

Figure 1. Rotor model: (**a**) perspective view; (**b**) top view.

Table 1. Calculation cases (model number and specification).

Model No.	0	1	2	3	4	5	6	7
L_{gap}	0	10	20	30	10	20	30	∞
L_{arm}	355	345	335	325	345	335	325	0
d	-	1	1	1	5	5	5	-

2.2. Settings of Numerical Analysis

STAR–CCM+ was used as the computational solver, and the 3D unsteady incompressible Reynolds–averaged Navier–Stokes (URANS) and continuity equations were used as the basic equations that were solved according to the SIMPLE algorithm. The space discretization procedure was the finite volume method, in which the second–order upwind difference scheme was used. The SST k–ω model [16] was adopted as the turbulence model. The entire computational domain was set as a cylinder with a diameter of $48D$ and a length of $64D$ (static region 1), and the center of the wind turbine model was placed at a position $24D$ from the inlet boundary (see Figure 2). The upstream and downstream boundaries of static region 1 were defined as constant wind speed ($U_\infty = 7$ m/s) and constant pressure ($P = 0$ Pa), respectively. A slip condition was applied to the side surface. Figure 3 illustrates the mesh around the wind turbine rotor. The entire rotor was placed in a cylindrical region (rotation region) with a diameter of $1.3D$, and this region was moved using the sliding mesh method to provide rotational motion. To adjust the mesh size of the wake region to a nearly constant size, static region 2 was set, which consisted of two semi–cylinders with a radius of $0.7D$ and a rectangular with a length of $3.6D$ and a height of $3H$. To gradually reduce the mesh size in the region approaching the object surface, a blade was surrounded by a cylindrical region (near–blade region) with a diameter of 2.0*c*, and an arm was surrounded by another cylindrical region (near–arm region) with a diameter of 1.0*c* (see Figure 4). An unstructured polyhedral mesh was adopted in most of the computational domain, and a 15–layer prism layer mesh was applied to the region where the boundary layer developed near the object. The minimum grid width was 2.3×10^{-6} m (blade surface), and the maximum y^+ was approximately 0.5. The number of cells in the static region is approximately 160,000 (common for all cases), whereas that in the rotation region is approximately 6.34 million and 3.57 million in the case of the gapless and armless rotors, respectively (total number of cells: approximately 6.5 million with arm and 3.7 million without arm). When there was a gap, the number of cells in the rotation region decreased at a rate of approximately 50,000/10 mm–gap. The time interval was 1.56×10^{-4} s, and the rotor rotated 0.5° per unit time step. Calculations were performed for up to eight revolutions, during which the calculated results, such as the torque, almost converged. The power coefficient was calculated from the average of the seventh rotation, and images were extracted at the eighth rotation. Although the mesh condition is slightly different from that in previous research [14], the calculation region and conditions of the numerical analysis in this research are almost similar to those in the previous research. The azimuth dependence of the one–blade torque of the armless rotor agrees approximately with the results of previous studies. Figure A1 in the Appendix A shows a comparison of the azimuth dependence of the torque coefficient. There is a slight difference at approximately 180°; however, at the other azimuth angles, the present and previous results [14] are almost the same. In addition, the present power coefficient of the rotor with arms at $\lambda = 3$ was $C_\mathrm{p} = 0.0995$, which is approximately 21% smaller than the experimental value of $C_\mathrm{p} = 0.126$ at TU Delft [17]. However, the details of the arm shape of the experimental turbine at TU Delft are unknown. The authors claim that this discrepancy is not an essential issue.

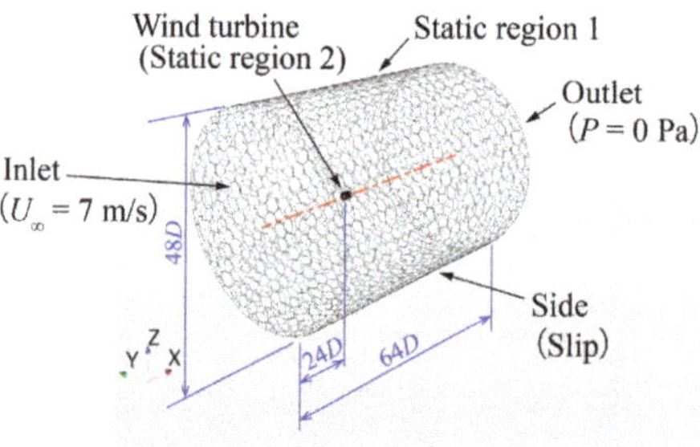

Figure 2. Computational domain of the present CFD.

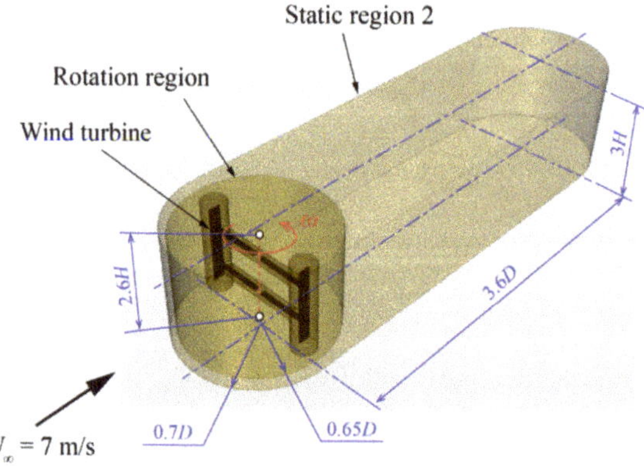

Figure 3. Computational mesh around the wind turbine rotor.

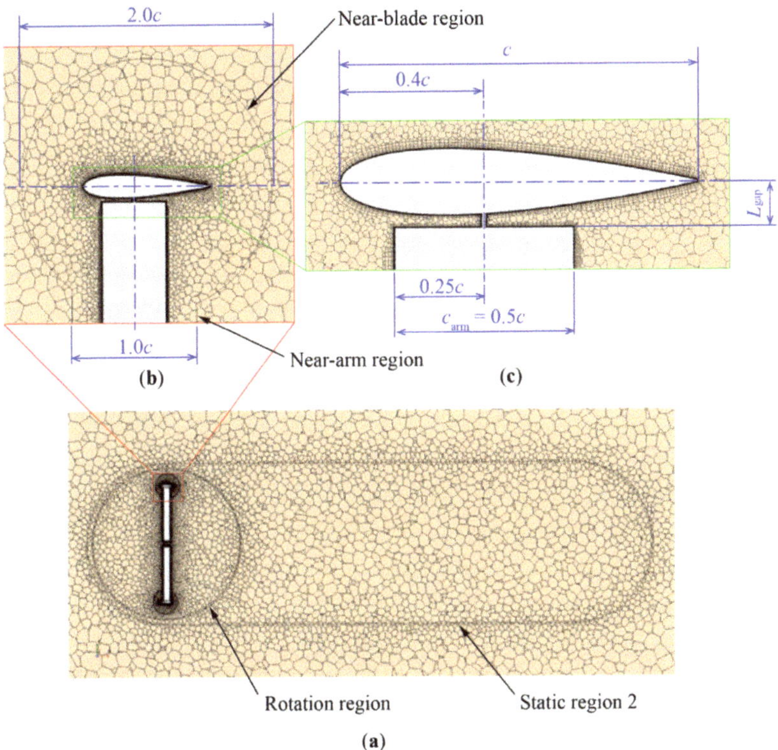

Figure 4. Details of the computational mesh around the wind turbine rotor and the definition of a gap at the connection portion: (**a**) rotation and wake regions (static region 2); (**b**) near–blade and near–arm regions; (**c**) mesh around a blade and a blade–arm connection gap.

3. Results and Discussion

3.1. Vortex Shedding from the Connection Part

To investigate the flow–field state near the connection between the blade and the arm, the Q–criterion isosurfaces [18] ($Q = 1000\ \text{s}^{-2}$) of model 0 (gapless rotor) and model 1 ($L_{gap} = 10$ mm, $d = 1$ mm) are shown in Figure 5a,b, respectively. The colors represent the strength of the vorticity on the Q–criterion isosurfaces. Figure 5 shows the conditions under which blade 1 of each rotor model is at azimuth angles $\psi = 0°, 90°, 180°$, and $270°$ for the two models. For both models, vortex shedding from the connection portion is evident for azimuth angles $\psi = 90°$ and $180°$; in particular, the shed vortex is large at the $180°$ position, where the blade enters from the upwind half to the downwind half. Furthermore, compared with model 1 with a gap, the vortices generated from the connection part in model 0 (gapless), in which the blade and arm are directly connected, appear to affect a wider area of the main blade above and below the connection.

3.2. Surface Pressure Distribution around the Connection Part

Figure 6 shows the surface pressure distribution in the vicinity of the connection part of each blade 1 in the four–rotor models, i.e., model 0 (gapless), model 7 (armless), model 1 ($L_{gap} = 10$ mm, $d = 1$ mm), and model 4 ($L_{gap} = 10$ mm, $d = 5$ mm). The states of azimuth angles $\psi = 90°$ and $180°$ are shown for each model in this figure. The surface pressure distribution of model 7 (armless) in Figure 6b is clearly different from that of the other rotors with arms. In the pressure distribution of azimuth $\psi = 90°$ in model 7, the inner surface (rotating shaft side) of the blade is entirely negative pressure (blue) to demonstrate that this surface becomes the suction side in the upstream region. The pressure near the trailing edge of the main blade is slightly higher (light blue), except in the vicinity of the blade tip. By contrast, in the other three models with arms, the pressure increase near the trailing edge of the main blade at the azimuth $\psi = 90°$ is suppressed around the arm–blade connection part. In particular, the suppressed area in model 0 (gapless) is wide. As shown in the pressure distribution at $\psi = 180°$ of model 7 (armless), the pressure near the center (equator level) of the blade span is high to indicate that the inner surface of the blade has already switched from the suction side to the pressure side. However, a negative pressure area is observed near the tip of the blade, which is presumed to be due to the influence of the blade tip; however, the cause is not clear. The surface pressure distributions at azimuth $\psi = 180°$ for the other three models with arms are roughly similar. However, unlike models 1 and 4 with gaps, a spot showing rather large positive pressure (yellow area) exists on the main blade in front of the connection portion for the pressure distribution at $\psi = 180°$ of model 0 (gapless rotor).

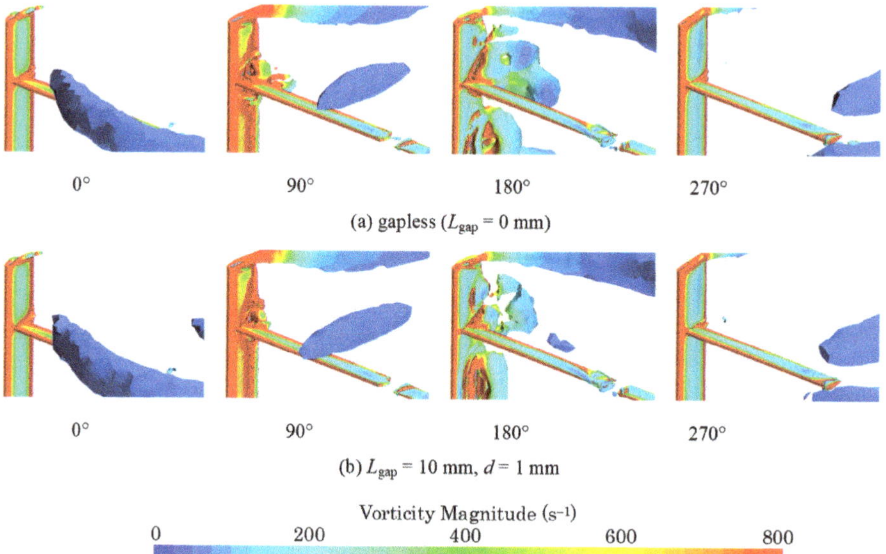

Figure 5. Azimuth dependency of vortex shedding (Q–criterion isosurfaces: $Q = 1000$ s^{-2}) and vorticity in the cases of (**a**) gapless and (**b**) 10 mm gap with $d = 1$ mm.

3.3. Wall Shear Stress Distribution around Connection Part

Figure 7 illustrates the wall shear stress distribution near the connection of the four-rotor models shown in Figure 6. In addition, the wall shear stress distribution of model 7 (armless) in Figure 7b is clearly different from those of the other rotors with arms. In particular, focusing on the distribution of azimuth $\psi = 90°$, the wall shear stress in model 7 (armless rotor) is linearly distributed along the leading edge of the main blade with positive and negative values. In the other models with arms, the linear distribution of wall shear stress along the leading edge of the blade is disturbed by the existence of the arms. In addition, in the vicinity of the arm connection portion, the area where the wall shear stress becomes negative spreads even near the trailing edge of the main blade. The degree of disturbance in the wall shear stress distribution around the connection part increases in the following order: models 1, 4, and 0.

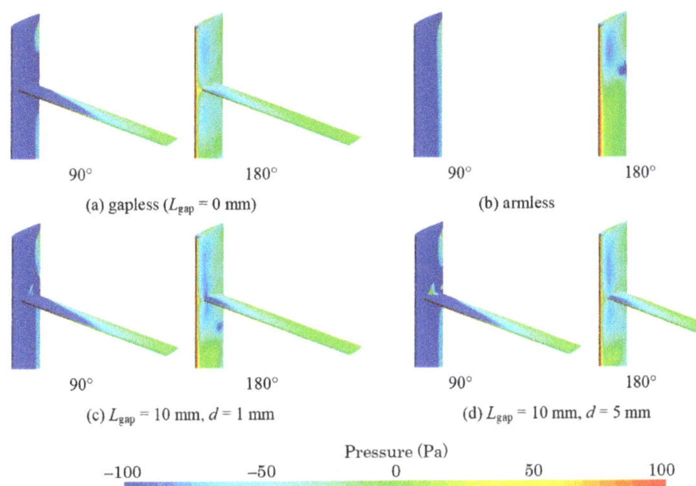

Figure 6. Surface pressure distribution in the cases of (**a**) gapless, (**b**) armless, (**c**) 10 mm gap with d = 1 mm, and (**d**) 10 mm gap with d = 5 mm.

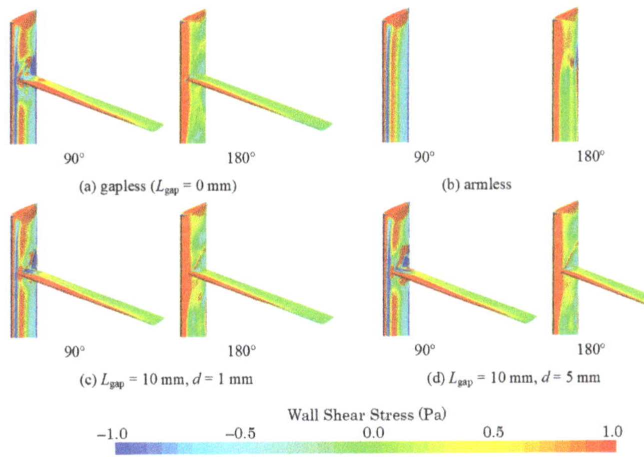

Figure 7. Wall shear stress distribution in the cases of (**a**) gapless, (**b**) armless, (**c**) 10 mm gap with d = 1 mm, and (**d**) 10 mm gap with d = 5 mm.

3.4. Gap Length Dependency of One-Blade Torque Variation in One Rotation

Figure 8a shows the gap–length dependence of the torque variation of one blade during one rotor rotation (seventh rotation) when the diameter of the cylindrical rod placed in the gap is d = 1 mm (models 1, 2, and 3). For comparison, the torque variations for the gapless rotor (model 0) and armless rotor (model 7) are also shown. The vertical axis in Figure 8a represents the torque coefficient C_q, which is defined by Equation (1).

$$C_q = \frac{Q_{\text{blade1}}}{0.5\rho U_\infty^2 (4HR^2)} \tag{1}$$

where Q_{blade1} is the rotational torque of one blade, ρ is the air density (ρ = 1.225 kg/m^3), and R is the rotor radius (R = $D/2$ = 0.375 m). Figure 8a shows that the effects of the gap length on the single–blade torque variation are small when the cylindrical rod diameter is

$d = 1$ mm. In addition, a large difference among the models is observed around azimuth $\psi = 90°$. Comparing at $\psi = 90°$, the armless rotor (model 7) has the largest torque coefficient ($C_q = 0.211$), whereas the cases with $d = 1$ mm (models 1, 2, and 3) have approximately 10% less ($C_q = 0.182, 0.189, 0.190$). The gapless rotor (model 0) has the smallest value ($C_q = 0.139$). In the downwind half ($180° < \psi < 360°$), the difference in torque coefficients among the models is small.

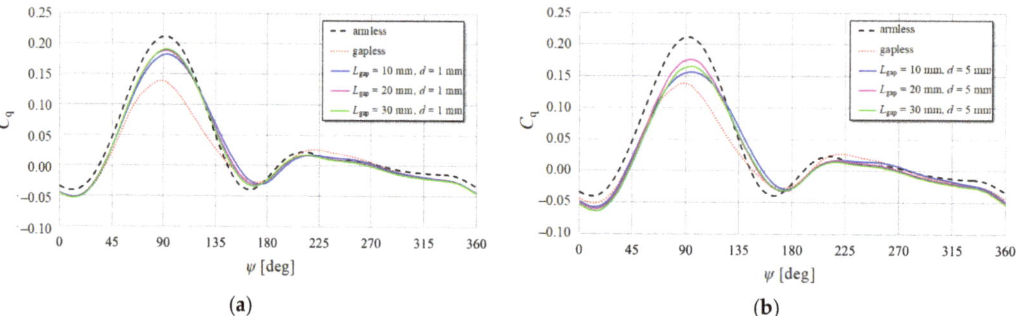

(**a**)   (**b**)

Figure 8. Variation in the torque coefficient of a blade: (**a**) $d = 1$ mm; (**b**) $d = 5$ mm.

Figure 8b depicts the variation in torque for one blade when the diameter of the cylinder is $d = 5$ mm (models 4, 5, and 6). The torque coefficients of the gapless and armless rotors are also shown in the figure. Similarly to Figure 8a, the difference in the torque coefficients among the models is small in the downwind half ($180° < \psi < 360°$). The difference among the models with $d = 5$ mm becomes large in the vicinity of $\psi = 90°$. Comparing at $\psi = 90°$, the C_q of models 4, 5, and 6 are 0.156, 0.175, and 0.164, respectively. There is no tendency to depend on the gap length, but all the torque coefficients at $\psi = 90°$ of models 4, 5, and 6 decrease compared to the models of $d = 1$ mm. However, in the vicinity of azimuth $\psi = 15°$, the difference among models 4, 5, and 6 is large. In this region, the torque coefficient of model 6 ($L_{gap} = 30$ mm, $d = 5$ mm) is smaller than those of the other two models.

3.5. Effects of Connection Gap on Power Coefficient

Power coefficients at $\lambda = 3$ of all eight models analyzed by CFD in this study are compared in Figure 9. The horizontal axis represents the gap length L_{gap}, and the vertical axis represents the power coefficient C_p defined by Equation (2).

$$C_p = \frac{P_{rotor}}{0.5\rho U_\infty^3 (4HR)} \tag{2}$$

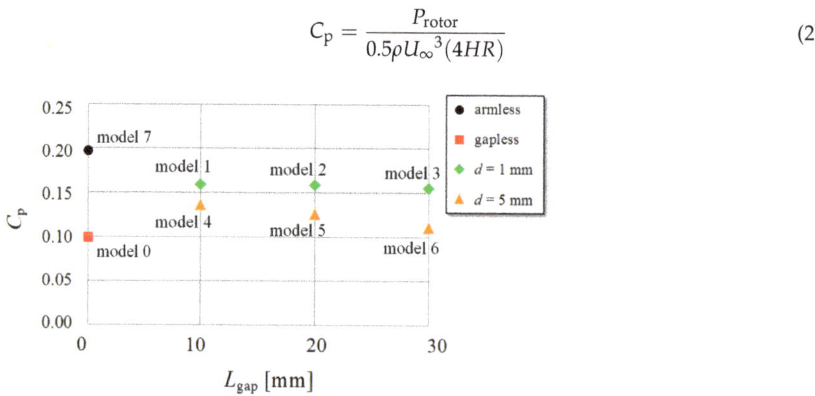

Figure 9. Dependence of power coefficient on gap length.

In the above equation, P_{rotor} is the average power output obtained from the product of the angular velocity and rotational torque generated by all the blades (including the arm) during the seventh rotor revolution.

As shown in Figure 9, the C_p without arms (armless) is approximately twice that with arms (gapless), and the existence of arms significantly degrades the VAWT performance. However, the power coefficient can be improved by creating a gap between the arm and the main blade. In addition, when the diameter of the cylindrical rod placed in the gap is small (d = 1 mm), the power output is higher than that when the diameter is large (d = 5 mm). The power is improved by the gap at the connection part because the fluid passing through the gap decreases the surface pressure on the inner surface of the blade (suction side), resulting in an increase in the lift force when the blade moves near the upstream azimuth (ψ = 90°). To support this conjecture, Figure 10 shows the blade surface pressure distribution in the cross section at z = 140 mm (15 mm above the arm neutral line) when blade 1 is at ψ = 90°. Figure 10 separately compares the pressure coefficients of model 0 without a gap, model 1 with a gap, and model 7 without arms (as a reference) for the outer surface (pressure side: broken line) and the inner surface (suction side: solid line). The pressure coefficient C_{pre} in Figure 10 is defined by Equation (3):

$$C_{pre} = \frac{\Delta p}{0.5\rho(R\omega)^2} \tag{3}$$

where Δp is the surface pressure (gauge pressure) of the blade.

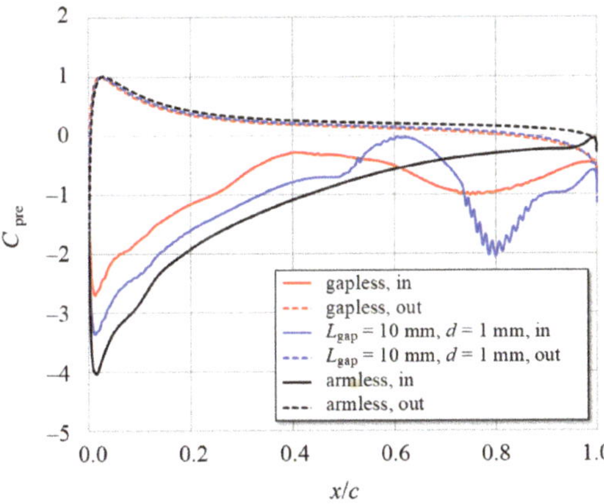

Figure 10. Pressure coefficient distribution on a blade at z = 140 mm when ψ = 90°.

As shown in Figure 10, there is no significant difference in the pressure distributions of the outer blade surface among the models. As a reference, Figures A2 and A3 in the Appendix A show the distribution of the surface pressure and wall shear stress on the outer surface of the blade; almost no difference is observed between the presence and absence of gaps. However, the pressure distributions on the inner surface of the blade differed considerably among the models, as shown in Figure 10. The area enclosed by the pressure coefficient curves on the outer and inner sides of the blade corresponds to the lift force. If the area of the armless model is assumed to be 100%, then those of models 1 and 2 are 87% and 43%, respectively. The figure shows that the provision of a gap improves the lift generated by the blade near the connection portion compared with the case without the gap.

As shown in Figure 9, the power coefficient decreases as the gap length increases, regardless of the diameter of the cylindrical rod installed in the gap. This can be explained by the replacement of the airfoil cross section with a small drag coefficient (2D–C_d =

approximately 0.02 [19] at Re_{arm} = 5.5 × 10^4) with the cylindrical cross section with a large drag coefficient (2D–C_d = approximately 1.2 [20]). To give a concrete example in the present numerical analysis, the surface pressure distribution in the cross section at the local radius r = 0.36 m (mid–gap) of the connecting rod (d = 5 mm) of model 6 is shown in Figure 11, compared with that of model 0 (airfoil arm) without gaps, when the arm 1 is at ψ = 0°. The pressure coefficient shown in Figure 11 is defined by Equation (4).

$$C_{pre} = \frac{\Delta p}{0.5\rho(U_\infty + r\omega)^2} \tag{4}$$

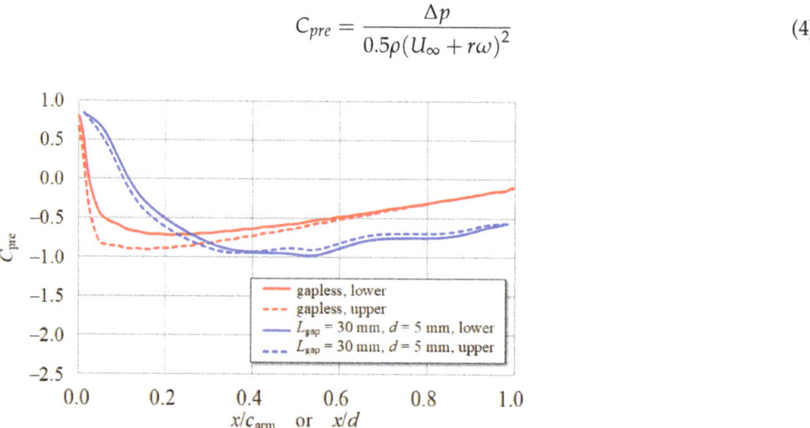

Figure 11. Pressure coefficient distribution on an arm or a cylindrical rod at r = 360 mm when ψ = 0°.

From the pressure distribution shown in Figure 11, the drag forces per unit length are calculated to be 0.55 N/m and 1.64 N/m for the airfoil arm (model 0) and cylindrical rod, respectively. The drag force of the connecting rod is approximately three times that of the arm. However, the effect of wall friction is neglected because it is relatively small (see [14]). In this case, the calculated drag coefficients of the airfoil arm and cylindrical rod are C_d = 0.17 and C_d = 0.73, respectively, which are somewhat different from the abovementioned two–dimensional values (2D–C_d) due to 3D effects.

The above results imply that it is desirable to create a short gap between the arm and the main blade of VAWT within the permissible range of structural strength to improve the power output. In addition, it is necessary to minimize the cross–sectional area of the connecting member in the gap. Figure 12 shows the schematic of the blade–arm connection part using a cylindrical rod of d = 5 mm when the L_{gap} is gradually shortened from 10 mm to 5 mm. Numerical analysis for cases with less than L_{gap} = 10 mm were not conducted in this study because the generation of reasonable mesh becomes difficult as the gap becomes shorter. As shown in Figure 12, when L_{gap} = 7.2 mm, the arm end surface touches the main blade surface, and as L_{gap} becomes even shorter, the cross–sectional area of the blade–arm connection increases. In this case, C_p shown in Figure 9 can be inferred to approach the value of C_p in model 0 from that in model 4. Ideally, when applied to an actual VAWT, connecting the arm and main blade with a member whose thickness does not exceed the arm thickness (7.2 mm in this model) within the range of L_{gap} = 7.5 to 10 mm (L_{gap}/R = 0.0200 to 0.0267) is recommended.

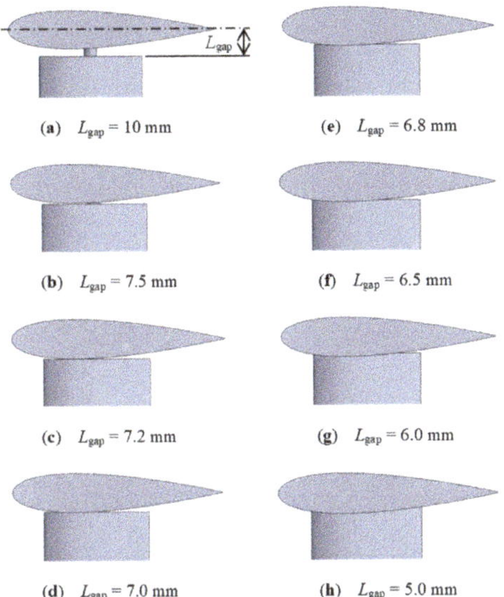

Figure 12. Schematic of the blade–arm connection part when d = 5 mm: (a) L_{gap} = 10 mm; (b) L_{gap} = 7.5 mm; (c) L_{gap} = 7.2 mm; (d) L_{gap} = 7.0 mm; (e) L_{gap} = 6.8 mm; (f) L_{gap} = 6.5 mm; (g) L_{gap} = 6.0 mm; (h) L_{gap} = 5.0 mm.

4. Conclusions

To investigate the effects of a gap that was placed between the blade and the arm of a vertical–axis wind turbine (VAWT), a computational fluid dynamics analysis was conducted for eight models of a two–bladed H–type VAWT (diameter: 0.75 m, height: 0.5 m) as the object, including the gap–length conditions of L_{gap}/R = 0.0267, 0.0533, and 0.0800. The visualization of vortex shedding using the Q–criterion isosurfaces showed that the size of the vortices generated from the connection portion was smaller with a gap than without. The surface pressure and wall shear stress distributions approached those of the armless rotor owing to the gap. Regarding the azimuth dependence of the single–blade torque, a significant difference in the effect of the gap was observed in the upstream range (near azimuth ψ = 90°). When assuming the power of the armless rotor at the tip speed ratio λ = 3 as 100%, those of the gapless rotor and the rotor, which had a connection rod (d = 10 mm, d/c = 0.125) in the shortest gap, became approximately 50% and 81%, respectively. However, the power coefficient decreased as the gap length increased. Moreover, the power coefficient decreased further as the diameter of the cylindrical rod placed in the gap increased. It is well known that applying an airfoil shape to the cross section of an arm effectively reduces the aerodynamic resistance of VAWTs when it is necessary to install the arm. Thus, providing a gap at the blade–arm connection part and reducing the cross–sectional area of the connecting member are effective in reducing the output loss caused by the connection portion.

Author Contributions: Conceptualization, Y.H.; methodology, Y.H.; software, Y.H.; validation, Y.H., A.M. and S.Y.; formal analysis, Y.H.; investigation, A.M.; resources, Y.H.; data curation, Y.H.; writing—original draft preparation, Y.H.; writing—review and editing, S.Y.; visualization, Y.H. and A.M.; supervision, Y.H.; project administration, Y.H.; funding acquisition, Y.H. All authors have read and agreed to the published version of the manuscript.

Funding: This research was supported by the Collaborative Research Program of the Research Institute for Applied Mechanics, Kyushu University, joint research with Nikkeikin Aluminium Core Technology Co., Ltd., the International Platform for Dryland Research and Education (IPDRE), Tottori University, and the 2021 Special Joint Project of the Faculty of Engineering, Tottori University.

Data Availability Statement: The data presented in this study are available on request from the corresponding author.

Acknowledgments: The corresponding author, Yutaka Hara, thanks the Advanced Mechanical and Electronic System Research Center (AMES) for supporting the article processing charge.

Conflicts of Interest: The authors declare no conflict of interest.

Appendix A

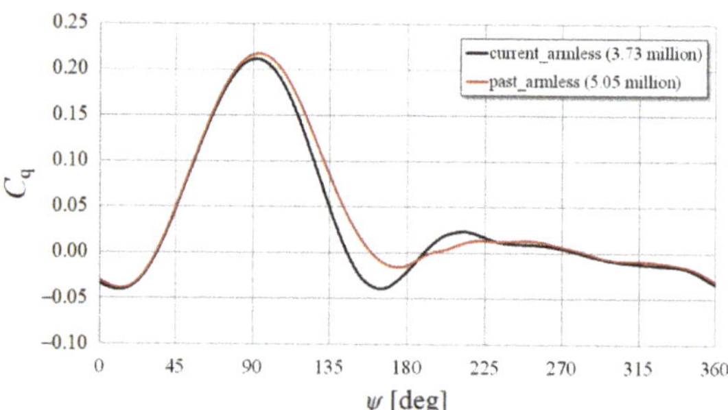

Figure A1. Comparison of azimuth dependence of single–blade torque for an armless rotor between the current analysis (rotational region: cylindrical) and the previous analysis [14] (rotational region: spherical).

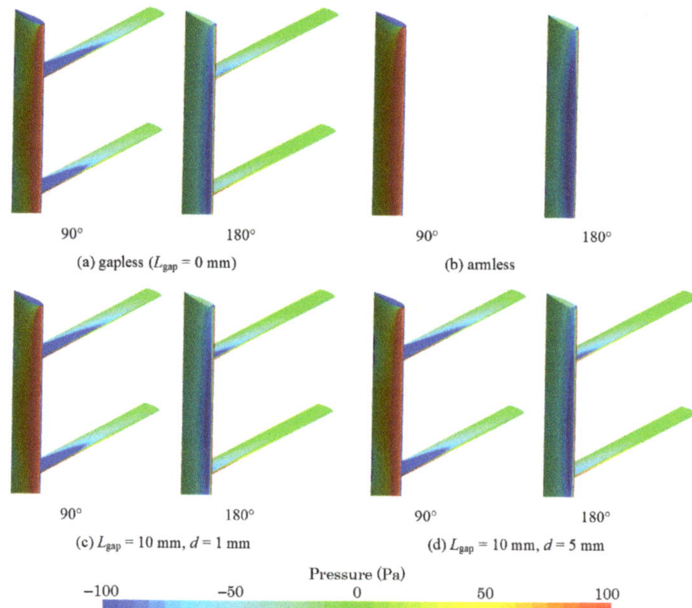

Figure A2. Pressure distribution on the outer surface of a blade in the cases of (**a**) gapless, (**b**) armless, (**c**) 10 mm gap with d = 1 mm, and (**d**) 10 mm gap with d = 5 mm.

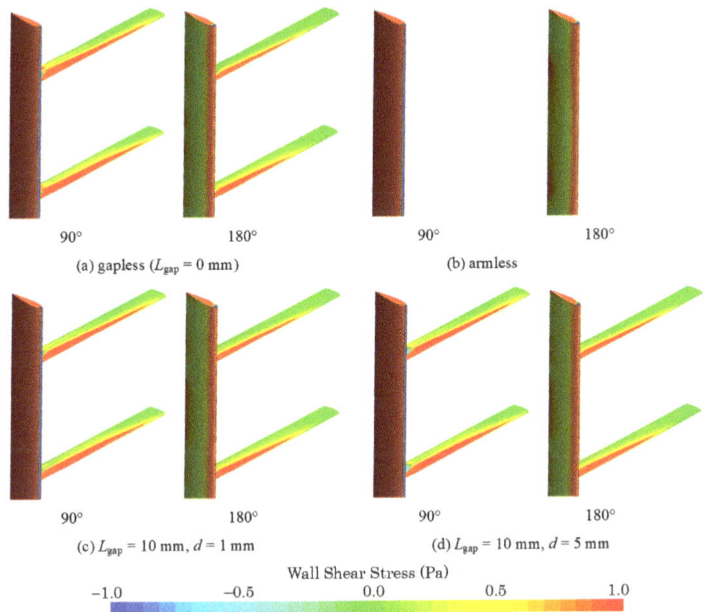

Figure A3. Wall shear stress distribution on the outer surface of a blade in the cases of (**a**) gapless, (**b**) armless, (**c**) 10 mm gap with $d = 1$ mm, and (**d**) 10 mm gap with $d = 5$ mm.

References

1. Ahmed, N.A.; Cameron, M. The challenges and possible solutions of horizontal axis wind turbines as a clean energy solution for the future. *Renew. Sustain. Energy Rev.* **2014**, *38*, 439–460. [CrossRef]
2. Luo, L.; Zhang, X.; Song, D.; Tang, W.; Yang, J.; Li, L.; Tian, X.; Wen, W. Optimal design of rated wind speed and rotor radius to minimizing the cost of energy for offshore wind turbines. *Energies* **2018**, *11*, 2728. [CrossRef]
3. Islam, M.; Fartaj, A.; Carriveau, R. Analysis of the design parameters related to a fixed-pitch straight-bladed vertical axis wind turbine. *Wind Eng.* **2008**, *32*, 491–507. [CrossRef]
4. Islam, M.; Ting, D.S.K.; Fartaj, A. Aerodynamic models for Darrieus-type straight-bladed vertical axis wind turbines. *Renew. Sustain. Energy Rev.* **2008**, *12*, 1087–1109. [CrossRef]
5. Kumar, R.; Raahemifar, K.; Fung, A.S. A critical review of vertical axis wind turbines for urban applications. *Renew. Sust. Energ. Rev.* **2018**, *89*, 281–291. [CrossRef]
6. Li, Y.; Calisal, S.M. Three-dimensional effects and arm effects on modeling a vertical axis tidal current turbine. *Renew. Sustain. Energy Rev.* **2010**, *35*, 2325–2334. [CrossRef]
7. Qin, N.; Howell, R.; Durrani, N.; Hamada, K.; Smith, T. Unsteady flow simulation and dynamic stall behaviour of vertical axis wind turbine blades. *Wind Eng.* **2011**, *35*, 511–527. [CrossRef]
8. De Marco, A.; Coiro, D.P.; Cucco, D.; Nicolosi, F. A numerical study on a vertical-axis wind turbine with inclined arms. *Int. J. Aerosp. Eng.* **2014**, *2014*, 180498. [CrossRef]
9. Marsh, P.; Ranmuthugala, D.; Penesis, I.; Thomas, G. Three-dimensional numerical simulations of straight-bladed vertical axis tidal turbines investigating power output, torque ripple and mounting forces. *Renew. Energy* **2015**, *83*, 67–77. [CrossRef]
10. Li, Q.A.; Maeda, T.; Kamada, Y.; Murata, J.; Shimizu, K.; Ogasawara, T.; Nakai, A.; Kasuya, T. Effect of solidity on aerodynamic forces around straight-bladed vertical axis wind turbine by wind tunnel experiments (depending on number of blades). *Renew. Energy* **2016**, *96*, 928–939. [CrossRef]
11. Aihara, A.; Mendoza, V.; Goude, A.; Bernhoff, H. A numerical study of strut and tower influence on the performance of vertical axis wind turbines using computational fluid dynamics simulation. *Wind Energy* **2022**, *25*, 897–913. [CrossRef]
12. Kirke, B.K.; Lazauskas, L. Limitations of fixed pitch Darrieus hydrokinetic turbines and the challenge of variable pitch. *Renew. Energy* **2011**, *36*, 893–897. [CrossRef]
13. The SeaTwirl Concept. Web Page. Available online: https://seatwirl.com/products/ (accessed on 19 September 2023).
14. Hara, Y.; Horita, N.; Yoshida, S.; Akimoto, H.; Sumi, T. Numerical analysis of effects of arms with different cross-sections on straight-bladed vertical axis wind turbine. *Energies* **2019**, *12*, 2106. [CrossRef]
15. Mertens, S.; van Kuik, G.; van Bussel, G. Performance of an H-Darrieus in the skewed flow on a roof. *J. Sol. Energy Eng.* **2003**, *125*, 433–440. [CrossRef]

16. Menter, F.R. Two-equation eddy-viscosity turbulence models for engineering applications. *AIAA J.* **1994**, *32*, 1598–1605. [CrossRef]
17. Van Bussel, G.J.W.; Polinder, H.; Sidler, H.F.A. TURBY®: Concept and realisation of a small VAWT for the built environment. In Proceedings of the EAWE/EWEA Special Topic Conference "The Science of Making Torque from Wind", Delft, The Netherlands, 19–21 April 2004; p. 8.
18. Fu, W.-S.; Lai, Y.-C.; Li, C.-G. Estimation of turbulent natural convection in horizontal parallel plates by the Q criterion. *Int. Commun. Heat Mass Transf.* **2013**, *45*, 41–46. [CrossRef]
19. Sheldahl, R.E.; Klimas, P.C. *Aerodynamic Characteristics of Seven Symmetrical Airfoil Sections through 180-Degree Angle of Attack for Use in Aerodynamic Analysis of Vertical Axis Wind Turbines*; SAND-80-2114; Sandia National Labs.: Albuquerque, NM, USA, 1981; pp. 43–44.
20. White, F.M. *Fluid Mechanics*, 4th ed.; McGraw-Hill: New York, NY, USA, 1999.

MDPI

St. Alban-Anlage 66

4052 Basel

Switzerland

www.mdpi.com

Education Sciences Editorial Office

E-mail: education@mdpi.com

www.mdpi.com/journal/education

www.ingramcontent.com/pod-product-compliance
Lightning Source LLC
LaVergne TN
LVHW071300170726
843515LV00006BA/1448